STORM WATER MANAGEMENT AND TECHNOLOGY

STORM WATER
MANAGEMENT AND
TECHNOLOGY

U.S. Environmental Protection Agency

NOYES DATA CORPORATION
Park Ridge, New Jersey, U.S.A.

Copyright © 1993 by Noyes Data Corporation
Library of Congress Catalog Card Number: 92–45153
ISBN: 0–8155–1327–5
Printed in the United States

Published in the United States of America by
Noyes Data Corporation
Mill Road, Park Ridge, New Jersey 07656

Library of Congress Cataloging–in–Publication Data

Storm water management and technology / U.S. Environmental Protection
    Agency.
              p.   cm.
        Includes bibliographical references and index.
        ISBN 0–8155–1327–5
          1. Factory and trade waste--Management. 2. Sewage--Purification.
        3. Runoff. 4. Water quality management. I. United States.
        Environmental Protection Agency.

Transferred to Digital Printing 2009

Foreword

Storm water runoff is part of a natural hydrologic process. However, human activities, particularly urbanization, can alter natural drainage patterns and add pollutants to the rainwater and snowmelt that runs off the surface and enters rivers, lakes, streams, and coastal waters. Recent studies have shown that storm water runoff is a source of water pollution, and results in declines in fisheries, and restrictions on swimming, among other concerns. In response to this problem, governmental bodies have been taking the initiative to regulate storm water runoff.

This book provides industrial facilities with comprehensive guidance on the development of storm water pollution prevention plans and identification of appropriate Best Management Practices (BMPs). It provides technical assistance and support to all facilities subject to pollution prevention requirements established under National Pollutant Discharge Elimination System (NPDES) permits for storm water point source discharges.

Owners and operators of industrial facilities will find that putting together a Storm Water Pollution Prevention Plan can be a straightforward process, accomplished by facility managers and employees.

In addition to providing guidance for facilities that are subject to storm water permit requirements, this book contains information that is generally useful for controlling storm water pollution from almost any type of developed site.

The book is presented as a user's guide to Storm Water Pollution Prevention Plan requirements. Step-by-step guidelines and accompanying worksheets will walk the reader through the process of developing and implementing a Storm Water Pollution Prevention Plan. This approach allows the reader to complete this process in the simplest and most efficient way. The worksheets are designed to help organize the required information.

The remainder of the manual is divided into three sections: Chapter 2 provides information on how to develop a plan; Chapter 3 serves as a resource for selecting activity-specific BMPs; and Chapter 4 discusses site-specific BMPs. The ten appendices included at the end of the manual provide additional information to help the reader make informed decisions on BMPs.

The information in the book is from *Storm Water Management for Industrial Activities—Developing Pollution Prevention Plans and Best Management Practices,* issued by the U.S. Environmental Protection Agency, September 1992.

The table of contents is organized in such a way as to serve as a subject index and provides easy access to the information contained in the book.

Advanced composition and production methods developed by Noyes Data Corporation are employed to bring this durably bound book to you in a minimum of time. Special techniques are used to close the gap between "manuscript" and "completed book." In order to keep the price of the book to a reasonable level, it has been partially reproduced by photo–offset directly from the original report and the cost saving passed on to the reader. Due to this method of publishing, certain portions of the book may be less legible than desired.

Notice

The materials in this book were prepared as accounts of work sponsored by the U.S. Environmental Protection Agency. On this basis the Publisher assumes no responsibility nor liability for errors or any consequences arising from the use of the information contained herein.

This document was issued in support of EPA regulations and policy initiatives involving the development and implementation of a national storm water program. This document is EPA guidance only. It does not establish or affect legal rights or obligations. EPA decisions in any particular case will be made applying the laws and regulations on the basis of specific facts when permits are issued or regulations promulgated. This manual is not intended in any way to substitute for binding legal requirements pursuant to National Pollutant Discharge Elimination System (NPDES) permits.

Mention of trade names or commercial products does not constitute endorsement or recommendation for use by the EPA or the Publisher. Final determination of the suitability of any information or product for use contemplated by any user, and the manner of that use, is the sole responsibility of the user. The book is intended for information purposes only. The reader is warned that caution must always be exercised with heavy storm water flows, and chemicals picked up by storm water. Expert advice should be obtained before implementation of any storm water control program.

All information pertaining to law and regulations is provided for background only. The reader must contact the appropriate legal sources and regulatory authorities for up–to–date regulatory requirements, and their interpretation and implementation.

The book is sold with the understanding that the Publisher is not engaged in rendering legal, engineering, or other professional service. If advice or other expert assistance is required, the service of a competent professional should be sought.

Contents and Subject Index

CHAPTER
1

INTRODUCTION

Storm water runoff is part of a natural hydrologic process. However, human activities, particularly urbanization, can alter natural drainage patterns and add pollutants to the rainwater and snowmelt that runs off the earth's surface and enters our Nation's rivers, lakes, streams, and coastal waters. A number of recent studies by the U.S. Environmental Protection Agency (EPA), State water pollution control authorities, and various universities have shown that storm water runoff is a major source of water pollution, declines in fisheries, restrictions on swimming, and these conditions limit our ability to enjoy many of the other benefits that the Nation's waters provide.

In response to this problem, the States and many municipalities have been taking the initiative to manage storm water more effectively. In acknowledgement of the importance of the storm water problem, the Congress has directed EPA to undertake a wide range of activities, including providing technical and financial assistance to States and other jurisdictions to help them improve their storm water management programs. In addition, through recent amendments to the Clean Water Act, the Congress has instructed EPA to develop a regulatory program for certain high priority storm water sources.

In carrying out its responsibilities, EPA is committed to promoting the concept and the practice of preventing pollution at the source, before it can cause environmental problems costing the public and private sector in terms of lost resources and the funding it takes to remediate or correct environmental damage.

1.1 PURPOSE OF THIS GUIDANCE MANUAL

This manual provides general guidance on developing and implementing a Storm Water Pollution Prevention Plan for industrial facilities. Owners and operators of industrial facilities will find that putting together a Storm Water Pollution Prevention Plan is a straightforward process that can be accomplished by facility managers and employees.

EPA is publishing this manual for several reasons. The primary purpose of this manual is to provide guidance for industrial facilities that are subject to requirements under EPA's General Permits for storm water discharges associated with industrial activity. Facilities located in the 12 nondelegated States or 6 Territories are subject to these requirements (see Section 1.6 for a list of States and Territories subject to EPA General Permit requirements). EPA anticipates that most storm water discharge permits issued under the Storm Water Program will require a pollution prevention plan. Throughout this manual, specific EPA General Permit pollution prevention requirements are given in the shaded boxes as seen below. Although the requirements for a Storm Water Pollution Prevention Plan may vary from one permit to another, and from State to State, EPA expects that most of the general concepts described in this manual are common to all plan requirements. Please also note that, although this manual presents EPA General Permit requirements that apply to facilities located in nondelegated States and Territories, some of the nondelegated States required modifications or additions to the pollution prevention plan requirements to ensure that the permit complies with State laws and standards. Therefore, it is important that all facilities located in delegated States, as well as nondelegated States, read their permits to determine whether there are

1

any special conditions. This manual is not intended in any way to substitute for binding legal requirements pursuant to National Pollutant Discharge Elimination System (NPDES) permits.

EPA GENERAL PERMIT REQUIREMENTS

Storm Water Pollution Prevention Plans

Part IV

A Storm Water Pollution Prevention Plan shall be developed for each facility covered by this permit. Storm Water Pollution Prevention Plans shall be prepared in accordance with good engineering practices. The plan shall identify potential sources of pollution which may reasonably be expected to affect the quality of storm water discharges associated with industrial activity from the facility. In addition, the plan shall describe and ensure the implementation of practices which are to be used to reduce the pollutants in storm water discharges associated with industrial activity at the facility and to assure compliance with the terms and conditions of this permit. Facilities must implement the provisions of the Storm Water Pollution Prevention Plan required under this part as a condition of this permit.

In addition to providing guidance for facilities that are subject to storm water permit requirements, this manual contains information that is generally useful for controlling storm water pollution from almost any type of developed site. EPA hopes this manual is widely used in furthering the prevention of pollution at its sources and the adoption of management practices that help us protect the overall quality of the environment.

EPA is also issuing a guidance manual on Best Management Practices (BMPs) for construction activities. If you are subject to requirements under the general permit for storm water discharges associated with construction activities, that manual is designed to help you comply with those somewhat different requirements.

1.2 ORGANIZATION OF THIS GUIDANCE MANUAL

This manual is presented as a user's guide to Storm Water Pollution Prevention Plan requirements. Step-by-step guidelines and accompanying worksheets will walk you through the process of developing and implementing a Storm Water Pollution Prevention Plan. This approach allows you to complete this process in the simplest and most efficient way. The worksheets are designed to help you organize the required information. The remainder of this manual is divided into three sections: Chapter 2 provides information on how to develop a plan; Chapter 3 serves as a resource for selecting activity-specific Best Management Practices (BMPs); and Chapter 4 discusses site-specific BMPs. As you complete each section, you will move through each of the following steps and end up with a fully developed Storm Water Pollution Prevention Plan. Each step is important and should be completed before moving on to the next step. The five major phases involved in developing and implementing your plan are as follows:

Phase 1	-	Planning and Organization
Phase 2	-	Assessment
Phase 3	-	BMP Identification
Phase 4	-	Plan Implementation
Phase 5	-	Evaluation

Chapter 2 provides step-by-step guidance for completing each of these phases. The Organization Phase starts the process by helping you to get organized and by identifying who is going to develop and implement the plan and by identifying site-specific pollution prevention objectives. The Assessment Phase involves gathering information about your site and identifying potential sources of storm water pollution. Using the information collected during the Assessment Phase, you can begin to design the storm water management program that best suits your site. During the BMP Identification Phase, you will evaluate the required baseline BMPs and select other preventive measures. The fourth stage of the Storm Water Pollution Prevention Planning process is the Implementation Phase, during which you put the plan into action. The final step, the Evaluation Phase, allows you to determine if your plan is actually accomplishing your pollution prevention objectives. Periodic reviews, inspections, and evaluations will allow you to keep the plan effective and up-to-date.

In Chapter 3, which details activity-specific BMPs, you will find a number of measures you can use to prevent or reduce the contamination of storm water caused by specific industrial activities. Chapter 4 describes site-specific BMPs. From the list of site-specific BMPs, you can select prevention and control measures that are most appropriate for the physical characteristics of your facility. A combination of these types of BMPs may be most appropriate for your site.

In addition, there are several appendices located at the end of this manual. Appendix A lists the references used to develop this manual. Appendix B includes a glossary of terms. Appendix C provides a model of what a pollution prevention plan might look like for a small industry. Appendix D provides State and Federal storm water and pollution prevention contacts and additional information on pollution prevention. Appendix E provides technical and design fact sheets for some of the storm water BMPs described in Chapter 4. Appendix F describes tests for non-storm water discharges. Appendix G compares Storm Water Pollution Prevention Plan requirements with plan requirements under other environmental programs. Appendix H is a list of reportable quantities for hazardous substances under 40 CFR Parts 117 and 302. Appendix I is the list of water priority chemicals under Emergency Planning and Community Right-to-Know Act (EPCRA), Section 313. Appendix J includes a table of the monitoring requirements that are contained in EPA's General Permits.

1.3 SCOPE OF THIS MANUAL

This manual provides useful information on many pollution prevention and best management practices which you can use to prevent or reduce the discharge of sediment and other pollutants in storm water runoff from your site. This manual describes the practices and controls, tells how, when, and where to use them, and how to maintain them. However, the effectiveness of these controls lies fully in your hands. Although specific recommendations will be offered in the following chapters, keep in mind that careful consideration must be given to selecting the most appropriate control measures based on site-specific features, and on properly installing the controls in a timely manner. Finally, although this manual provides guidelines for maintenance, it is up to you to make sure that your controls are carefully maintained or they will prove to be ineffective.

This manual describes the EPA General Permit requirements for pollution prevention plans. However, requirements may vary from permit to permit. You should read your permit to determine the required components of your pollution prevention plan. Although this manual describes "typical" permit requirements, do not assume that the typical permit requirements described in this manual are the same as your permit requirements even if you are included under an NPDES general permit for storm water discharges associated with industrial activities. Permit conditions may vary between different permits and/or different versions of the permit.

EPA has issued a number of regulations addressing pollution control practices for different environmental media (i.e., land, water, air, and ground water). However, this manual focuses on identifying pollution prevention measures and BMPs specifically for industrial storm water discharges and provides guidance to industrial facilities on how to comply with storm water permits.

Although Storm Water Pollution Prevention Plans primarily focus on storm water, it is important to consider the impacts of selected storm water management measures on other environmental media (i.e., land, air, and ground water). For example, if the water table is unusually high in your area, a retention pond for contaminated storm water may also lead to contamination of a ground water source unless special preventive measures are taken. Permittees must take these issues into consideration in selecting appropriate pollution prevention measures and should make certain that adoption of storm water measures is consistent with other Federal, State, and local environmental laws. For instance, under EPA's July 1991 Ground Water Protection Strategy, States are encouraged to develop Comprehensive State Ground Water Protection Programs. Your facility's efforts to control storm water should be compatible with the ground water protection objectives reflected in your State's program.

1.4 DEFINITIONS

As you use this manual to select pollution prevention approaches, you will see two key phrases used frequently: "pollution prevention plan" and "best management practice." A solid understanding of these terms is very important in meeting the goals of storm water management discussed above.

> **Pollution Prevention Plan**

The first term of importance is "storm water pollution prevention plan." As mentioned in Section 1.1, this manual is designed to help you to prepare and implement a Storm Water Pollution Prevention Plan. As you will learn in Chapter 2, Storm Water Pollution Prevention Plans consist of a series of steps and activities to, first, identify sources of pollution or contamination on your site, and, second, select and carry out actions which prevent or control the pollution of storm water discharges.

> **Best Management Practice**

The other concept used throughout this manual is "Best Management Practice" or BMP. BMPs are measures or practices used to reduce the amount of pollution entering surface water, air, land, or ground waters. BMPs may take the form of a process, activity, or physical structure. Some BMPs are simple and can be put into place immediately, while others are more complicated and require extensive planning or space. They may be inexpensive or costly to implement. Although BMPs are used in many environmental programs, the BMPs presented in this manual are specifically designed to reduce or eliminate pollutants in storm water discharges. Chapter 2 describes the baseline BMP requirements of EPA's General Permit for storm water discharges associated with industrial activity. Chapters 3 and 4 describe numerous specific BMPs that will help you comply with these requirements.

1.5 GOALS OF STORM WATER MANAGEMENT

Federal, State, and local storm water management programs have a common goal:

> **To Improve Water Quality By Reducing the Pollutants Contained In Storm Water Discharges**

Meeting this goal is a difficult challenge for many reasons. For example, the original sources of the pollutants transported in storm water can be diffuse or spread out over a wide area. So, small oil and grease spills at hundreds of different facilities within a single city can collectively represent a major pollution problem. In addition, the nature of storm water is such that the amount of pollutants that enter receiving waters will vary in accordance with the frequency, intensity and duration of rainfall and the nature of local drainage patterns. Considering the wide variety of types of industries in the United States and the wide range of materials and chemical compounds that are used as part of different industrial activities, a site-specific pollution prevention plan tailored for each facility is considered the most effective, flexible, and economically practical approach to achieve effective storm water management.

The pollution prevention plan approach required by EPA gives facilities flexibility to establish a site-specific storm water management program to meet Best Available Technology/Best Control Technology (BAT/BCT) standards required by the Clean Water Act instead of imposing numerical discharge limitations. Yet, the BMP framework established by the pollution prevention plan requirements must be fully implemented to meet these standards.

1.6 SUMMARY OF THE STORM WATER PROGRAM

Storm water discharges have been increasingly identified as a significant source of water pollution in numerous nationwide studies on water quality. To address this problem, the Clean Water Act Amendments of 1987 required EPA to publish regulations to control storm water discharges under NPDES. EPA published storm water regulations on November 16, 1990, which require certain dischargers of storm water to waters of the United States to apply for NPDES permits. "Waters of the United States" is generally defined as surface waters, including lakes, rivers, streams, wetlands, and coastal waters. NPDES storm water discharge permits will allow the States and EPA to track and monitor sources of storm water pollution. According to the November 16, 1990, final rule, facilities with a "storm water discharge associated with industrial activity" are required to apply for a storm water permit. EPA has defined this phrase in terms of 11 categories of industrial activity that include: (1) facilities subject to storm water effluent limitations guidelines, new source performance standards, or toxic pollutant effluent standards under 40 CFR Subchapter N; (2) "heavy" manufacturing facilities; (3) mining and oil and gas operations with "contaminated" storm water discharges; (4) hazardous waste treatment, storage, or disposal facilities; (5) landfills, land application sites, and open dumps; (6) recycling facilities; (7) steam electric generating facilities; (8) transportation facilities, including airports; (9) sewage treatment plants; (10) construction operations disturbing 5 or more acres*; and (11) other industrial facilities where materials are exposed to storm water*. Operators of industrial facilities that are Federally, State, or municipally owned or operated that meet the above description must also submit applications. If you have questions about whether or not your facility needs to seek permit coverage, contact the EPA Storm Water Hotline at (703) 821-4823.

Storm water discharges associated with industrial activity that reach waters of the United States through Municipal Separate Storm Sewer Systems (MS4s) are also required to obtain NPDES storm water permit coverage. Discharges of storm water to a combined sewer system or to a Publicly Owned Treatment Works (POTW) are excluded.

The storm water regulation presents three permit application options for storm water discharges associated with industrial activity. The first option is to submit an individual application consisting of Forms 1 and 2F. The second option is to participate in a group application. The third option is to file a Notice of Intent (NOI) to be covered under a general permit in accordance with the

---

*On June 4, 1992, the United States Court of Appeals for the Ninth Circuit remanded the exemptions for manufacturing facilities which do not have materials or activities exposed to storm water and for construction sites of less than five acres to the EPA for further rulemaking.

requirements of an issued general permit. Regardless of the permit application option a facility selects, the resulting storm water discharge permit will most likely contain a requirement to develop and implement a Storm Water Pollution Prevention Plan.

NPDES permits are issued by the State for States that have been delegated NPDES permitting authority or by EPA for States that have not been delegated NPDES permitting authority. Therefore, the specific EPA General Permit requirements discussed in this guidance manual apply only to facilities located in one of the 12 nondelegated States or Territories (Alaska; Arizona; Idaho; Louisiana; Maine; Massachusetts; New Hampshire; New Mexico; Oklahoma; South Dakota; Texas; the District of Columbia; Puerto Rico; Guam; American Samoa; Northern Mariana Islands; Trust Territory of the Pacific Islands; Indian lands in Alabama, California, Georgia, Kentucky, Michigan, Minnesota, Mississippi, Montana, North Carolina, North Dakota, New York, Nevada, South Carolina, Tennessee, Utah, Wisconsin, Wyoming; located within Federal facilities or Indian lands in Colorado and Washington; and located within Federal facilities in Delaware). EPA expects, however, that the Federal general permit will be used as a model by NPDES-authorized States, tailored to meet State-specific conditions. Even though storm water permit requirements will vary from State to State depending on water quality concerns and permitting priorities for the permitting authority, EPA expects that most NPDES storm water discharge permits will contain Storm Water Pollution Prevention Plan requirements similar to the requirements presented in this manual.

CHAPTER
2

STORM WATER POLLUTION PREVENTION PLAN

Chapter 2 presents a step-by-step guide to help you develop a Storm Water Pollution Prevention Plan for your facility. Figure 2.1 is a flowchart showing each step involved in developing and implementing a successful plan. As shown in this flowchart, the steps have been grouped into five general phases, which are: (1) planning and organization; (2) assessment; (3) BMP identification; (4) implementation; and (5) evaluation/monitoring. In addition, Storm Water Pollution Prevention Plans also must address a number of general requirements, including developing a schedule or deadlines for the accomplishment of tasks, and an identification of signature authority, where required by Federal regulations. Some types of facilities will also have to meet other special requirements. For example, special requirements apply to facilities that discharge through municipal separate storm water systems as well as those facilities that are subject to reporting requirements under EPCRA, Section 313 for water priority chemicals.

Figure 2.1 also identifies a number of worksheets that can help walk you through the planning process. These worksheets are located at the end of Chapter 2. You can pull them out, photocopy them, and simply incorporate the completed forms in your plan.

The five planning phases, general requirements, and special requirements are discussed in turn in the remainder of this chapter. To help you follow along, a simplified version of the flowchart for the entire planning process is shown at the beginning of each section, with a highlighted box showing the particular phase that is being discussed. So, for example, you will find that the Planning and Organization Phase is highlighted on the flowchart at the top of page 2-3, signaling the beginning of our detailed discussion of this first step.

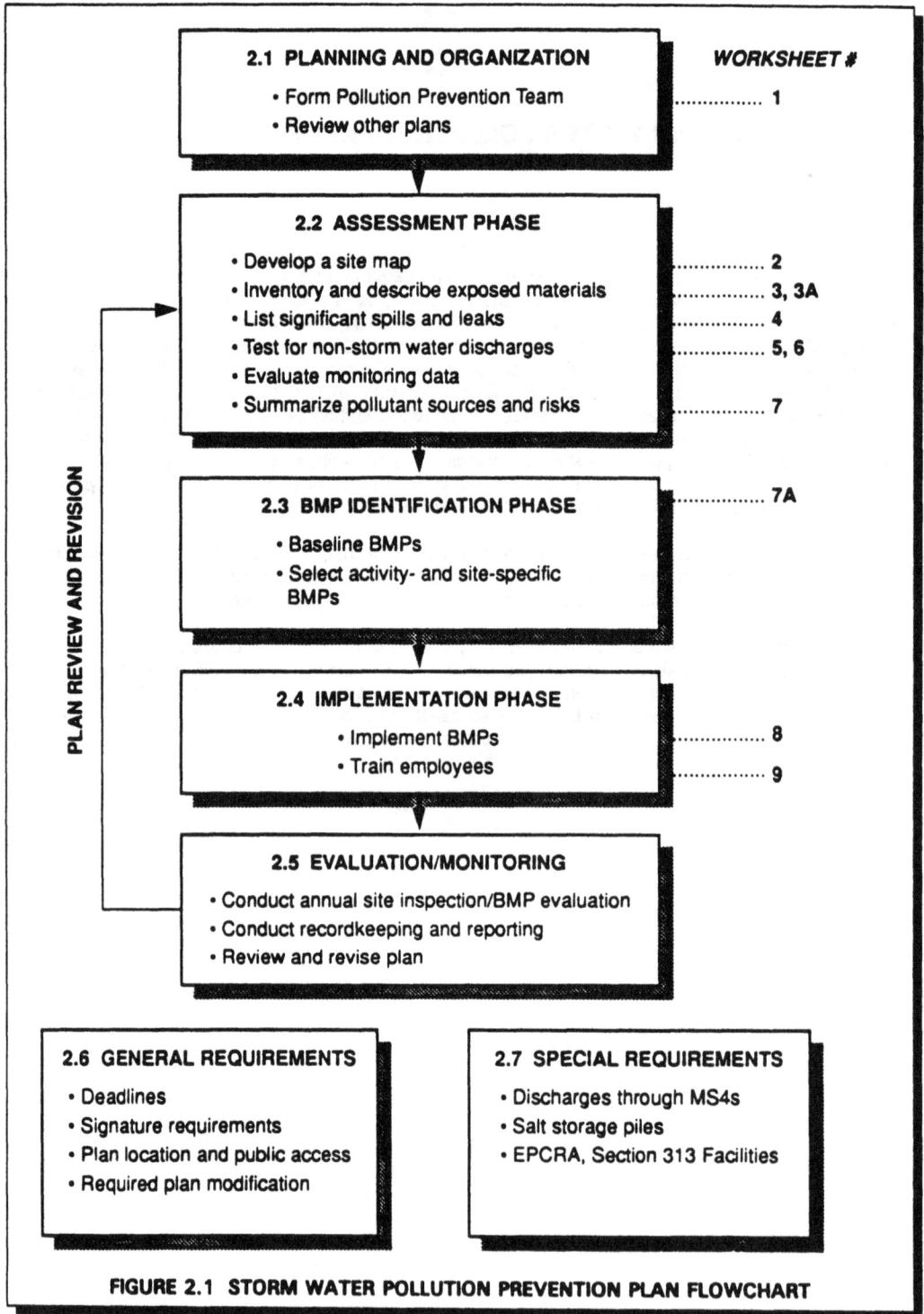

2.1 PLANNING AND ORGANIZATION

- Form Pollution Prevention Team
- Review other plans

WORKSHEET #

.... 1

2.2 ASSESSMENT PHASE

- Develop a site map
- Inventory and describe exposed materials
- List significant spills and leaks
- Test for non-storm water discharges
- Evaluate monitoring data
- Summarize pollutant sources and risks

.... 2
.... 3, 3A
.... 4
.... 5, 6

.... 7

2.3 BMP IDENTIFICATION PHASE

- Baseline BMPs
- Select activity- and site-specific BMPs

.... 7A

2.4 IMPLEMENTATION PHASE

- Implement BMPs
- Train employees

.... 8
.... 9

2.5 EVALUATION/MONITORING

- Conduct annual site inspection/BMP evaluation
- Conduct recordkeeping and reporting
- Review and revise plan

PLAN REVIEW AND REVISION

2.6 GENERAL REQUIREMENTS

- Deadlines
- Signature requirements
- Plan location and public access
- Required plan modification

2.7 SPECIAL REQUIREMENTS

- Discharges through MS4s
- Salt storage piles
- EPCRA, Section 313 Facilities

FIGURE 2.1 STORM WATER POLLUTION PREVENTION PLAN FLOWCHART

Figure

```
┌─────────────────────────────────────┐
│  2.1  PLANNING AND ORGANIZATION      │
│   • Form Pollution Prevention Team   │
│   • Review other plans               │
└─────────────────────────────────────┘
              │
┌─────────────────────────────────────┐
│       2.2  ASSESSMENT PHASE          │
└─────────────────────────────────────┘
              │
┌─────────────────────────────────────┐
│    2.3  BMP IDENTIFICATION PHASE     │
└─────────────────────────────────────┘
              │
┌─────────────────────────────────────┐
│    2.4  IMPLEMENTATION PHASE         │
└─────────────────────────────────────┘
              │
┌─────────────────────────────────────┐
│    2.5  EVALUATION/MONITORING        │
└─────────────────────────────────────┘

┌──────────────────┐ ┌──────────────────┐
│  2.6  GENERAL    │ │  2.7  SPECIAL    │
│  REQUIREMENTS    │ │  REQUIREMENTS    │
└──────────────────┘ └──────────────────┘
```

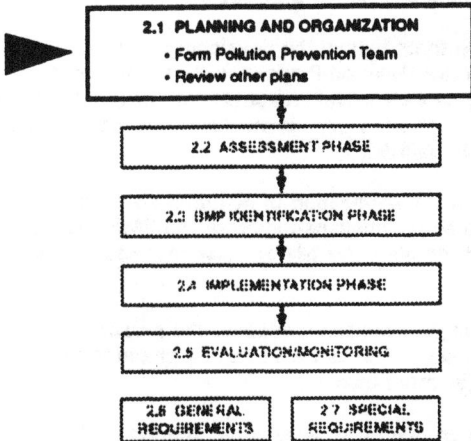

2.1 PLANNING AND ORGANIZATION PHASE

Before you start putting your Storm Water Pollution Prevention Plan together, there are two tasks to complete to make developing the plan easier. These steps are designed to help you organize your staff and make preliminary decisions:

- Decide who will be responsible for developing and implementing your Storm Water Pollution Prevention Plan

- Look at other existing environmental facility plans for consistency and overlap.

2.1.1 Who Will Develop and Implement Your Plan?

EPA GENERAL PERMIT REQUIREMENTS

Pollution Prevention Team

Part IV.D.1.

Each plan shall identify a specific individual or individuals within the facility organization as members of a storm water pollution prevention team that are responsible for developing the Storm Water Pollution Prevention Plan and assisting the facility or plant manager in its implementation, maintenance, and revision. The plan shall clearly identify the responsibilities of each team member. The activities and responsibilities of the team shall address all aspects of the facility's Storm Water Pollution Prevention Plan.

What is the Purpose of Designating an Individual or a Team?

Designating a specific individual or team who will develop and implement your pollution prevention plan serves several purposes. Naming the individual or team members makes it clear that part of that person's job is to prevent storm water pollution. Identifying a specific individual also provides a point of contact for those outside the facility who may need to discuss aspects of the facility's pollution prevention plan (i.e., regulatory officials, etc.).

Where setting up a pollution prevention team is appropriate, it is important to identify the key people onsite who are most familiar with the facility and its operations, and to provide adequate structure and direction to the facility's entire storm water management program. The pollution prevention team concept is flexible and should be molded to conform to the resources and specific conditions of the facility. Specific activities of the pollution prevention team, the number of members, and their background and experience will vary for each facility.

Effective organization of the pollution prevention team is important in order for the team to be able to accomplish the task of developing and implementing a comprehensive Storm Water Pollution Prevention Plan. There are two important features in organizing a team of this nature: (1) selecting the right individuals to serve on the team; and (2) establishing good channels of communication.

What are the Roles and Responsibilities of the Designated Individual or Team?

The designated individual or team will be the driving force behind the development, implementation, maintenance, and revision of the facility's Storm Water Pollution Prevention Plan. One of the first tasks of those responsible is to define and agree upon a clear and reasonable set of goals for the facility's overall storm water management program. Where a team is involved, the responsibilities or duties of specific team members should be clearly defined.

Areas of responsibilities include initial site assessment, identification of pollutant sources and risks, decision making on appropriate BMPs, directing the actual implementation of the BMPs, and then, regular evaluations to measure the effectiveness of the plan. Details of these procedures are described in the latter part of this chapter.

To ensure that the Storm Water Pollution Prevention Plan remains effective, the person or team responsible for maintaining the pollution prevention plan must be aware of any changes that are made in plant operations to determine if any changes must be made.

While a designated individual or a pollution prevention team can be assigned the job of developing and implementing a Storm Water Pollution Prevention Plan, plant management is ultimately responsible for the implementation of the plan and for compliance with all applicable storm water requirements. Accordingly, the designated individual or team must have a clear line of communication with plant management to ensure that they are able to function in a cooperative partnership.

Who Should be on a Storm Water Pollution Prevention Team?

Any team, by definition, involves decision making and planning in a group setting. This allows for people with different ideas and areas of expertise to share knowledge and collectively figure out what works best for a particular facility. To broaden the base of involvement in the facility's storm water pollution prevention program, team members should represent all phases of the facility's operations.

For example, at a large facility, a team may be comprised of representatives from plant management, all aspects of production operations, engineering, waste handling and treatment (environmental department), and, if applicable, research and development. See Figure 2.2 for an illustration of an example team organizational chart.

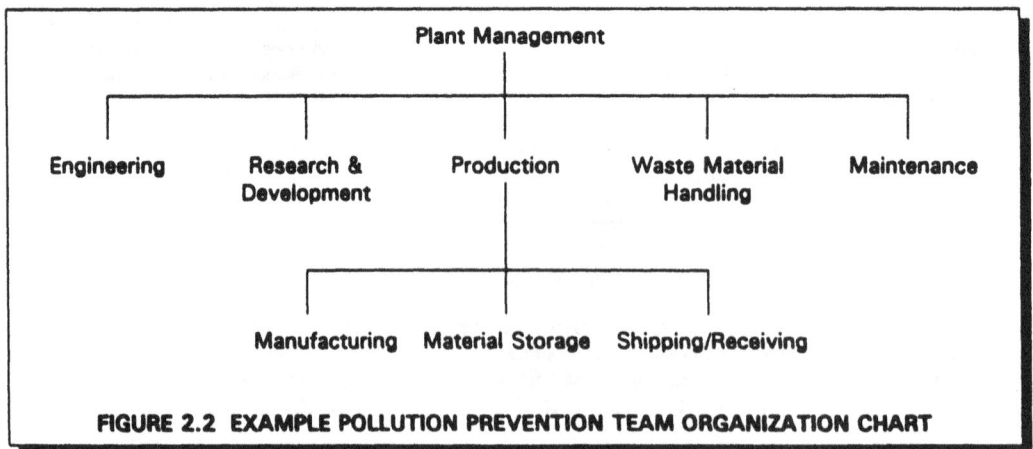

FIGURE 2.2 EXAMPLE POLLUTION PREVENTION TEAM ORGANIZATION CHART

Not all facilities will have or require all of these "team" positions. As mentioned above, team membership depends on the type of operations occurring at a facility. For example, a small trucking operation may find it appropriate to designate a single individual or a very small pollution prevention team with experience in key types of facility operations, such as vehicle maintenance, vehicle washing, fueling, and materials handling.

For a facility that has already designated a spill prevention and response team, the facility may use some of these personnel on the storm water pollution prevention team, thus overlapping the two groups to a certain extent. However, the roles and responsibilities of the pollution prevention team reach beyond the activities of a spill prevention and response team, and consequently, it would not be appropriate for a facility simply to substitute the spill response team for the pollution prevention team without clearly examining the roles and requirements related to storm water management (see Section 2.1.2).

Worksheet #1 (located at the end of Chapter 2) is an example of an appropriate form on which to list the team members. To complete this worksheet, list the pollution prevention team members by name, facility position (title), phone number, and include a brief description of each member's specific responsibilities. This list can be directly incorporated into the Storm Water Pollution Prevention Plan, but it should also be displayed or posted within the facility so that other plant employees are aware of who is responsible for storm water management.

EPCRA, Section 313 Facility Team Requirements

EPA's General Permit contains more specific pollution prevention team requirements for facilities subject to reporting under EPCRA, Section 313 for water priority chemicals [Part IV.D.7.b.(9).]. The team must designate a person who will be accountable for spill prevention at the facility and identify this person in the plan. The designated person is responsible for setting up necessary spill emergency procedures and reporting requirements to isolate, contain, and clean up spills and emergency releases of Section 313 water priority chemicals before a discharge can occur.

2.1.2 Building on Existing Environmental Management Plans

EPA GENERAL PERMIT REQUIREMENTS
Consistency with Other Plans
Part IV.D.6.
Storm Water Pollution Prevention Plans may reflect requirements for Spill Prevention Control and Countermeasure (SPCC) plans developed for the facility under Section 311 of the Clean Water Act or BMP programs otherwise required by an NPDES permit for the facility as long as such requirement is incorporated into the Storm Water Pollution Prevention Plan.

Many industrial facilities may have already incorporated storm water management practices into day-to-day operations as a part of an environmental management plan required by other regulations. Potentially relevant elements of a number of different types of plans are listed in Appendix G at the end of this manual. The plans addressed include: the Preparedness, Prevention and Contingency Plan [40 Code of Federal Regulations (CFR) 264 and 265], the Spill Control and Countermeasures requirements (40 CFR 112), the National Pollutant Discharge Elimination System Toxic Organic Management Plan (40 CFR 413, 433, 469), and the Occupational Safety and Health Administration (OSHA) Emergency Action Plan (29 CFR 1910). It is the responsibility of the pollution prevention team to evaluate these other plans to determine which, if any, provisions may be incorporated into the Storm Water Pollution Prevention Plan.

In some cases, it may be possible to build on elements of these plans that are relevant to storm water pollution prevention. For example, if your facility already has in place an effective spill prevention and response plan, elements of that spill prevention strategy may be relevant to your approach for storm water pollution prevention. More specifically, lists of potential pollutants or constituents of concern may provide a starting point for your list of potential storm water pollutants. Although you should build on relevant portions of other environmental plans as appropriate, it is important to note that your Storm Water Pollution Prevention Plan must be a comprehensive, stand-alone document.

```
┌──────────────────────────────────────┐
│  2.1  PLANNING AND ORGANIZATION        │
└──────────────────────────────────────┘
                  │
┌──────────────────────────────────────┐
│      2.2  ASSESSMENT PHASE             │
│                                        │
│  • Develop a site map                  │
│  • Inventory and describe exposed materials │
│  • List significant spills and leaks   │
│  • Test for non-storm water discharges │
│  • Evaluate monitoring data            │
│  • Summarize pollutant sources and risks │
└──────────────────────────────────────┘
                  │
┌──────────────────────────────────────┐
│  2.3  BMP IDENTIFICATION PHASE         │
└──────────────────────────────────────┘
                  │
┌──────────────────────────────────────┐
│  2.4  IMPLEMENTATION PHASE             │
└──────────────────────────────────────┘
                  │
┌──────────────────────────────────────┐
│  2.5  EVALUATION/MONITORING            │
└──────────────────────────────────────┘

┌──────────────┐    ┌──────────────┐
│ 2.6 GENERAL  │    │ 2.7 SPECIAL  │
│ REQUIREMENTS │    │ REQUIREMENTS │
└──────────────┘    └──────────────┘
```

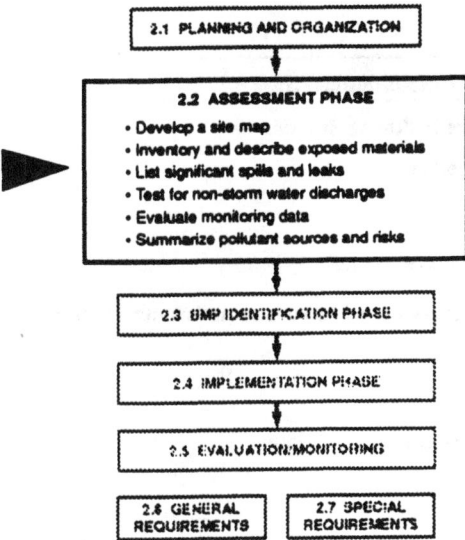

2.2 ASSESSMENT PHASE - DESCRIPTION OF POTENTIAL POLLUTANT SOURCES

After identifying who is responsible for developing and implementing your plan and organizing your planning process, you should proceed to this next step—a pollutant source assessment. This is where you take a look at your facility and site and determine what materials or practices are or may be a source of contaminants to the storm water running off your site. To complete this phase, you will:

- Assess the potential sources of storm water pollution at your facility

- Create a map of the facility site to locate pollutant sources and determine storm water management opportunities

- Conduct a material inventory

- Evaluate past spills and leaks

- Identify non-storm water discharges and illicit connections

- Collect or evaluate storm water quality data

- Summarize the findings of this assessment.

EPA GENERAL PERMIT REQUIREMENTS

Description of Potential Pollutant Sources

Part IV.D.2.

Each plan should provide a description of potential sources which may be reasonably expected to add significant amounts of pollutants to storm water discharges or which may result in the discharge of pollutants during dry weather from separate storm sewers draining the facility. Each plan shall identify all activities and significant materials which may potentially be significant pollutant sources.

This phase is designed to help you to target the most important pollutant sources for corrective and/or preventive action, thus using a "risk-based" approach to environmental protection. Details on how to complete this assessment are provided in the next six subsections of this chapter (see 2.2.1-2.2.6). These sections of the manual will help you discover areas at your facility that have the potential for contributing pollutants to storm water. Within each of the following sections, you will find helpful worksheets and suggestions for accomplishing a complete and accurate assessment of existing and potential problems. Each of the required components builds on the others; therefore, it is very important to perform each step thoroughly.

2.2.1 Developing a Site Map

```
┌─────────────────────────────────────────────────────────────────────┐
│                   EPA GENERAL PERMIT REQUIREMENTS                     │
├─────────────────────────────────────────────────────────────────────┤
│                                                                       │
│              Site Drainage and Potential Pollutant Sources            │
│                          Part IV.D.2.a.(1).                           │
│                                                                       │
│  The facility site map must include:                                  │
│                                                                       │
│    • An outline of the drainage area of each storm water outfall      │
│                                                                       │
│    • Location of any existing structural control measures used to     │
│      reduce pollutants in storm water runoff                          │
│                                                                       │
│    • Surface water bodies                                             │
│                                                                       │
│    • Locations where significant materials are exposed to             │
│      precipitation                                                    │
│                                                                       │
│    • Locations where major spills or leaks have occurred              │
│                                                                       │
│    • Locations for each of the following activities (where exposed    │
│      to storm water):                                                 │
│                                                                       │
│        - Fueling stations                                             │
│        - Vehicle and equipment maintenance and/or cleaning areas      │
│        - Loading/unloading areas                                      │
│        - Treatment, storage, or waste disposal areas                  │
│        - Liquid storage tanks                                         │
│        - Processing areas                                             │
│        - Storage areas.                                               │
│                                                                       │
└─────────────────────────────────────────────────────────────────────┘
```

The facility site map is basically an illustration of the overall site and location, and should indicate property boundaries, buildings and operation or process areas, as well as provide information on drainage, storm water control structures, and receiving streams. Locating these features on the map will help you assess where potential storm water pollutants are located on your site, where they mix with storm water, and where storm water leaves your site. All of this information is essential in identifying the best opportunities for storm water pollution prevention or control. Worksheet #2 (located at the end of Chapter 2) is designed to help you develop an appropriate and useful site map.

Figures 2.3 and 2.4 are good examples of site maps with different layers of information to help locate sources of pollution on your site. When properly drafted, your site map will be a very useful tool to assist in designing the proper pollution prevention controls, thereby preventing further degradation of water quality by reducing additional water pollution.

Outfalls and Drainage Areas

Once boundaries and facility structures have been shown on your site map, you should identify all of the storm water outfalls (also called "discharge points") on your site. A storm water outfall is the point where storm water enters a natural waterway or a separate storm sewer system. If your facility has its own storm water conveyance system, locate where the pipes or conveyances discharge to a stream, river, lake, or other water body. If your facility discharges to a municipal separate storm sewer system, your onsite drainage point into the system is an outfall. However, on many sites, storm water is simply collected in ditches. The discharge points may not be so obvious, particularly when it is not raining. In these cases, it may be necessary to inspect your site

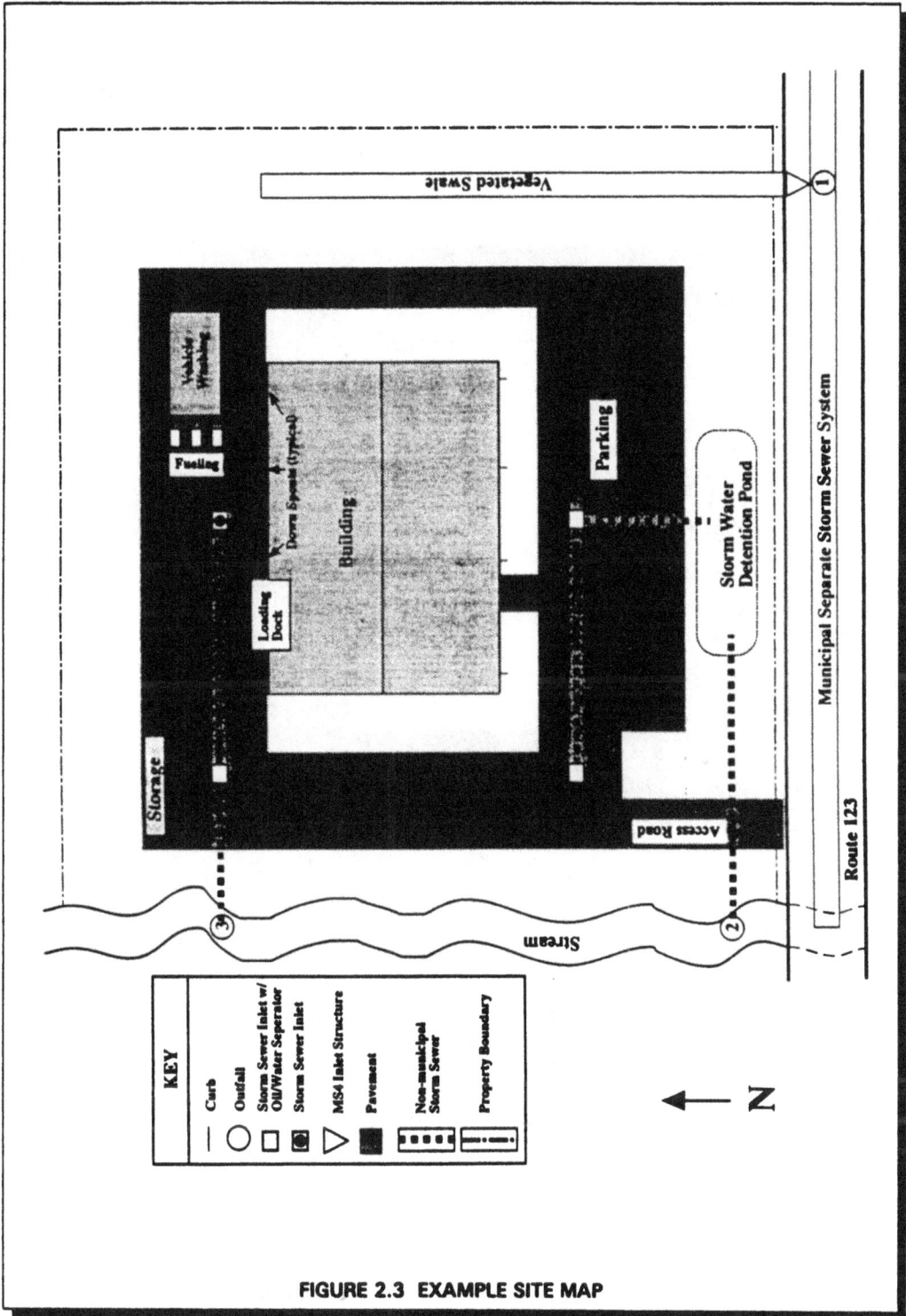

FIGURE 2.3 EXAMPLE SITE MAP

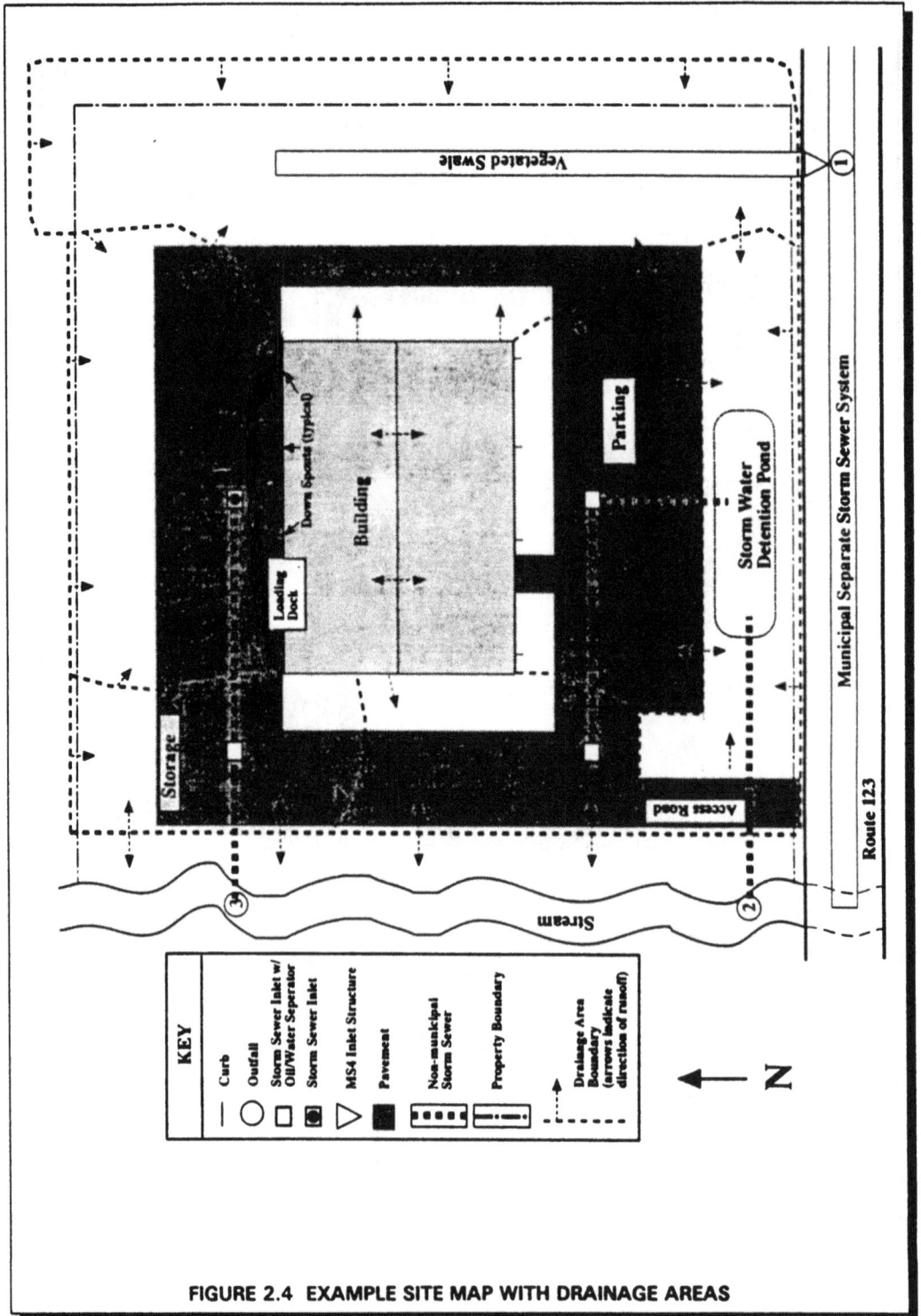

FIGURE 2.4 EXAMPLE SITE MAP WITH DRAINAGE AREAS

during a rain storm to identify your discharge points. Clearly label each outfall either with letters (A, B, C, etc.) or numbers (1, 2, 3, etc.) so that you can easily reference these discharge points in other sections of your Storm Water Pollution Prevention Plan.

Working back from the storm water outfalls you have identified, now determine the drainage areas for each outfall (see Figure 2.4). A topographic map can help with this task if one with the suitable scale is readily available. For larger facilities (greater than 25 acres), 7.5 minute topographic maps, available from the United States Geological Survey (USGS), probably have the level of detail necessary to determine site drainage patterns.

Maps may be purchased from local commercial dealers or directly from USGS information offices. Check your local yellow pages for commercial dealers. Topographic maps may also be purchased by mail. Standard topographic quadrangles cost $2.50. You can order maps from the following locations:

USGS Map Sales   for Alaska maps:
Box 25286   or   USGS Map Sales
Denver, CO 80225   101 12th Ave., #12
                        Fairbanks, AK 99701

For smaller sites, examination of a topographic map may not reveal very much about the drainage patterns of the site. A simple alternative is to examine the contours of your site. A visual observation of flows or the use of small floatables or dyes in concentrated flows are simple methods to determine drainage patterns on your facility. Drainage patterns may be very obvious in some cases, such as drainage down a particular hill on the site. In areas where the site appears to be relatively flat, a rough study of storm water flow during a rain event should provide you with a sufficient sense of the flow patterns.

Structural Storm Water Controls

Other features to include on the site map are the locations and identification of any existing structural control measures already in place that are used to control or direct storm water runoff. A structural control measure is any physically constructed feature you have onsite that is used specifically to change the way that storm water flows or that is used to remove pollutants from storm water. Examples of structural controls include: retention/detention ponds, flow diversion structures (including ditches and culverts), vegetative swales, porous pavement, sediment traps, and any soil stabilization or erosion control practices. See Chapter 4 for a more complete description and illustrations of these structures. Each structure should be clearly identified on the site map, as illustrated in Figure 2.3.

Surface Waters

On the site map, you should label all surface water bodies on or next to the site. This includes any stream, river, lake, or other water body (see Figure 2.3 as an example). Each water body should be identified by name. If you do not know the name of the water body, you can check the USGS topographical maps discussed above for the legal name. Your municipal government may also have municipal maps that identify small streams by name. If your storm water runoff flows into a small, unnamed tributary, the name of the downstream water body will be sufficient.

Potential Pollutant Sources

To develop a useful site map for your facility's Storm Water Pollution Prevention Plan, you must also indicate other items on the map so that you understand what activities are taking place in each drainage area, and therefore, what types of pollutants may be present in storm water from these areas. These features include:

- Topography of site (discussed above)

- Location of exposed significant materials (see Section 2.2.2)

- Locations of past spills and leaks (see Section 2.2.3)

- High-risk waste generating areas and activities common on industrial sites, such as:

    - Fueling stations
    - Vehicle and equipment maintenance
    - Vehicle and equipment washing
    - Loading and unloading areas
    - Above-ground liquid storage tanks
    - Industrial waste management areas and outside manufacturing
    - Outside storage of raw materials, by-products, or finished products.

You will notice that specific BMPs may be applied to control the amount of pollutants in storm water discharges from these areas (see Chapter 3). Now is the time to determine if any of these activities take place onsite, and in which drainage areas they take place.

EPA GENERAL PERMIT REQUIREMENTS

Types of Pollutants and Flow Direction

Part IV.D.2.a.(2).

For each area of the facility that generates storm water discharges associated with industrial activity with a reasonable potential for containing significant amounts of pollutants, include a prediction of the direction of flow and identify the types of pollutants which are likely to be present in storm water discharges associated with industrial activity. Factors to consider include the toxicity of the chemical; quantity of chemicals used, produced, or discharged; the likelihood of contact with storm water; and the history of significant leaks or spills of toxic or hazardous pollutants. Flows with a significant potential for causing erosion shall be identified.

2.2.2 Material Inventory

```
┌─────────────────────────────────────────────────────────────────────┐
│                   EPA GENERAL PERMIT REQUIREMENTS                     │
├─────────────────────────────────────────────────────────────────────┤
│                     Inventory of Exposed Materials                    │
│                                                                       │
│                           Part IV.D.2.b.                              │
│                                                                       │
│  Conduct an inventory of materials that may be exposed to storm water │
│  at your site, and include a narrative description of:                │
│                                                                       │
│   • Significant materials that have been handled, treated, stored, or │
│     disposed in a manner to allow exposure to storm water between the │
│     time of three years prior to the date of permit issuance and the  │
│     present                                                           │
│                                                                       │
│   • Method(s) and location of onsite storage or disposal              │
│                                                                       │
│   • Materials management practices employed to minimize contact of    │
│     these materials with storm water runoff between the time of three │
│     years prior to the date of the issuance of the permit and the     │
│     present                                                           │
│                                                                       │
│   • Existing structural and nonstructural control measures to reduce  │
│     pollutants in storm water runoff, including their locations       │
│                                                                       │
│   • Any treatment of storm water runoff.                              │
└─────────────────────────────────────────────────────────────────────┘
```

The next step in the Assessment Phase is to conduct a material inventory at your site, specifically looking for materials that have been exposed to storm water and measures you have taken to prevent the contact of these materials with storm water. Maintaining an up-to-date material inventory is an efficient way to identify what materials are handled onsite and which may contribute to storm water contamination problems. As discussed above, these potential pollutant sources should be identified on your facility's site map.

Worksheet #3 (located at the end of Chapter 2) will help guide you through the process of conducting a material inventory for your Storm Water Pollution Prevention Plan. Although an inventory of all materials (exposed and not exposed) is required as part of EPA's General Permits, conducting such an inventory is a good first step in compiling a list of exposed materials. If any of the significant materials on your site have been exposed to storm water in the three years prior to the effective date of your permit, fill out Worksheet #3A and include it in your plan.

Inventory of Exposed Significant Materials

"Significant materials," as defined in 40 CFR 122.26(b)(12), are substances related to industrial activities such as process chemicals, raw materials, fuels, pesticides, and fertilizers (see Glossary in Appendix B for exact definition). When these substances are exposed to storm water runoff, they may be carried to a receiving stream with the storm water flow. Therefore, identification of these materials helps to determine where a potential for contamination exists and is the first step in identifying appropriate BMPs to address this contamination potential.

To inventory the materials on your site, inspect your site carefully. You may wish to use the site checklist (page 2-14) to help you identify exposed materials. Focus on areas where you store, process, transport or transfer any materials used or produced during your industrial processes. Check any storage tanks, pipes or pumping areas and note any leaks or spills. Observe any loading or unloading operations and indicate whether any industrial materials are exposed to storm water during those processes. Look at any unsealed dumpsters or disposal units/areas where you deposit wastes from your industrial activities and document instances where waste materials are exposed to rain. Also pay attention to material handling equipment, including everything from vehicles to pallets, where raw and waste materials from your industrial activities are exposed. Finally, consider areas such as the roof where particles are emitted from air vents and are likely to fall within your drainage areas.

Site Checklist

☐ Does your facility show signs of poor housekeeping (cluttered walkways, unswept floors, uncovered materials, etc.)?

☐ Are there spots, pools, puddles, or other traces of oil, grease, or other chemicals on the ground?

☐ Is there discoloration, residue, or corrosion on the roof or around vents or pipes that ventilate or drain work areas?

☐ Do you see leaking equipment, pipes, containers, or lines?

☐ Are there areas where absorbent materials (kitty litter, saw dust, etc.) are regularly used?

☐ Do you notice signs such as smoke, dirt, or fumes that indicate material losses?

☐ Do you smell strange odors, or experience eye, nose, or throat irritation when you first enter the work area? These are indications of equipment leaks.

☐ Do storage containers show signs of corrosion or leaks?

☐ Are there open containers, stacked drums, shelving too small to properly handle inventory, or other indications of poor storage procedures?

☐ Are containers properly labeled?

These are some basic guidelines meant to help you determine what kinds of things to look out for. This list does not necessarily cover every possible source of pollutants. As the site operator, you are responsible for knowing the particular concerns associated with your activity. Be as detailed as you can in your description of the significant materials exposed at your facility. Discuss what you found in your assessment, the amounts present and their location. Update this inventory whenever new, significant materials are introduced and exposed onsite so that your management practices can be modified to suit any changes.

Next, you should give closer scrutiny to areas where you store or dispose of industrial materials. Inspect your various containers carefully and note whether there are any openings, holes or leaks that allow storm water to contact significant materials in those containers.

Existing Management Measures and Treatment of Storm Water Runoff

Now that you have described the potential pollutant sources in storm water runoff from your site, you should describe what management practices you currently use. Management practices can be as simple as scheduled sweeping of the material transfer area. In this section of your plan you must describe both structural and nonstructural management practices. Structural management practices are those practices that entail construction of manmade structures such as berms, detention ponds, or grassed swales, whereas nonstructural management practices involve regularly scheduled actions (such as sweeping, inspections, or improved materials handling and management practices).

Remember that the purpose of BMPs is to keep the pollutants out of storm water runoff by reducing material exposure to storm water, directing the storm water away from contaminated areas, or reducing the volume of potentially polluting materials on the site.

Finally, you must describe any treatment that you provide for the storm water discharges from your site. The treatment of storm water is often accomplished through holding in a detention pond which allows for settling of inorganic solids and partial removal of organic contaminants. In the case of detention ponds, you should describe the size and average depth of each pond on your site (storage volume). You should also provide any design criteria (i.e., design flow rates, etc.) for the pond that may be available to you from engineering design reports or diagrams. Your site may also direct some of your storm water into your process water treatment system. If so, you should identify what type of treatment is provided, and whether this is allowed under your NPDES or other discharge permit. In any case, be sure to specify areas from which the treated storm water drains.

2.2.3 Identifying Past Spills and Leaks

EPA GENERAL PERMIT REQUIREMENTS

Spills and Leaks

Part IV.D.2.c.

Include a list of significant spills and significant leaks of toxic or hazardous pollutants that occurred at areas that are exposed to precipitation or that otherwise drain to a storm water conveyance at the facility after the date of three years prior to the effective date of this permit. Such list shall be updated as appropriate during the term of this permit.

The next component of the assessment phase of your pollution prevention plan is a list of significant spills and significant leaks of toxic or hazardous materials that have occurred at your facility. This list provides information on potential sources of storm water contamination. The first question that comes to mind is "What is a significant spill or leak?"

EPA has defined "significant spills" to include releases within a 24-hour period of hazardous substances in excess of reportable quantities under Section 311 of the Clean Water Act and Section 102 of the Comprehensive Environmental Response, Compensation, and Liability Act (CERCLA). Reportable quantities are set amounts of substances in pounds, gallons, or other units and are listed in 40 CFR Part 117 and 40 CFR Part 302. This list is included as Appendix H in this manual. If your facility releases these listed hazardous substances to the environment in excess of these amounts, you are required to notify the National Response Center at (800) 424-8802 as soon as possible. Releases are defined to include any spilling, leaking, pumping, pouring, emitting, emptying, discharging, injecting, escaping, leaching, dumping, or disposing into the environment.

Worksheet #4 (located at the end of Chapter 2) can help you organize this list of leaks and spills. The areas on your site where significant leaks or spills have occurred are areas on which you should focus very closely when selecting activity-specific or site-specific BMPs.

If several of these events have occurred at your facility, pay special attention to Section 2.3.1, which discusses spill prevention and response procedures. Adequate spill prevention and response procedures are one of the BMPs that should be included in your pollution prevention plan. Using the proper procedures will reduce the likelihood of spills or releases in the future, thus reducing the opportunity for spilled pollutants to come into contact with storm water.

The above list of significant leaks and spills, together with the other information gathered to identify pollutants and sources, provides the necessary focus for the BMP Identification Phase of your facility's Storm Water Pollution Prevention Plan. This information is used to target pollution prevention activities such as preventive maintenance, good housekeeping, spill prevention and response procedures, employee training, and storm water management controls such as covering, flow diversion, erosion control and treatment that ultimately will reduce pollutant loadings in storm water discharges.

2.2.4 Identifying Non-Storm Water Discharges

EPA GENERAL PERMIT REQUIREMENTS

Non-Storm Water Discharges

Part IV.D.3.g.(1).

The plan must include a certification that all storm water outfalls have been tested or evaluated for the presence of non-storm water discharges. The certification shall include:

- Identification of potential non-storm water discharges

- A description of the results of any test and/or evaluation for the presence of non-storm water discharges

- The evaluation criteria or test method used

- The date of testing and/or evaluation

- The onsite drainage points that were directly observed during the test and/or evaluation.

This certification shall be signed in accordance with Section 2.6.2 in this manual and must be included in your Storm Water Pollution Prevention Plan. An example certification form is provided as Worksheet #5.

If this certification is not feasible because you do not have access to an outfall, manhole, or other point of access to the final storm water discharge point(s), you should describe why the certification was infeasible. You also must notify the permitting authority by October 1, 1993 [or 180 days after submitting the Notice of Intent (NOI) for facilities that begin industrial activities after October 1, 1992], of any potential sources of non-storm water discharges to the storm water discharge and why you could not perform the test for non-storm water discharges. This certification must be signed in accordance with Section 2.6.2 of this manual and submitted to the permitting authority. An example Failure to Certify form is provided as Worksheet #6.

Examples of non-storm water discharges include any water used directly in the manufacturing process (process water), air conditioner condensate, non-contact cooling water, vehicle wash water, or sanitary wastes. Connections of non-storm water discharges to a storm water collection system are common yet are often unidentified. Those types of discharges are significant sources of water quality problems. Unless permitted by an NPDES permit, such discharges are illegal. If such connections are discovered, disconnect them or submit an NPDES permit application (Form 2C for process wastewater or 2E for nonprocess wastewater) to your permitting authority. Such interconnections must be disconnected or covered by an NPDES permit.

To check for non-storm water discharges, you may elect to use one of four common dry weather tests described below and in more detail in Appendix F: (1) visual inspection; (2) plant schematic review; and (3) dye testing.

Visual Inspection

The easiest method for detecting non-storm water connections into the storm water collection system is simply to observe all discharge points during dry weather. Inspect each discharge point on three separate occasions. As a rule, the discharge point should be dry during a period of extended dry weather since a storm water collection system should only collect storm water. Keep in mind, however, that drainage of a particular rain event can continue for three days or more after the rain has stopped. In addition, infiltration of ground water into the underground collection system is also common. To be sure about the source of any flow during dry weather, you may need to perform one of the additional tests described below.

Sewer Map

A review of a plant schematic is another simple way to determine if there are any interconnections into the onsite storm water collection system. A sewer map or plant schematic is a map of pipes and drainage systems used to carry process wastewater, non-contact cooling water, air conditioner condensate, and sanitary wastes (bathrooms, sinks, etc.). A common problem, however, is that sites often do not have accurate, up-to-date schematics. If you do have an accurate and reliable plant schematic, you can simply examine the pathways of the different water circuits listed above. Be sure also to investigate where the floor drains discharge. These are commonly connected to the storm sewer system, especially in older buildings.

Dye Testing

Another method for detecting improper connections to the storm water collection system is dye testing. A dye test can be performed by simply releasing a dye into either your sanitary or process wastewater system and examining the discharge points from the storm water collection system for discoloration. A detailed description of the equipment needed and proper procedures for a dye test is included in Appendix F.

Non-Storm Water Discharges

As noted above, unless covered by an NPDES permit, non-storm water discharges are illegal. Generally, non-storm water discharges are issued individual NPDES permits based on application Form 2C (for process wastewater) or Form 2E (for nonprocess wastewater). However, EPA's General Permit authorizes the following types of non-storm water discharges:

- Discharges from fire fighting activities

- Fire hydrant flushings

- Potable water sources including waterline flushings

- Irrigation drainage

- Lawn watering

- Uncontaminated ground water

- Foundation or footing drains where flows are not contaminated with process materials

- Discharges from springs

- Routine exterior building washdown which does not use detergents or other compounds

- Pavement wash waters where spills or leaks of toxic or hazardous materials have not occurred and where detergents are not used

- Air conditioning condensate.

Be sure to examine your facility's storm water permit to determine whether it authorizes any of these or other non-storm water discharges. If your permit does not authorize non-storm water discharges occurring at your facility, you should contact your permitting authority or the Storm Water Hotline for more information about how to address these discharges.

EPA GENERAL PERMIT REQUIREMENTS

Non-Storm Water Discharges

Part IV.D.3.g.(2).

Except for flows from fire fighting activities, sources of non-storm water that are authorized by this permit must be identified in the plan. The plan shall identify and ensure the implementation of appropriate pollution prevention measures for the non-storm water component of the discharge.

Generally, except for flows from fire fighting activities, all non-storm water connections that are identified and that are authorized by your storm water discharge permit should be identified in the Storm Water Pollution Prevention Plan. Where necessary to minimize pollutants in these discharges, pollution prevention measures should be adopted and implemented. The pollution potential from these sources can be significantly reduced where a conscious effort is taken to control them.

2.2.5 Storm Water Monitoring Data

EPA GENERAL PERMIT REQUIREMENTS

Sampling Data

Part IV.D.2.d.

Include a summary of any existing discharge sampling data describing pollutants in storm water discharges from the facility and a summary of sampling data collected during the term of this permit.

Storm water sampling data provide information that describes the quality of storm water discharges. These data are valuable because they indicate the potential environmental risk of the discharge by identifying the types and amounts of pollutants present. In addition, these data can be used to identify potential sources of storm water pollution.

During the site assessment phase, permittees should collect and summarize any storm water sampling data that were collected in the past. Historical storm water monitoring data may be very useful in locating areas which have previously contributed pollutants to storm water discharges and identifying what the problem pollutants are. In your summary of these data, describe the sample collection procedures used. Be sure to cross-reference the particular storm water outfall sampled to one of the outfalls designated on your site map.

Although some permittees may not have to conduct storm water sampling under the permit that is issued to that facility, incorporation of these data into the Storm Water Pollution Prevention Plan as it is collected will provide a basis for evaluating the effectiveness of the plan. Under EPA's General Permit, certain classes of facilities are required to conduct storm water sampling either annually or semiannually throughout the term of the permit. Appendix J contains a table summarizing these sampling requirements, including the parameters for which analysis is required and the sampling frequency. State-issued storm water general permits may include similar provisions. Generally, where sampling is required, facilities must collect and analyze grab and composite samples in accordance with the protocol established in 40 CFR Part 136. EPA has published a guidance manual addressing storm water sampling requirements and procedures for NPDES storm water discharge permit applications. Although directed toward application requirements, the guidance manual contains information that would be of assistance to facilities required to sample under a storm water general permit. To obtain a copy of the manual, call the Storm Water Hotline at (703) 821-4823.

2.2.6 Assessment Summary

```
┌─────────────────────────────────────────────────────────────────────┐
│                    EPA GENERAL PERMIT REQUIREMENTS                    │
├─────────────────────────────────────────────────────────────────────┤
│          Risk Identification and Summary of Potential Pollutant Sources │
│                             Part IV.D.2.e.                            │
│                                                                       │
│  Include in your plan a narrative description of the potential pollutant sources and identify any │
│  pollutant of concern that may be generated by the following activities at your facility: │
│                                                                       │
│    • Loading and unloading operations                                 │
│                                                                       │
│    • Outdoor storage activities                                       │
│                                                                       │
│    • Outdoor manufacturing or processing activities                   │
│                                                                       │
│    • Significant dust or particulate generating activities            │
│                                                                       │
│    • Onsite waste disposal practices.                                 │
└─────────────────────────────────────────────────────────────────────┘
```

Once you have completed the above steps in your pollutant source assessment, you should have enough information to determine which areas, activities or materials may contribute pollutants to storm water runoff from your site. With this information, you can select the most appropriate BMPs to prevent or control pollutants from these areas.

The following paragraph is an example of how you can analyze the information you have gathered and start to figure out what you can do to correct these problems:

> In a particular drainage area, you have a vehicle maintenance facility area where oil filters are stored outdoors. You found that no material management practices were currently being used to protect the used filters from contact with storm water. You would then suspect that the storm water draining from that area would most likely contain a significant amount of oil and grease. Therefore, you have concluded that you need to do something to reduce the possibility of oil and grease mixing with storm water.

EPA's General Permit requires this type of narrative description summarizing any potential source of storm water pollutants, and what types of pollutants have already been or may be found in storm water runoff from the site.

Worksheet #7 (located at the end of Chapter 2) will help you organize the pollutant sources that you identified during the site assessment phase, relate them to management practices that you already have in place, and list potential new BMP options to address remaining pollutant sources.

```
┌─────────────────────────────────┐
│   2.1  PLANNING AND ORGANIZATION │
└─────────────────────────────────┘
                │
                ▼
┌─────────────────────────────────┐
│   2.2  ASSESSMENT PHASE          │
└─────────────────────────────────┘
                │
                ▼
┌─────────────────────────────────┐
│   2.3  BMP IDENTIFICATION PHASE  │
│   • Baseline BMPs                │
│   • Select activity- and site-specific │
│     BMPs (Ch. 3 and Ch. 4)       │
└─────────────────────────────────┘
                │
                ▼
┌─────────────────────────────────┐
│   2.4  IMPLEMENTATION PHASE      │
└─────────────────────────────────┘
                │
                ▼
┌─────────────────────────────────┐
│   2.5  EVALUATION/MONITORING     │
└─────────────────────────────────┘

┌──────────────┐   ┌──────────────┐
│ 2.6 GENERAL  │   │ 2.7 SPECIAL  │
│ REQUIREMENTS │   │ REQUIREMENTS │
└──────────────┘   └──────────────┘
```

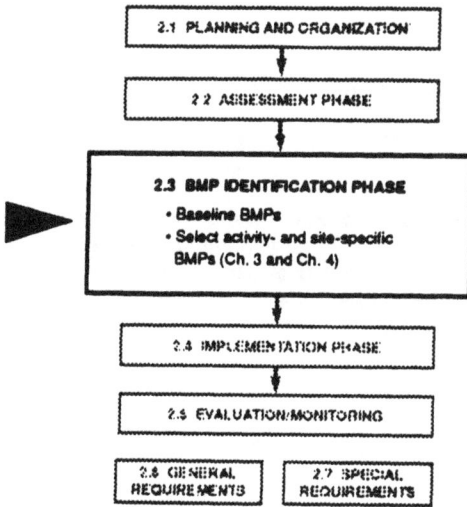

2.3 BMP IDENTIFICATION PHASE

Once you have identified and assessed potential and existing sources of contamination to storm water at your facility, the next step is to select the proper measures or BMPs that will eliminate or reduce pollutant loadings in storm water discharges from your facility site. Specifically, your plan design will include the following BMPs:

- Good housekeeping

- Preventive maintenance

- Visual inspections

- Spill prevention and response

- Sediment and erosion control

- Management of runoff

- Employee training

- Recordkeeping and reporting

- Other BMPs as appropriate

BMPs are measures used to prevent or mitigate pollution from any type of activity. BMPs are a very broad class of measures and may include processes, procedures, schedules of activities, prohibitions on practices, and other management practices to prevent or reduce water pollution. In essence, they are anything a plant manager, department foreman, environmental specialist, consultant or employee may identify as a method, short of actual treatment, to curb water pollution. They may be inexpensive or costly. BMPs can be just about anything that "does the job" of preventing toxic or hazardous substances from entering the environment.

The purpose of this section is to describe the "baseline" BMPs that you must include in your facility's storm water pollution prevention program and offer some guidelines about how to select more "advanced" BMPs that are tailored to the specific pollutant sources on your particular site. With this information, you should be able to design a storm water management program that best addresses any problems with runoff from your facility's site.

2.3.1 Baseline Best Management Practices

<table>
<tr><td>

EPA GENERAL PERMIT REQUIREMENTS

Measures and Controls

Part IV.D.3.

Each facility covered by this permit shall develop a description of storm water management controls appropriate for the facility and implement such controls. The appropriateness and priorities of controls in a plan shall reflect identified potential sources of pollutants at the facility. The description of storm water management controls shall address the following minimum components, including a schedule for implementing such controls:

- Good Housekeeping

- Preventive Maintenance

- Visual Inspections

- Spill Prevention and Response

- Sediment and Erosion Control

- Management of Runoff

- Employee Training (see Section 2.4.2)

- Recordkeeping and Reporting (see Section 2.5.2)

</td></tr>
</table>

"Baseline" BMPs are practices that are inexpensive, relatively simple, and applicable to a wide variety of industries and activities. Most industrial facilities already have these measures in place for product loss prevention, accident and fire prevention, worker health and safety, or to comply with other environmental regulations. The purpose of this section is to highlight how these common practices can be improved and tailored to prevent storm water pollution. EPA's Storm Water Program is emphasizing these generic measures because they can be effective, are cost-effective, and because they emphasize prevention over treatment.

Industrial facilities must implement, at a minimum, the above-listed eight baseline BMPs, where appropriate. How each of these BMPs can prevent storm water pollution is described in detail below.

Worksheet #7a (located at the end of Chapter 2) is designed to help you list the specific activities or practices that you select to include in your plan for each of the baseline BMPs.

Good Housekeeping

EPA GENERAL PERMIT REQUIREMENTS

Good Housekeeping

Part IV.D.3.a.

Good housekeeping requires the maintenance of areas which may contribute pollutants to storm water discharges in a clean, orderly manner.

Good housekeeping practices are designed to maintain a clean and orderly work environment. Often the most effective first step towards preventing pollution in storm water from industrial sites simply involves using good common sense to improve the facility's basic housekeeping methods. Poor housekeeping can result in more waste being generated than necessary and an increased potential for storm water contamination. A clean and orderly work area reduces the possibility of accidental spills caused by mishandling of chemicals and equipment and should reduce safety hazards to plant personnel. Well maintained material and chemical storage areas will reduce the possibility of storm water mixing with pollutants.

There are some simple procedures a facility can use to promote good housekeeping, including improved operation and maintenance of industrial machinery and processes, material storage practices, material inventory controls, routine and regular clean-up schedules, maintaining well organized work areas, and educational programs for employees about all of these practices. The following sections describe these good housekeeping procedures and provide a checklist that you can use to evaluate and improve your facility's storm water pollution prevention program.

Operation and Maintenance

These practices ensure that processes and equipment are working well. Improved operation and maintenance practices are easy to implement. Here are a few examples of basic operation and maintenance BMPs that should be incorporated in your good housekeeping program:

- Maintain dry and clean floors and ground surfaces by using brooms, shovels, vacuum cleaners, or cleaning machines

- Regularly pickup and dispose of garbage and waste material

- Make sure equipment is working properly (see Section 2.3.4 on preventive maintenance)

- Routinely inspect for leaks or conditions that could lead to discharges of chemicals or contact of storm water with raw materials, intermediate materials, waste materials, or products (see Visual Inspection BMP below)

- Ensure that spill cleanup procedures are understood by employees (see Spill Prevention and Response BMP below).

Material Storage Practices

Improper storage can result in the release of materials and chemicals that can cause storm water runoff pollution. Proper storage techniques include:

- Providing adequate aisle space to facilitate material transfer and easy access for inspections

- Storing containers, drums, and bags away from direct traffic routes to prevent accidental spills (see Spill Prevention and Response BMP below)

- Stacking containers according to manufacturers' instructions to avoid damaging the containers from improper weight distribution

- Storing containers on pallets or similar devices to prevent corrosion of the containers which can result when containers come in contact with moisture on the ground

- Assigning the responsibility of hazardous material inventory to a limited number of people who are trained to handle hazardous materials.

Material Inventory Procedures

Keeping an up-to-date inventory of all materials (hazardous and non-hazardous) present on your site will help to keep material costs down caused by overstocking, track how materials are stored and handled onsite, and identify which materials and activities pose the most risk to the environment. The following instructions explain the basic steps to completing a material inventory. **Worksheets #3 and 3A** provide an example of the types of information you should collect while conducting the inventory.

- **Identify all chemical substances present in the workplace.** Walk through the facility and review the purchase orders for the previous year. List all of the chemical substances used in the workplace, and then obtain the Material Safety Data Sheet (MSDS) for each.

- **Label all containers** to show the name and type of substance, stock number, expiration date, health hazards, suggestions for handling, and first aid information. This information can usually be found on the MSDS. Unlabeled chemicals and chemicals with deteriorated labels are often disposed of unnecessarily or improperly.

- **Clearly mark** on the inventory hazardous materials that require special handling, storage, use, and disposal considerations.

Improved material tracking and inventory practices, such as instituting a shelf-life program, can reduce the waste that results from overstocking and the disposal of out-dated materials. Careful tracking of all materials ordered may also result in more efficient materials use.

Decisions on the amount of hazardous materials the facility stores should include an evaluation of your emergency control systems. Ensure that storage areas are designed to contain spills.

Employee Participation

Frequent and proper training of employees in good housekeeping techniques reduces the possibility that the chemicals or equipment will be mishandled. Motivating employees to reduce waste generation is another important pollution prevention technique. Section 2.4.2 provides more information on employee training programs. Here are some suggestions for involving employees in good housekeeping practices:

- Incorporate information sessions on good housekeeping practices into the facility's employee training program

- Discuss good housekeeping at employee meetings

- Publicize pollution prevention concepts through posters

- Post bulletin boards with updated good housekeeping procedures, tips and reminders.

Good Housekeeping Checklist
☐ Is good housekeeping included in the storm water pollution prevention program?
☐ Are outside areas kept in a neat and orderly condition?
☐ Is there evidence of drips or leaks from equipment or machinery onsite?
☐ Is the facility orderly and neat? Is there adequate space in work areas?
☐ Is garbage removed regularly?
☐ Are walkways and passageways easily accessible, safe, and free of protruding objects, materials or equipment?
☐ Is there evidence of dust on the ground from industrial operations or processes?
☐ Are cleanup procedures used for spilled solids?
☐ Is good housekeeping included in the employee program?
☐ Are good housekeeping procedures and reminders posted in appropriate locations around the workplace?
☐ Are there regular housekeeping inspections?

Preventive Maintenance

EPA GENERAL PERMIT REQUIREMENTS

Preventive Maintenance

Part IV.D.3.b.

Your preventive maintenance program must include:

- Timely inspection and maintenance of storm water management devices (e.g., cleaning oil/water separators, catch basins)

- Inspection and testing of facility equipment and systems to uncover conditions that could cause breakdowns or failures resulting in discharges of pollutants to surface waters

- Proper maintenance of facility equipment and systems.

Most plants already have preventive maintenance programs that provide some degree of environmental protection. The program you undertake as part of the Storm Water Pollution Prevention Plan should not just duplicate previous efforts, but should expand the current preventive maintenance programs to include storm water considerations, especially the upkeep and maintenance of storm water management devices. The pollution prevention team should evaluate the existing plant preventive maintenance program and recommend any necessary changes.

Preventive maintenance involves the regular inspection and testing of plant equipment and operational systems (see Visual Inspections description below). These inspections should uncover conditions such as cracks or slow leaks which could cause breakdowns or failures that result in discharges of chemicals to storm sewers and surface waters. The program should prevent breakdowns and failures by adjustment, repair or replacement of equipment. An effective preventive maintenance program should therefore include the following elements:

- Identification of equipment, systems, and facility areas that should be inspected

- Schedule for periodic inspections or tests of these equipment and systems

- Appropriate and timely adjustment, repair or replacement of equipment and systems

- Maintenance of complete records on inspections, equipment, and systems.

Identification of Equipment to Inspect

The first step is to identify which systems or equipment may malfunction and cause spills, leaks, or other situations that could lead to storm water runoff contamination. Look back at what sources of potential storm water contamination were identified during the pollutant source assessment phase. The following list identifies some types of equipment to include in your preventive maintenance inspection and testing program:

```
┌─────────────────────────────────────────────┐
│            Equipment to Inspect              │
│                                              │
│   •  Pipes                                   │
│                                              │
│   •  Pumps                                   │
│                                              │
│   •  Storage tanks and bins                  │
│                                              │
│   •  Pressure vessels                        │
│                                              │
│   •  Pressure release valves                 │
│                                              │
│   •  Process and material handling equipment │
│                                              │
│   •  Storm water management devices (oil/    │
│      water separators, catch basins, or other│
│      structural or treatment BMPs).          │
└─────────────────────────────────────────────┘
```

Schedule Routine Preventive Maintenance Inspections

Once you have identified which equipment and areas to inspect at your facility, set schedules for routine inspections. Include examination for leaks, corrosion, support or foundation failure, or other forms of deterioration or leaks in your inspection. Look for spots or puddles of chemicals and document any detection of smoke, fumes, or other signs of leaks. Periodic testing of plant equipment for structural soundness is a key element of preventive maintenance. This can be done by making sure storage tanks are solid and strong enough to hold materials. Another important consideration is when and how often preventive maintenance inspections should be conducted to ensure that this practice is effective. Smaller facilities with little equipment and few systems may still find it necessary to conduct frequent inspections if the equipment is older and more susceptible to leaks or other discharges. Preventive maintenance inspections may be conducted as part of your regular visual inspections.

Equipment Repair or Replacement

Promptly repair or replace defective equipment found during inspections and testings. Keeping spare parts for equipment that needs frequent repair is another simple practice that can help avoid problems and equipment down-time.

Records on Preventive Maintenance

Include a suitable records system for scheduling tests and documenting inspections in the preventive maintenance program. Record test results and follow up with corrective action. Make sure records are complete and detailed. These records should be kept with other visual inspection records.

EPCRA, Section 313 Facility Preventive Maintenance Inspection Requirements

EPA's General Permit contains additional preventive maintenance inspection requirements for facilities subject to reporting under EPCRA, Section 313 for water priority chemicals [Part IV.D.7.b.(7).]. For these facilities, all areas of the facility must be inspected for the following at appropriate intervals as specified in the plan:

- Leaks or conditions that would lead to discharges of Section 313 water priority chemicals

- Conditions that could lead to direct contact of storm water with raw materials, intermediate materials, waste materials or products

- Examine piping, pumps, storage tanks and bins, pressure vessels, process and material handling equipment, and material bulk storage areas for leaks, wind blowing, corrosion, support or foundation failure, or other deterioration or noncontainment.

These inspections must occur at intervals based on facility design and operational experience, and the timing must be specified in the plan.

When a leak or other threatening condition is found, corrective action must be taken immediately or the facility unit or process must be shut down until the problem is repaired.

Visual Inspections

EPA GENERAL PERMIT REQUIREMENTS

Visual Inspections

Part IV.D.3.d.

- Identify qualified plant personnel who will inspect plant equipment and areas at appropriate intervals in the plan

- Track results of inspections to ensure that appropriate actions are taken

- Maintain records of all inspections.

Preventing pollution of storm water runoff from your facility requires good housekeeping in areas where materials are handled, stored, or transferred and preventive maintenance of process equipment and systems. Such practices are described in detail above and should be outlined in your Storm Water Pollution Prevention Plan. Regular visual inspections are your means to ensure that all of the elements of the plan are in place and working properly.

Routine visual inspections are not meant to be a comprehensive evaluation of the entire storm water pollution prevention program—that is the function of the Annual Site Inspection and Site Evaluation described in Section 2.5.1 below. Rather, they are meant to be a routine look-over of the facility to identify conditions which may give rise to contamination of storm water runoff with pollutants from your facility.

Every facility is different, so it is up to the facility owner/operator to determine what areas of your facility could potentially contribute pollutants to storm water runoff, and to devise and implement a visual inspection program based on this information. The visual inspection is simply a way to confirm that the measures chosen are in place and working and should periodically take place during storm events. The frequency of visual inspection should be determined by the types and amounts of materials handled at the facility, existing BMPs at the facility, and any other factors that may be relevant, such as the age of the facility (in general, older facilities should be inspected at more frequent intervals than new facilities). The following lists identify some types of equipment and plant areas to include in your Visual Inspections and preventive maintenance plan:

Areas to Inspect

- Areas around all of equipment listed in Preventive Maintenance box

- Areas where spills and leaks have occurred in the past

- Material storage areas (tank farms, drum storage)

- Outdoor material processing areas

- Material handling areas (e.g., loading, unloading, transfer)

- Waste generation, storage, treatment and disposal areas.

Implementation of a Visual Inspection Plan

The best plan is a simple one, and this includes the visual inspection plan - there is no reason for it to be highly technical, complicated or labor-intensive. If your facility already has a routine surveillance program in place, consider expanding it to include the visual inspection element of your Storm Water Pollution Prevention Plan. For example, if your facility has a security surveillance program, you might consider training facility security personnel to perform the visual inspection program. If your facility has no routine surveillance or inspection program already in place, then a plan must be developed and people must be assigned the responsibility for carrying the inspections out. It is important to remember that the employees carrying out the visual inspection program should be properly trained, familiar with the storm water pollution prevention program, and knowledgeable about proper recordkeeping and reporting procedures.

Records of Inspections

The most important thing for you to remember here is to document all inspections. Inspection records should note when inspections were done, who conducted the inspection, what areas were inspected, what problems were found, and steps taken to correct any problems, including who has been notified. Many industrial facilities will already have some sort of incident reporting procedure in place — existing incident reporting and security surveillance procedures could easily be incorporated into the Storm Water Pollution Prevention Plan. These records should be kept with the plan. EPA's General Permit requires that records be kept until at least one year after coverage under the permit expires.

Visual Inspection Checklist

Do you see:

☐ Corroded drums or drums without plugs or covers

☐ Corroded or damaged tanks, tank supports, and tank drain valves

☐ Torn bags or bags exposed to rain water

☐ Corroded or leaking pipes

☐ Leaking or improperly closed valves and valve fittings

☐ Leaking pumps and/or hose connections

☐ Broken or cracked dikes, walls or other physical barriers designed to prevent storm water from reaching stored materials

☐ Windblown dry chemicals

☐ Improperly maintained or overloaded dry chemical conveying systems.

Spill Prevention and Response

EPA GENERAL PERMIT REQUIREMENTS

Spill Prevention and Response

Part IV.D.3.c.

- Identify areas where spills can occur onsite and their drainage points

- Specify material handling procedures, storage requirements, and use of equipment such as diversion valves, where appropriate

- Identify procedures used for cleaning up spills and inform personnel about these procedures

- Provide the appropriate spill clean-up equipment to personnel.

Spills and leaks together are one of the largest industrial sources of storm water pollutants, and in most cases are avoidable. Establishing standard operating procedures such as safety and spill prevention procedures along with proper employee training can reduce these accidental releases. Avoiding spills and leaks is preferable to cleaning them up after they occur, not only from an environmental standpoint, but also because spills cause increased operating costs and lower productivity.

Development of spill prevention and response procedures is a very important element of an effective Storm Water Pollution Prevention Plan. A spill prevention and response plan may have already been developed in response to other environmental regulatory requirements. If your facility already has a spill prevention and response plan, it should be evaluated and revised if necessary to address the objectives of the Storm Water Pollution Prevention Plan.

The next section outlines the steps you should take to identify and characterize potential spills, to eliminate or reduce spill potential, and how to respond when spills occur.

Identify Potential Spill Areas

As part of the Assessment Phase of developing the Storm Water Pollution Prevention Plan, you should have created a list or inventory of materials handled, used, and disposed of. A site map indicating the drainage area of each storm water outfall was also created. Now overlay the drainage area map with the locations of areas and activities with high material spill potential to determine where spills will most likely occur. Spill potential also depends on how materials are handled, the types and volumes of materials handled, and how materials are stored on your site. You must describe these factors in your plan.

The activities and areas where spills are likely to occur on your site are listed and described below:

- Loading and unloading areas

- Storage areas

- Process activities

- Dust or particulate generating processes

- Waste disposal activities.

Loading and unloading areas have a high spill potential because the nature of the activity involves transfer of materials from one container to another. The spill potential is affected by the integrity of the container, the form of the chemical being transferred, the design of the transfer area (bermed vs. direct connection to the storm water collection system), the proximity of this area to the storage area, and procedures for loading and unloading. Evaluate the spill potential from all loading and unloading equipment, such as barges, railroad cars, tank trucks, and front end loaders, as well as storage and vehicle wash areas.

Storage areas, both indoor and outdoor, are potential spill areas. Outdoor storage areas are exposed to storm water runoff and may provide direct contact between potential pollutants and storm water. Indoor storage areas may contaminate storm water if the drains in the storage area are connected to the storm sewer or if improper clean up procedures in the event of a spill are used. This evaluation should consider the type, age, and condition of storage containers and structures (including tanks, drums, bags, bottles). An evaluation of the spill potential of storage areas should also focus on how employees handle materials.

All process areas are potential sources of storm water contamination if the floor drains in these areas are connected to storm sewers (see Section 2.2.4). If these drains cannot be sealed, the process area should be evaluated for the adequacy of spill control structures such as secondary containment, if necessary. One should also consider normal housekeeping procedures. Some process areas are hosed down periodically and the resulting wash water contains pollutants. Outdoor process activities may contaminate storm water if spills are diverted to the storm sewer.

Also, evaluate spill potential from the following stationary facilities:

- Manufacturing areas

- Warehouses

- Chemical processing and or blending areas

- Temporary and permanent storage sites

- Power generating facilities

- Food processing areas

- Tank farms

- Service stations

- Parking lots

- Access roads.

Also evaluate the possibility of storm water contamination from underground sources, such as tanks and pipes. Leaking underground storage tanks are often a source of storm water contamination.

In addition to identifying these and other potential spill areas, projecting possible spill volume and type of material is critical to developing the correct response procedures for a particular area.

Specify Material Handling Procedures and Storage Requirements

Through the process of developing various spill scenarios, ideas for eliminating or minimizing the spill or its impact will emerge. These solutions should be prioritized and adopted according to conditions of effectiveness, cost, feasibility, and ease of implementation. Following is a list of some suggested activities or alterations that may be made to reduce the potential that spills will occur or impact storm water quality:

- Develop ways to recycle, reclaim and/or reuse process materials to reduce the volume brought into the facility

- Install leak detection devices, overflow controls, and diversion berms

- Disconnect drains from processing areas that lead to the storm sewer (however, be sure that any such action would not create a health hazard within your facility)

- Adopt effective housekeeping practices

- Adopt a materials flow/plant layout plan (i.e., do not store bags that are easily punctured near high-traffic areas where they may be hit by moving equipment or personnel)

- Perform regular visual inspections to identify signs of wear on tanks, drums, containers, storage shelves, and berms and to identify sloppy housekeeping or other clues that could lead to potential spills

- Perform preventive maintenance on storage tanks, valves, pumps, pipes, and other equipment

- Use filling procedures for tanks and other equipment that minimize spills

- Use material transfer procedures that reduce the chance of leaks or spills

- Substitute less or non-toxic materials for toxic materials

- Ensure appropriate security.

Identify Spill Response Procedures and Equipment

In the event that spill prevention measures fail, a swiftly executed response may prevent contamination of storm water. Spill response plans are required by numerous programs for various reasons. However, this may be the first time that a spill response plan specifically addresses protection of storm water quality.

Past experience has shown that the single most important obstacle to an effective spill response plan is its implementation. Develop the plan with its ease of implementation in mind. The spill response procedures should be clear, concise, step-by-step instructions for responding to the spill events at a particular facility. Organize the plan to facilitate rapid identification of the appropriate set of procedures. For example, you may find that the plan works best for your facility when organized by spill location. Another possible method of organization is by spilled material. The key component to implementation is the ability of employees to use the plan quickly and effectively. The specific approach you take will depend on the specific conditions at your facility such as size, number of employees and the spill potential of the site.

The spill response plan is developed based on the spill potential scenarios identified. It reflects a consideration of the potential magnitude and frequency of spills, of the types of materials spilled, and of the variety of potential spill locations. Specific procedures may be needed to correspond to particular chemicals onsite. At all times during the operation of a facility, personnel with appropriate training and authority should be available to respond to spills.

The spill response plan should describe:

- Identification of spill response "team" responsible for implementing the spill response plan.

- Safety measures.

- Procedures to notify appropriate authorities providing assistance [police, fire, hospital, Publicly Owned Treatment Works (POTW), etc.].

- Spill containment, diversion, isolation, cleanup.

- Spill response equipment including:

  - Safety equipment such as respirators, eye guards, protective clothing, fire extinguisher, and two-way radios.

  - Cleanup equipment such as booms, barriers, sweeps, adsorbents, containers, etc.

Following any spills, evaluate how the prevention plan was successful or unsuccessful in responding and how it can be improved.

EPCRA, Section 313, Facility Spill Prevention and Response Requirements

EPA's General Permit sets forth more specific requirements for facilities subject to reporting under EPCRA, Section 313 for water priority chemicals [Part IV.D.7.b.(7).]. When a leak or spill of a Section 313 water priority chemical has occurred, the contaminated soil, material, or debris must be removed promptly and disposed of in accordance with Federal, State, and local requirements and as described in the Storm Water Pollution Prevention Plan.

These facilities are also required to designate a person responsible for spill prevention, response, and reporting procedures (see Section 2.1.1, Pollution Prevention Team).

Sediment and Erosion Control

EPA GENERAL PERMIT REQUIREMENTS

Sediment and Erosion Control

Part IV.D.3.h.

Identify areas which, due to topography, activities, or other factors, have a high potential for significant soil erosion, and identify structural, vegetative, and/or stabilization measures to be used to limit erosion.

There may be certain areas on your site which, due to construction activities, steep slopes, sandy soils, or other reasons, are prone to soil erosion. Construction activities typically remove grass and other protective ground covers resulting in the exposure of underlying soil to wind and rain. Similarly, steep slopes or sandy soils may not be able to hold plant life so that soils are exposed. Because the soil surface is unprotected, dirt and sand particles are easily picked up by wind and/or washed away by rain. This process is called erosion. Erosion can be controlled or prevented with the use of certain BMPs. A number of these measures are described in Chapter 4.

Management of Runoff

EPA GENERAL PERMIT REQUIREMENTS

Management of Runoff

Part IV.D.3.i.

The plan shall contain a narrative consideration of the appropriateness of traditional storm water management practices (practices other than those which control the source of pollutants) used to divert, infiltrate, reuse, or otherwise manage storm water runoff in a manner that reduces pollutants in storm water discharges from the site. The plan shall provide that measures determined to be reasonable and appropriate shall be implemented and maintained. The potential of various sources at the facility to contribute pollutants to storm water discharges associated with industrial activity (see Part IV.D.2. (description of potential pollutant sources) of this permit) shall be considered when determining reasonable and appropriate measures. Appropriate measures may include: vegetative swales and practices, reuse of collected storm water (such as for a process or as an irrigation source), inlet controls (such as oil/water separators), snow management activities, infiltration devices, and wet detention/retention devices.

Many BMPs discussed in this chapter are measures to reduce pollutants at the source before they have an opportunity to contaminate storm water runoff. Traditional storm water management practices also can be used to direct storm water away from areas of exposed materials or potential pollutants. Further, traditional storm water management practices can be used to direct storm water that contains pollutants to natural or other types of treatment locations. For example, using an oil/water separator on storm water that has oil and grease in it will take out some of the oil and grease before the storm water leaves the site. Permits will generally not require specific storm water management practices since these practices must be selected on a case-by-case basis depending on the activities at your site and the amount of space you have available.

Chapter 4 provides descriptions of several traditional storm water management practices. Additional sources of information are listed in Appendix A.

2.3.2 Advanced Best Management Practices

In addition to those BMPs that should be routinely incorporated into your storm water prevention pollution plan, you may need to implement some "advanced" BMPs that are specifically directed to address particular pollutant sources or activities on your site. As discussed in Chapters 3 and 4, these BMPs must be tailored to address specific problems.

In determining which BMPs represent the Best Available Technology Economically Achievable (BAT), the following factors are considered: (1) the age of equipment and facilities involved; (2) the process employed; (3) the engineering aspects of the application of various types of control techniques; (4) process changes; (5) the cost of achieving effluent reduction; and (6) non-water quality environmental impact (including energy requirements).

BMP Cost and Effectiveness

The costs of implementing the BMPs described in this manual vary depending upon many factors and site-specific conditions. In general, the required baseline BMPs are relatively low in cost when compared with more traditional storm water treatment or highly engineered controls. Costs also vary depending upon the size of the facility, the number of employees, the types of chemicals or raw materials stored or used, and the nature of plant operations. However, because many of the baseline practices are widely accepted and considered "common sense" or standard good operating practices, many facilities have them in place.

Because BMP effectiveness is also site-specific, this manual does not attempt to provide specific guidance on this matter.

Reduce, Reuse, Recycle

As described in Chapter 1, EPA encourages industrial facilities to choose practices that prevent the contamination of storm water rather than treat it once it is polluted. Use of the Storm Water Management Hierarchy (see Table 2.1) as a tool to help select BMPs for your program will help you discover how to prevent pollution and avoid its associated costs and liabilities while meeting the environmental goals of EPA's Storm Water Program.

When selecting a BMP for your storm water management program, EPA recommends that you choose practices that eliminate or reduce the amount of pollutants generated on your site. This practice is referred to as "source reduction." When it is impossible, select options that recycle or reuse the storm water in your industrial processes, or those that reduce the need to store and expose more hazardous materials to storm water by recycling or recovering used materials. Treating storm water to remove pollutants before they leave the site is the next best option, although this often just transfers the problem from one place or medium to another. Table 2.1, below, provides examples of BMPs that are representative of the different types of storm water management.

TABLE 2.1 CLASSIFICATION OF STORM WATER BMPs

Storm Water Management Hierarchy	Example BMPs
Source Reduction	• Preventive maintenance • Spill prevention • Chemical substitution • Housekeeping • Training • Materials management practices
Containment/Diversion	• Segregating the activity of concern • Covering the activity • Berming the activity • Diverting flow to grassed area • Dust control
Recycling	• Recycling
Treatment	• Oil/water separator • Vegetated swale • Storm water detention pond

2.3.3 Completing the BMP Identification Phase

When you started designing your pollution prevention plan, you assembled certain crucial pieces of information:

* A list of actual and potential storm water discharge problems

* The location of each outfall on a site map showing the drainage route from your property

* A list of the management plans and practices that are already in place at your facility

* Information contained in this manual on "baseline" BMPs and "advanced" BMPs for resolving storm water problems.

At the completion of the BMP identification phase, you should have accomplished the following:

* Reviewed your current management plans and practices to assess their effectiveness in addressing storm water discharges on your site.

* Scheduled the implementation of "baseline" BMPs and whatever "advanced" BMPs were necessary to effectively eliminate storm water pollution problems at your site.

* Determined what to do about any identified, unpermitted connections of non-storm water discharges to separate storm sewers. Your options were to:

  - Discontinue any connections of non-storm water discharges to a separate storm sewer system

  - Obtain an NPDES permit for the non-storm water discharge.

- Identified options for addressing any unresolved storm water discharge problems.

- Gained management approval and acceptance of the plan.

2.1 PLANNING AND ORGANIZATION
2.2 ASSESSMENT PHASE
2.3 BMP IDENTIFICATION PHASE
2.4 IMPLEMENTATION PHASE • Implement BMPs • Train employees
2.5 EVALUATION/MONITORING

2.6 GENERAL REQUIREMENTS	2.7 SPECIAL REQUIREMENTS

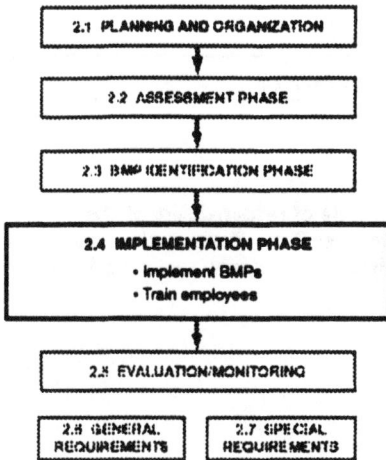

2.4 IMPLEMENTATION PHASE

At this point, you have designed your Storm Water Pollution Prevention Plan and the plan has been approved by facility management. This next section of the manual will guide you through the next major phase in the planning process—implementation. Specifically, you will:

• Implement the selected storm water BMPs

• Train all employees to carry out the goals of the plan.

2.4.1 Implement Appropriate Controls

EPA GENERAL PERMIT REQUIREMENTS
Implementation
Part IV.D.
Facilities must implement the provisions of the storm water pollution prevention plan as a condition of EPA's general permit. The plan shall include a schedule for implementing identified storm water management controls.

Implementing your plan will involve several steps:

• Develop a schedule for implementation. For example, your schedule might include a deadline for putting improved housekeeping measures into practice. Should implementation involve certain types of modifications to your site (e.g., any construction), you will need to account for the time required to secure any necessary local or State permits.

• Assign specific individuals with responsibility for implementing aspects of the plan and/or monitoring implementation.

• Ensure that management approves of your implementation schedule and strategy and schedule regular times for reporting progress to management.

Worksheet #8 (located at the end of Chapter 2) will help you list the schedule for implementation of your facility's plan.

2.4.2 Employee Training

EPA GENERAL PERMIT REQUIREMENTS

Employee Training

Part IV.D.3.e.

Employee training programs must inform personnel at all levels of responsibility of the components and goals of the Storm Water Pollution Prevention Plan. Training should address each component of your pollution prevention plan, including how and why tasks are to be implemented. Topics will include:

- Spill prevention and response

- Good housekeeping

- Material management practices.

The pollution prevention plan must specify how often training is conducted.

Employee training is essential to effective implementation of the Storm Water Pollution Prevention Plan. The purpose of a training program is to teach personnel at all levels of responsibility the components and goals of the Storm Water Pollution Prevention Plan. When properly trained, personnel are more capable of preventing spills, responding safely and effectively to an accident when one occurs, and recognizing situations that could lead to storm water contamination.

The following sections include ideas about how to create an effective storm water pollution prevention training program for your facility.

Worksheet #9 (located at the end of Chapter 2) is designed to help you organize your employee training program.

Spill Prevention and Response

Spill prevention and response procedures are described in detail in Section 2.3.1. Discuss these procedures or plans in the training program in order to ensure all plant employees, not just those on the spill response teams, are aware of what to do if a spill occurs. Specifically, all employees involved in the industrial activities of your facility should be trained about the following measures:

- Identifying potential spill areas and drainage routes, including information on past spills and causes

- Reporting spills to appropriate individuals, without penalty (e.g., employees should be provided "amnesty" when they report such instances)

- Specifying material handling procedures and storage requirements

- Implementing spill response procedures.

Onsite contractors and temporary personnel should also be informed of the plant operations and design features in order to help prevent accidental discharges or spills from occurring.

Good Housekeeping

Also, teach facility personnel how to maintain a clean and orderly work environment. Section 2.3.1 above outlines the steps for practicing good housekeeping. Emphasize these points in the good housekeeping portion of your training program:

- Require regular vacuuming and/or sweeping

- Promptly clean up spilled materials to prevent polluted runoff

- Identify places where brooms, vacuums, sorbents, foams, neutralizing agents, and other good housekeeping and spill response equipment are located

- Display signs reminding employees of the importance and procedures of good housekeeping

- Discuss updated procedures and report on the progress of practicing good housekeeping at every meeting

- Provide instruction on securing drums and containers and frequently checking for leaks and spills

- Outline a regular schedule for housekeeping activities to allow you to determine that the job is being done.

Materials Management Practices

- Neatly organize materials for storage

- Identify all toxic and hazardous substances stored, handled, and produced onsite

- Discuss handling procedures for these materials.

Tools For a Successful Training Program

Here are some suggestions of training tools that you can include in your facility's training program:

- Employee handbooks

- Films and slide presentations

- Drills

- Routine employee meetings

- Bulletin boards

- Suggestion boxes

- Newsletters

- Environmental excellence awards or other employee incentive programs.

Providing employees with incentives, such as awards for practicing pollution prevention, is a good way to motivate personnel in working to achieve the goals of the Storm Water Pollution Prevention Plan.

How Often to Conduct Training

You should examine your plan to determine how often you should train the employees at your facility. Frequency should take into account the complexity of your management practices and the nature of your staff, including staff turnover and changes in job assignments. Facilities are required to specify a schedule for periodic training activities in their plan. In any case, you should regularly evaluate the effectiveness of your training efforts. In many cases, this will simply involve speaking with your employees to verify that information has been communicated effectively.

EPCRA, Section 313 Facility Requirements

EPA's General Permit contains additional training requirements for employees and contractor personnel that work in areas where EPCRA, Section 313 water priority chemicals are used or stored [Part IV.D.7.b.(9).]. These individuals must be trained in the following areas at least once per year:

- Preventive measures, including spill prevention and response and preventive maintenance

- Pollution control laws and regulations

- The facility's Storm Water Pollution Prevention Plan

- Features and operations of the facility which are designed to minimize discharges of Section 313 water priority chemicals, particularly spill prevention procedures.

| 2.1 PLANNING AND ORGANIZATION |
| 2.2 ASSESSMENT PHASE |
| 2.3 BMP IDENTIFICATION PHASE |
| 2.4 IMPLEMENTATION PHASE |

2.5 EVALUATION MONITORING
- Annual site inspection/BMP evaluation
- Recordkeeping and reporting
- Review and revise plan

| 2.6 GENERAL REQUIREMENTS | 2.7 SPECIAL REQUIREMENTS |

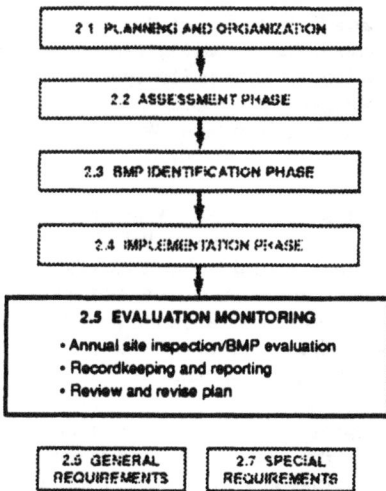

2.5 EVALUATION PHASE

Now that your Storm Water Pollution Prevention Plan has been put to action, you must keep it up-to-date by regularly evaluating the information you collected in the Assessment Phase and the controls you selected in the BMP Identification Phase. Specifically, you will:

- **Conduct site evaluations**

- **Keep records of all inspections and reports**

- **Revise the plan as needed.**

2.5.1 Annual Site Compliance Evaluation

EPA GENERAL PERMIT REQUIREMENTS

Comprehensive Site Compliance Evaluation

Part IV.D.4.

Qualified personnel must conduct site compliance evaluations at appropriate intervals specified in the plan at least once a year (at least once in three years for inactive mining sites). As part of your compliance evaluations, you are required to:

- Inspect storm water drainage areas for evidence of pollutants entering the drainage system

- Evaluate the effectiveness of measures to reduce pollutant loadings and whether additional measures are needed

- Observe structural measures, sediment controls, and other storm water BMPs to ensure proper operation

- Inspect any equipment needed to implement the plan, such as spill response equipment

- Revise the plan as needed within two weeks of inspection (potential pollutant source description and description of measures and controls)

- Implement any necessary changes in a timely manner, but at least within 12 weeks of the inspection

- Prepare a report summarizing inspection results and follow up actions, the date of inspection and personnel who conducted the inspection; identify any incidents of noncompliance or certify that the facility is in compliance with the plan.

- All incidents of noncompliance must be documented in the inspection report. Where there are no incidents of noncompliance, the inspection report must contain a certification that the facility is in compliance with the plan.

- Sign the report in accordance with Section 2.6.2 and keep it with the plan.

Annual site compliance evaluations are comprehensive inspections performed by individuals specifically designated in the Storm Water Pollution Prevention Plan as having responsibility for conducting such inspections. These employees should be familiar with all facility industrial operations and Storm Water Pollution Prevention Plan goals and requirements. Furthermore, inspectors should be able to make necessary management decisions or have direct access to management.

This annual evaluation provides a basis for evaluating the overall effectiveness of your Storm Water Pollution Prevention Plan. In particular, the annual site compliance evaluation will allow you to verify that the description of potential pollutant sources contained in the plan is accurate, that the plan drainage map is accurate or has been updated to reflect current conditions, and that controls identified in the plan to reduce pollutants in storm water discharges are accurately identified, in place and working. The annual site compliance evaluation will also identify where new controls are needed so that you may implement them and incorporate them into the plan.

The scope of the annual site compliance evaluation will depend on various factors, including the scope of the Storm Water Pollution Prevention Plan and the size and nature of the activities occurring at the facility. The process for conducting the evaluation should follow these steps:

- Review the Storm Water Pollution Prevention Plan and draw up a list of those items which are part of material handling, storage, and transfer areas covered by the plan

- List all equipment and containment in these areas covered in the plan

- Review facility operations for the past year to determine if any more areas should be included in the original plan, or if any existing areas were modified so as to require plan modification; change plan as appropriate

- Conduct inspection to determine (1) if all storm water pollution prevention measures are accurately identified in the plan, and (2) are in place and working properly

- Document findings

- Modify Storm Water Pollution Prevention Plan as appropriate.

As each facility and Storm Water Pollution Prevention Plan is unique, so the exact inspection format will vary from facility to facility. All documentation regarding conditions necessitating modification to the Storm Water Pollution Prevention Plan should be kept on file as part of the plan until one year after coverage under the permit expires.

2.5.2 Recordkeeping and Internal Reporting

EPA GENERAL PERMIT REQUIREMENTS
Keeping Records Part IV.D.3.f.
Incidents such as spills or other discharges, along with other information describing the quality and quantity of storm water discharges must be included in the records. Inspections and maintenance activities shall be documented and recorded in the plan. Records must be maintained for one year after the permit expires.

Keeping records of and reporting events that occur onsite is an effective way of tracking the progress of pollution prevention efforts and waste minimization. Analyzing records of past spills, for example, can provide useful information for developing improved BMPs to prevent future spills of the same kind. Recordkeeping and internal reporting represent good operating practices because they can increase the efficiency of the facility and effectiveness of BMPs.

Recordkeeping and Reporting Procedures for Spills, Leaks, and Other Discharges

A recordkeeping system set up for documenting spills, leaks, and other discharges, including discharges of hazardous substances in reportable quantities (for a discussion of reportable quantities, see Section 2.2.3 and Appendix H), could help your facility minimize incident recurrence, correctly respond with appropriate cleanup activities, and comply with legal requirements. The system for recordkeeping and reporting could also include any other information that would enhance the effectiveness of the Storm Water Pollution Prevention Plan. You should make a point of keeping track of reported incidents and following up on results of inspections and reported spills, leaks, or other discharges.

Records should include the following, as appropriate:

- The date and time of the incident, weather conditions, duration, cause, environmental problems, response procedures, parties notified, recommended revisions of the BMP program, operating procedures, and/or equipment needed to prevent recurrence.

- Formal written reports. These are helpful in reviewing and evaluating the discharges and making revisions to improve the BMP program. Document all reports you call in to the National Response Center in the event of a reportable quantity discharge. For more information on reporting spills or other discharges, refer to Section 2.2.3 and 40 CFR 117.3 and 40 CFR 302.4.

- A list of the procedures for notifying the appropriate plant personnel and the names and telephone numbers of responsible employees. This enables more rapid reporting of and response to spills and other incidents.

Recordkeeping and Reporting Procedures for Inspections and Maintenance Activities

Maintaining records for all inspections is an important element of any Storm Water Pollution Prevention Plan. Documenting all inspections, whether routine or detailed, is a good preventive maintenance technique, because analysis of inspection records allows for early detection of any potential problems. Recordkeeping also helps to devise improvements in the BMP program after inspection records have been analyzed. Recordkeeping and reporting for maintenance activities should also be a part of the plan as another preventive maintenance measure. Keeping a log of all maintenance activities, such as the cleaning of oil and grit separators or catch basins, will enable the facility to evaluate the effectiveness of the BMP program, equipment, and operation.

There are various simple techniques used to accurately document and report inspection results including the following:

- Field notebooks

- Timed and dated photographs

- Video tapes

- Drawings and maps.

Keeping Records Updated

It is important to keep all records updated on:

- The correct name and address of facility

- The correct name and location of receiving waters

- The number and location of discharge points

- Principal products and production rates (where appropriate).

Records Retention

Records of spills, leaks, or other discharges, inspections, and maintenance activities must be retained for at least one year after coverage under the permit expires.

2.5.3 Plan Revisions

EPA GENERAL PERMIT REQUIREMENTS
Keeping Plans Current
Part IV.C.
You must amend your plan whenever there is a change in design, construction, operation, or maintenance, which may impact the potential for pollutants to be discharged or if the Storm Water Pollution Prevention Plan proves to be ineffective in controlling the discharge of pollutants. Facilities are not required to submit a notice to the Director each time the pollution prevention plan is modified unless the Director specifically requests changes to be made to the plan.

For your Storm Water Pollution Prevention Plan to be effective, you should ensure that your plan complies with any permit conditions that apply to your facility and that you have accurately represented facility features and operations. Should either of these conditions not be met by the plan, you must make the necessary changes. Either the managers of facilities or the permitting authority may recommend changes to the plan (see Section 2.6.4 for requirements).

Storm Water Pollution Prevention Plans are developed based on site-specific features. When there are changes in design, construction, operation, or maintenance, and that change will have a significant effect on the potential for discharging pollutants in storm water at a facility, your Storm Water Pollution Prevention Plan should be modified to reflect the changes and new conditions. For example, if your facility begins to use a new chemical in its production operations, proper handling procedures for this chemical should be incorporated into the facility plan.

You may also decide to change the plan because it has proven to be ineffective in controlling storm water contamination based on the results of routine visual inspections (see Section 2.3.1) or more comprehensive site evaluations (see Section 2.5.1).

2.1 PLANNING AND ORGANIZATION

2.2 ASSESSMENT PHASE

2.3 BMP IDENTIFICATION PHASE

2.4 IMPLEMENTATION PHASE

2.5 EVALUATION MONITORING

2.7 SPECIAL REQUIREMENTS

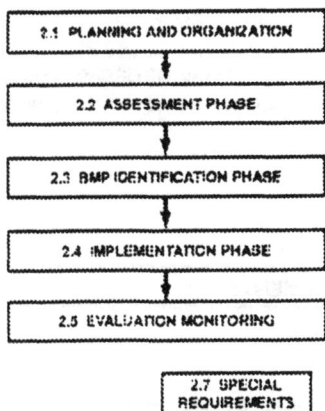

2.6 GENERAL REQUIREMENTS
• Deadlines
• Signature requirements
• Plan location and public access
• Required plan modification

2.6 GENERAL REQUIREMENTS

This Section provides guidance on some of the administrative requirements related to organizing and developing your Storm Water Pollution Prevention Plan. This information should be reviewed prior to beginning to develop your facility's Storm Water Pollution Prevention Plan. These requirements include:

- Deadlines for plan development and implementation

- Who must sign the plan

- Where to keep the plan

- How to make changes to the plan that are required by the Director.

2.6.1 Schedule for Plan Development and Implementation

EPA GENERAL PERMIT REQUIREMENTS		
Schedule for Plan Development and Implementation Part IV.A.		
Type of Facility	Deadline for Plan Completion	Deadline for Plan Compliance
Facilities with industrial activities existing on or before October 1, 1992	April 1, 1993	October 1, 1993
Facilities commencing industrial activities after October 1, 1992, but on or before December 31, 1992	60 days after commencement of discharge	60 days after commencement of discharge
Facilities commencing industrial activities on or after January 1, 1993	48 hours prior to commencement of discharge (upon submittal of NOI)	48 hours prior to commencement of discharge (upon submittal of NOI)
Oil and gas exploration, production, processing or treatment operations discharging a reportable quantity release in storm water after October 1, 1992	60 days after release	60 days after release
Industrial facilities that are owned or operated by a municipality that are rejected or denied from the group application process	365 days after date of rejection or denial	545 days after date of rejection or denial
Note: The Director may grant a written extension for plan preparation and compliance for new dischargers (after October 1, 1992) upon showing of good cause.		

The deadlines to complete and comply with or implement your facility's Storm Water Pollution Prevention Plan may depend on the type of permit under which your facility is covered. Be sure to read your permit carefully so that you know what the deadlines are. Many NPDES-delegated States may issue general permits for storm water that contain deadlines similar to the deadlines in EPA's General Permits.

2.6.2 Required Signatures

EPA GENERAL PERMIT REQUIREMENTS

Signature Requirements

Part VII.G.1.

Where your facility is subject to storm water permit requirements, all reports, certifications, or information either submitted to the permitting authority or to the operator of a large or medium municipal separate storm sewer system, or required to be maintained by the permittee onsite should be signed as follows:

- For a corporation, the plan must be signed by a "responsible corporate officer." A responsible corporate officer may be <u>any one of the following</u>:

  - A president, secretary, treasurer, or vice-president of the corporation in charge of a principal business function, or any other person who performs similar policy or decision-making functions for the corporation

  - The manager of one or more manufacturing, production, or operating facilities employing more than 250 persons or having gross annual sales or expenditures exceeding $25,000,000 (in second quarter 1980 dollars) if authority to sign documents has been assigned or delegated to the manager in accordance with corporate procedure.

- For a partnership or sole proprietorship, the plan must be signed by a general partner or the proprietor, respectively.

- For a municipality, State, Federal, or other public agency, the plan must be signed by either:

  - The principal executive officer or ranking official, which includes the chief executive officer of the agency, or

  - The senior officer having responsibility for the overall operations of a principal geographic unit of the agency.

Designating Signatory Authority

Part VII.G.2.

Any of the above persons may designate a duly authorized representative to sign for them. The representative should either have overall responsibility for the operation of the facility or environmental matters for the company. If an authorized representative is appointed, the authorization must be put in writing by the responsible signatory and submitted to the Director. Any change in an authorized individual or an authorized position must be made in writing and submitted to the permitting authority.

EPA GENERAL PERMIT REQUIREMENTS

Certification

Part VII.G.2.d.

Any person signing documents under this permit shall make the following certification: "I certify under penalty of law that this document and all attachments were prepared under my direction or supervision in accordance with a system designed to assure that qualified personnel properly gathered and evaluated the information submitted. Based on my inquiry of the person or persons who manage the system, or those persons directly responsible for gathering the information, the information submitted is, to the best of my knowledge and belief, true, accurate, and complete. I am aware that there are significant penalties for submitting false information, including the possibility of fine and imprisonment for knowing violations."

To ensure that your facility's Storm Water Pollution Prevention Plan is completely developed and adequately implemented, your NPDES permit will generally require that an authorized facility representative sign and certify the plan. The authorized facility representative should be someone at or near the top of your facility's management chain, such as the president, vice president, or a production manager who has been delegated the authority to sign and certify this type of document. In signing the plan, the corporate officer is attesting that the information is true. This signature provides a basis for an enforcement action to be taken against the person signing the plan and related reports. The permittee should be aware that Section 309 of the Clean Water Act provides for significant penalties where information is false or the permittee violates, either knowingly or negligently, its permit requirements. In some cases, your general permit may require certification of the plan by a professional engineer. Specific signatory requirements will be listed in your NPDES permit.

EPCRA, Section 313 Facility Plan Certification Requirements

EPA's General Permit contains additional certification requirements for facilities subject to reporting under EPCRA, Section 313 for water priority chemicals [Part IV.D.7.b.(10).]. The plan must be reviewed and certified by a Registered Professional Engineer and recertified every three years or as soon as practicable after significant modifications are made to the facility. This certification that the plan was prepared in accordance with good engineering practices does not relieve the facility owner or operator of responsibility to prepare and implement the plan, however.

2.6.3 Plan Location and Public Access

EPA GENERAL PERMIT REQUIREMENTS

Where and How Long to Keep the Plan

Parts IV.B. and VI.E.

Plans are required to be maintained onsite of the facility unless the Director, or authorized representative, or the operator of a large or medium municipal separate storm sewer system, requests that the plan be submitted. Plans and all required records must be kept until at least one year after coverage under the permit expires.

Although all plans are to be maintained onsite, some NPDES storm water permits may require that facilities submit copies of their Storm Water Pollution Prevention Plans to the Director for review. Examine your permit carefully to determine what submittal requirements apply to your facility. Even if your permit does not require you automatically to submit your plan to your permitting authority, you must provide copies of the plan to your permitting authority or to your municipal operator upon request. Plans and associated records are available to the public by request through the permitting authority.

2.6.4 Director-Required Plan Modifications

EPA GENERAL PERMIT REQUIREMENTS
Required Changes
Part IV.B.3.
Any changes required by the permitting authority shall be made within 30 days, unless otherwise provided by the notification, and the facility must submit a certification signed in accordance with Section 2.6.2 to the Director that the requested changes have been made.

Upon reviewing your plan, the permitting authority may find that it does not meet one or more of the minimum standards established by the pollution prevention plan requirements. In this case, the permitting authority will notify you of changes needed to improve the plan.

For example, where a facility has not addressed spill response procedures for a toxic chemical to the extent that the permitting authority believes is necessary, the facility will be required to revise the procedures. The permitting authority retains the authority to make this type of request at any time during the effective period of the plan. In the notification, the permitting authority will establish a deadline for the incorporation of the required changes, unless the permit specifies a deadline. Permittees may or may not have to certify that the requested changes have been implemented depending on their specific permit conditions. You should examine your permit for such details.

```
┌─────────────────────────────────┐
│  2.1  PLANNING AND ORGANIZATION │
└─────────────────────────────────┘
              │
              ▼
┌─────────────────────────────────┐
│  2.2  ASSESSMENT PHASE          │
└─────────────────────────────────┘
              │
              ▼
┌─────────────────────────────────┐
│  2.3  BMP IDENTIFICATION PHASE  │
└─────────────────────────────────┘
              │
              ▼
┌─────────────────────────────────┐
│  2.4  IMPLEMENTATION PHASE      │
└─────────────────────────────────┘
              │
              ▼
┌─────────────────────────────────┐
│  2.5  EVALUATION/MONITORING     │
└─────────────────────────────────┘

┌───────────────────┐
│  2.6  GENERAL      │
│  REQUIREMENTS      │
└───────────────────┘

┌───────────────────────────────┐
│  2.7  SPECIAL REQUIREMENTS     │
│  • Discharges through MS4s     │
│  • Salt storage piles          │
│  • EPCRA, Section 313 Facilities│
└───────────────────────────────┘
```

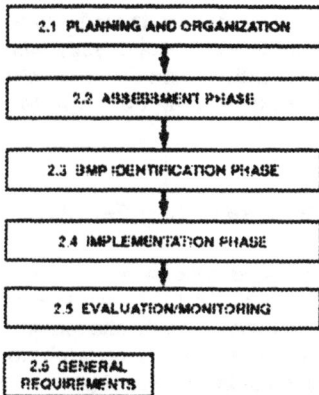

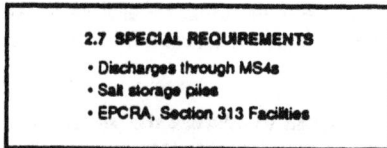

2.7 SPECIAL REQUIREMENTS

In addition to the minimum "baseline" BMPs discussed in previous sections, facilities may be subject to additional "special" requirements. Not all facilities will have to include these special requirements in their Storm Water Pollution Prevention Plan. Be sure to check your permit closely for these conditions. In particular, EPA's General Permit includes special requirements for:

- Facilities that discharge storm water through municipal separate storm sewer systems

- Facilities subject to EPCRA, Section 313 reporting requirements

- Facilities with salt storage piles.

2.7.1 Special Requirements for Discharges Through Municipal Separate Storm Sewer Systems

EPA GENERAL PERMIT REQUIREMENTS
Discharges Through Large or Medium Municipal Separate Storm Sewer Systems (MS4s)
Part IV.D.5.
Permittees must comply with conditions in municipal storm water management programs developed under the NPDES permit issued for that system to which the industrial facility discharges, provided that the facility was directly notified of the applicable requirements by the municipal operator. The facility must be in compliance with these conditions by the deadlines specified in the pollution prevention plan listed in Section 2.6.1.

The November 16, 1990, storm water discharge permit application regulations require large and medium municipal separate storm sewer systems (systems serving a population of 100,000 or more) to develop storm water management programs in order to control pollutants discharged through the municipal systems. These management programs will address discharges of industrial storm water through the systems to the extent that they are harmful to the water quality of receiving streams. Municipalities should be aware of the facilities with storm water discharges associated with industrial activity that discharge into their separate storm sewer system because the November 16, 1990, final rule required these facilities to notify the municipal operator. In addition, facilities covered by general permits will typically be required to submit a copy of their NOI to the municipal operator. EPA emphasizes that it is the facility's responsibility to inform the municipality of all storm water discharges associated with industrial activity to the separate storm sewer system. Facilities with such discharges that have not yet contacted the appropriate municipal authority should do so immediately.

Although facility-specific Storm Water Pollution Prevention Plans for industries are designed to prevent pollutants from entering storm water discharges, the municipal operator may find it necessary to impose specific requirements on a particular industrial facility or class of industrial facilities in some situations. One way to ensure that facilities comply with these requirements is to include a provision in the facility's NPDES storm water discharge permit that directly requires compliance. This mechanism provides a basis for enforcement action to be directed, where necessary, against the owner or operator of the facility with a storm water discharge associated with industrial activity.

2.7.2 Special Requirements for EPCRA, Section 313 Reporting Facilities

Section 313 of EPCRA requires operators of manufacturing facilities that handle toxic chemicals in amounts exceeding threshold levels (listed at 40 CFR 372.25) to report to the government on an annual basis. Because these types of facilities handle large amounts of toxic chemicals, EPA concluded that they have an increased potential to degrade the water quality of receiving streams. To address this risk, EPA established specific control requirements in its general permit. In particular, these requirements apply to Section 313 facilities that report for "water priority chemicals" that include any of over 200 chemicals that have been identified by EPA as especially toxic to water ecosystems. For reference, Appendix I contains a list of Section 313 water priority chemicals.

Many of the requirements outlined below are specifically designed to address the water quality concerns that toxic chemicals present. Incorporation of these requirements into site-specific Storm Water Pollution Prevention Plans will prevent spills and leaks of water priority chemicals and eliminate or reduce other opportunities for exposure of toxic chemicals to storm water, thus protecting receiving streams from toxic discharges.

Specific Requirements

The following specific control requirements must be practiced in areas where Section 313 water priority chemicals are stored, handled, processed, or transferred:

- Provide containment, drainage control, and/or diversionary structures:

  - Prevent or minimize runon by installing curbing, culverting, gutters, sewers, or other controls, and/or

  - Prevent or minimize exposure by covering storage piles.

- Prevent discharges from all areas:

  - Use manually activated valves with drainage controls in all areas, and/or

  - Equip the plant with a drainage system to return spilled material to the facility.

- Prevent discharges from liquid storage areas:

  - Store liquid materials in compatible storage containers

  - Provide secondary containment designed to hold the volume of the largest storage tank plus precipitation.

- Prevent discharges from loading/unloading areas:

  - Use drip pans and/or

  - Implement a strong spill contingency and integrity testing plan.

- Prevent discharges from handling/processing/transferring areas:

  - Use covers, guards, overhangs, door skirts

  - Conduct visual inspections or leak tests for overhead piping.

- Introduce facility security programs to prevent spills:

  - Use fencing, lighting, traffic control, and/or secure equipment and buildings.

Additional requirements are baseline BMPs that have been enhanced to address specific storm water concerns associated with the handling of toxic chemicals. These additional requirements are highlighted in previous sections on the pages indicated below:

Pollution Prevention Team	p. 2-5
Preventive Maintenance	p. 2-27
Spill Prevention Response	p. 2-34
Employee Training	p. 2-42
Professional Engineer Certification	p. 2-49

2.7.3 Special Requirements for Salt Storage Piles

EPA GENERAL PERMIT REQUIREMENTS

Salt Storage Piles

Part IV.D.8.

Where storm water from a salt storage pile is discharged to waters of the United States, the pile must be covered or enclosed to prevent exposure to precipitation, except when salt is being added to or taken from the pile. Discharges shall comply with this provision as expeditiously as practicable, but in no event later than October 1, 1995.

Facilities may use salt for de-icing purposes or part of their industrial processes. Since exposed salt piles will easily contaminate storm water runoff, an obvious BMP for these piles is to cover them with a tarp or other covering or enclose them in a shed or building. This requirement may not be applicable to all Storm Water Pollution Prevention Plans, however. Where runoff from the salt pile is not discharged to waters of the United States, then this requirement would not apply since the pollutants will not reach a waterbody. Since it may not be feasible to maintain cover over a salt pile when adding to it or taking salt from it, permits will generally incorporate some flexibility, as does EPA's General Permit.

STORM WATER POLLUTION PREVENTION PLAN WORKSHEETS

POLLUTION PREVENTION TEAM (Section 2.1.1) MEMBER ROSTER	Worksheet #1 Completed by: _____ Title: _____ Date: _____

Leader: _____   Title: _____

Office Phone: _____

Responsibilities: _____

_____

_____

Members:

(1) _____   Title: _____

Office Phone: _____

Responsibilities: _____

_____

_____

(2) _____   Title: _____

Office Phone: _____

Responsibilities: _____

_____

_____

(3) _____   Title: _____

Office Phone: _____

Responsibilities: _____

_____

_____

(4) _____   Title: _____

Office Phone: _____

Responsibilities: _____

_____

_____

DEVELOPING A SITE MAP (Section 2.2.1)	Worksheet #2 Completed by: _____ Title: _____ Date: _____

Instructions:	Draw a map of your site including a footprint of all buildings, structures, paved areas, and parking lots. The information below describes additional elements required by EPA's General Permit (see example maps in Figures 2.3 and 2.4).

EPA's General Permit requires that you indicate the following features on your site map:

- All outfalls and storm water discharges

- Drainage areas of each storm water outfall

- Structural storm water pollution control measures, such as:

  - Flow diversion structures
  - Retention/detention ponds
  - Vegetative swales
  - Sediment traps

- Name of receiving waters (or if through a Municipal Separate Storm Sewer System)

- Locations of exposed significant materials (see Section 2.2.2)

- Locations of past spills and leaks (see Section 2.2.3)

- Locations of high-risk, waste-generating areas and activities common on industrial sites such as:

  - Fueling stations
  - Vehicle/equipment washing and maintenance areas
  - Area for unloading/loading materials
  - Above-ground tanks for liquid storage
  - Industrial waste management areas (landfills, waste piles, treatment plants, disposal areas)
  - Outside storage areas for raw materials, by-products, and finished products
  - Outside manufacturing areas
  - Other areas of concern (specify:_____)

Worksheet #3
Completed by: _____
Title: _____
Date: _____

MATERIAL INVENTORY
(Section 2.2.2)

Instructions: List all materials used, stored, or produced onsite. Assess and evaluate these materials for their potential to contribute pollutants to storm water runoff. Also complete Worksheet 3A if the material has been exposed during the last three years.

Material	Purpose/Location	Quantity (units) Used	Quantity (units) Produced	Quantity (units) Stored	Quantity Exposed in Last 3 Years	Likelihood of contact with storm water. If yes, describe reason.	Past Significant Spill or Leak Yes	Past Significant Spill or Leak No

DESCRIPTION OF EXPOSED SIGNIFICANT MATERIAL
(Section 2.2.2)

Worksheet #3A
Completed by: _____
Title: _____
Date: _____

Instructions: Based on your material inventory, describe the significant materials that were exposed to storm water during the past three years and/or are currently exposed. For the definition of "significant materials" see Appendix B of the manual.

Description of Exposed Significant Material	Period of Exposure	Quantity Exposed (units)	Location (as indicated on the site map)	Method of Storage or Disposal (e.g., pile, drum, tank)	Description of Material Management Practice (e.g., pile covered, drum sealed)

LIST OF SIGNIFICANT SPILLS AND LEAKS
(Section 2.2.3)

Worksheet #4
Completed by: _____
Title: _____
Date: _____

Directions: Record below all significant spills and significant leaks of toxic or hazardous pollutants that have occurred at the facility in the three years prior to the effective date of the permit.

Definitions: Significant spills include, but are not limited to, releases of oil or hazardous substances in excess of reportable quantities.

1st Year Prior

Date (month/day/year)	Spill	Leak	Location (as indicated on site map)	Description				Response Procedure		Preventive Measures Taken
				Type of Material	Quantity	Source, If Known	Reason	Amount of Material Recovered	Material No Longer Exposed to Storm Water (True/False)	

2nd Year Prior

Date (month/day/year)	Spill	Leak	Location (as indicated on site map)	Description				Response Procedure		Preventive Measures Taken
				Type of Material	Quantity	Source, If Known	Reason	Amount of Material Recovered	Material No Longer Exposed to Storm Water (True/False)	

3rd Year Prior

Date (month/day/year)	Spill	Leak	Location (as indicated on site map)	Description				Response Procedure		Preventive Measures Taken
				Type of Material	Quantity	Source, If Known	Reason	Amount of Material Recovered	Material No Longer Exposed to Storm Water (True/False)	

NON-STORM WATER DISCHARGE ASSESSMENT AND CERTIFICATION (Section 2.2.4)

Worksheet #5
Completed by: _____
Title: _____
Date: _____

Date of Test or Evaluation	Outfall Directly Observed During the Test (identify as indicated on the site map)	Method Used to Test or Evaluate Discharge	Describe Results from Test for the Presence of Non-Storm Water Discharge	Identify Potential Significant Sources	Name of Person Who Conducted the Test or Evaluation

CERTIFICATION

I, _____ (responsible corporate official), certify under penalty of law that this document and all attachments were prepared under my direction or supervision in accordance with a system designed to assure that qualified personnel properly gather and evaluate the information submitted. Based on my inquiry of the person or persons who manage the system or those persons directly responsible for gathering the information, the information submitted is, to the best of my knowledge and belief, true, accurate, and complete. I am aware that there are significant penalties for submitting false information, including the possibility of fine and imprisonment for knowing violations.

A. Name & Official Title (type or print)	B. Area Code and Telephone No.
C. Signature	D. Date Signed

NON-STORM WATER DISCHARGE ASSESSMENT AND FAILURE TO CERTIFY NOTIFICATION
(Section 2.2.4)

Worksheet #6
Completed by: _____
Title: _____
Date: _____

Directions: If you cannot feasibly test or evaluate an outfall due to one of the following reasons, fill in the table below with the appropriate information and sign this form to certify the accuracy of the included information.

List all outfalls not tested or evaluated, describe any potential sources of non-storm water pollution from listed outfalls, and state the reason(s) why certification is not possible. Use the key from your site map to identify each outfall.

Important Notice: A copy of this notification must be signed and submitted to the Director within 180 days of the effective date of this permit.

Identify Outfall Not Tested/Evaluated	Description of Why Certification Is Infeasible	Description of Potential Sources of Non-Storm Water Pollution

CERTIFICATION

I certify under penalty of law that this document and all attachments were prepared under my direction or supervision in accordance with a system designed to assure that qualified personnel properly gather and evaluate the information submitted. Based on my inquiry of the person or persons who manage the system or those persons directly responsible for gathering the information, the information submitted is, to the best of my knowledge and belief, true, accurate, and complete. I am aware that there are significant penalties for submitting false information, including the possibility of fine and imprisonment for knowing violations, and that such notification has been made to the Director within 180 days of _____ (date permit was issued), the effective date of this permit.

A. Name & Official Title (type or print)

B. Area Code and Telephone No.

C. Signature

D. Date Signed

Worksheet #7
Completed by: _____
Title: _____
Date: _____

POLLUTANT SOURCE IDENTIFICATION
(Section 2.2.6)

Instructions: List all identified storm water pollutant sources and describe existing management practices that address those sources. In the third column, list BMP options that can be incorporated into the plan to address remaining sources of pollutants.

Storm Water Pollutant Sources	Existing Management Practices	Description of New BMP Options
1.		
2.		
3.		
4.		
5.		
6.		
7.		
8.		
9.		
10.		

Worksheet #7a
Completed by: _____
Title: _____
Date: _____

BMP IDENTIFICATION
(Section 2.3.1)

Instructions: Describe the Best Management Practices that you have selected to include in your plan. For each of the baseline BMPs, describe actions that will be incorporated into facility operations. Also describe any additional BMPs [activity-specific (Chapter 3) and site-specific BMPs (Chapter 4)] that you have selected. Attach additional sheets if necessary.

BMPs	Brief Description of Activities
Good Housekeeping	
Preventive Maintenance	
Inspections	
Spill Prevention Response	
Sediment and Erosion Control	
Management of Runoff	
Additional BMPs (Activity-specific and Site-specific)	

Worksheet #8
Completed by: _____
Title: _____
Date: _____

IMPLEMENTATION
(Section 2.4.1)

Instructions: Develop a schedule for implementing each BMP. Provide a brief description of each BMP, the steps necessary to implement the BMP (i.e., any construction or design), the schedule for completing those steps (list dates) and the person(s) responsible for implementation.

BMPs	Description of Action(s) Required for Implementation	Scheduled Completion Date(s) for Req'd. Action	Person Responsible for Action	Notes
Good Housekeeping	1.			
	2.			
	3.			
Preventive Maintenance	1.			
	2.			
	3.			
Inspections	1.			
	2.			
	3.			
Spill Prevention and Response	1.			
	2.			
	3.			
Sediment and Erosion Control	1.			
	2.			
	3.			
Management of Runoff	1.			
	2.			
	3.			
Additional BMPs (Actively-specific and site-specific)	1.			
	2.			
	3.			

Worksheet #9
Completed by: _____
Title: _____
Date: _____

EMPLOYEE TRAINING
(Section 2.4.2)

Instructions: Describe the employee training program for your facility below. The program should, at a minimum, address spill prevention and response, good housekeeping, and material management practices. Provide a schedule for the training program and list the employees who attend training sessions.

Training Topics	Brief Description of Training Program/Materials (e.g., film, newsletter course)	Schedule for Training (list dates)	Attendees
Spill Prevention and Response			
Good Housekeeping			
Material Management Practices			
Other Topics			

CHAPTER

3

ACTIVITY-SPECIFIC SOURCE CONTROL BMPs

This chapter describes specific BMPs for common industrial activities that may contaminate storm water. Chapter 2 led you through the steps of identifying activities at your facility that can contaminate storm water. At this point, you should be ready to choose the BMPs that best fill your facility's need. You should read this chapter if any of the activities listed below take place at your facility. BMPs for each of these activities are provided in the sections listed below:

Activity	Section
Fueling	3.1
Maintaining Vehicles and Equipment	3.2
Painting Vehicles and Equipment	3.3
Washing Vehicles and Equipment	3.4
Loading and Unloading Materials	3.5
Liquid Storage in Above-Ground Tanks	3.6
Industrial Waste Management and Outside Manufacturing	3.7
Outside Storage of Raw Materials, By-Products, or Finished Products	3.8
Salt Storage	3.9

Each section is presented in a question and answer format. By answering these questions, you will be able to quickly identify source controls or recycling BMPs that are suitable for your facility. The BMPs suggested are relatively easy to use, are inexpensive, and often are effective in removing the source of storm water contaminants. This is not a complete list of BMPs for every industrial activity; rather, it is meant to help you think about ways you can reduce storm water contamination on your site. You may want to contact one of the State or Federal pollution prevention assistance offices listed in Appendix D for suggestions or help in choosing or using these and other BMP options.

3.1 BMPs FOR FUELING STATIONS

When storm water mixes with fuel spilled or leaked onto the ground, it becomes polluted with chemicals that are harmful to humans and to fish and wildlife. The following questions will help you identify activities that can contaminate storm water and suggest BMPs to reduce or eliminate storm water contamination from fueling stations. Read this section if your facility has outdoor fueling operations or if fueling occurs in areas where leaks or spills could contaminate storm water. Also refer to the BMPs listed in Section 4.2 on Exposure Minimization.

Q. Have you installed spill and overfill prevention equipment?

Fuel overflows during storage tank filling are a major source of spills. Overflows can be prevented. Watch the transfer constantly to prevent overfilling and spilling. Overfill prevention equipment automatically shuts off flow, restricts flow, or sounds an alarm when the tank is almost full. Federal regulations require overfill prevention equipment on all Underground Storage Tanks (USTs) installed after December 1988. For USTs installed before December 1988, overfill prevention equipment is required by 1998. State or local regulations may be stricter, so contact your State and/or local government for details. Consider installing overflow prevention equipment sooner than the required deadline as part of your pollution prevention plan.

> **FUEL STATION ACTIVITIES THAT CAN CONTAMINATE STORM WATER:**
> - Spills and leaks that happen during fuel or oil delivery
> - Spills caused by "topping off" fuel tanks
> - Allowing rainfall on the fuel area or storm water to run onto the fuel area
> - Hosing or washing down the fuel area
> - Leaking storage tanks

Q. Are vehicle fuel tanks often "topped off"?

Gas pumps automatically shut off when the vehicle fuel tank is almost full to prevent spills. Trying to completely fill the tanks or topping off the tank often results in overfilling the tank and spilling fuel. Discourage topping off by training employees and posting signs.

Q. Have you taken steps to protect fueling areas from rain?

Fueling areas can be designed to minimize spills, leaks, and incidental losses of fuel, such as vapor loss, from coming into contact with rain water:

- Build a roof over the fuel area.

- Pave the fuel area with concrete instead of asphalt. Asphalt soaks up fuel or can be slowly dissolved by fuel, engine fluids, and other organic liquids. Over time, the asphalt itself can become a source of storm water contamination.

Q. Is runon to the fueling area minimized?

Runon is storm water generated from other areas that flows or "runs on" to your property or site. Runon flowing across fueling areas can wash contaminants into storm drains. Runon can be minimized by:

- Grading, berming, or curbing the area around the fuel site to direct runon away from the fuel area

- Locating roof downspouts so storm water is directed away from fueling areas

- Using valley gutters to route storm water around fueling area.

Q. Are oil/water separators or oil and grease traps installed in storm drains in the fueling area?

Oil/water separators and oil and grease traps are devices that reduce the amount of oil entering storm drains. These devices should be installed and routinely inspected, cleaned, and maintained.

Q. Is the fueling area cleaned by hosing or washing?

Cleaning the fueling area with running water should be avoided because the wash water will pick up fuel, oil, and grease and make it storm water. Consider using a damp cloth on the pumps and a damp mop on the pavement rather than a hose. Check with your local sewer authority about any treatment required before discharging the mop water or wash water to the sanitary sewer.

Q. Do you control petroleum spills?

Spills should be controlled immediately. Small spills can be contained using sorbent material such as kitty litter, straw, or sawdust. Do not wash petroleum spills into the storm drain or sanitary sewer. For more information on spill control measures, see sections on Containment Diking and Curbing in Chapter 4.

Q. Are employees aware of ways to reduce contamination of storm water at fueling stations?

Storm water contamination from fueling operations often occurs from small actions such as topping off fuel tanks, dripping engine fluids, and hosing down fuel areas. Inform employees about ways to eliminate or reduce storm water contamination.

EMPLOYEE INVOLVEMENT IS THE KEY:

Getting employees interested in reducing waste generation is the key to a successful storm water pollution prevention plan. Discuss pollution prevention with your employees. They are most familiar with the operations that generate wastes and may have helpful waste reduction suggestions. Consider setting up an employee reward program to promote pollution prevention.

Q. Where does the water drain from your fueling area?

In many cases, wash water and storm water in fueling areas drain directly to the storm sewer without adequate treatment. Some types of oil/water separators installed at these locations can provide treatment to discharges from oil contaminated pavements, but this equipment is only effective when properly maintained (i.e., cleaned frequently). Some States require that these discharges be tied in to a sanitary sewer system or process wastewater treatment system. If discharges from fueling or other high risk areas at your facility drain to a sanitary sewer system, you should inform your local POTW.

SUMMARY OF FUELING STATION BMPs

- Consider installing spill and overflow protection.

- Discourage topping off of fuel tanks.

- Reduce exposure of the fuel area to storm water.

- Use dry cleanup methods for the fuel area.

- Use proper petroleum spill control.

- Encourage employee participation.

3.2 BMPs FOR VEHICLE AND EQUIPMENT MAINTENANCE

Many vehicle and equipment maintenance operations use materials or create wastes that are harmful to humans and the environment. Storm water runoff from areas where these activities occur can become polluted by a variety of contaminants such as solvents and degreasing products, waste automotive fluids, oils and greases, acids, and caustic wastes. These and other harmful substances in storm water can enter water bodies through storm drains or through small streams where they can harm fish and wildlife.

The following questions will help you find sources of storm water contamination from vehicle and equipment maintenance operations on your site and to help you choose BMPs that can reduce or eliminate these sources.

Q. Are parts cleaned at your facility?

Parts are often cleaned using solvents such as trichloroethylene, 1,1,1-trichloroethane or methylene chloride. Many of these cleaners are harmful and must be disposed of as a hazardous waste. Cleaning without using liquid cleaners whenever possible reduces waste. Scrape parts with a wire brush, or use a bake oven if one is available. Prevent spills and drips of solvents and cleansers to the shop floor. Do all liquid cleaning at a centralized station so the solvents and residues stay in one area. If you dip parts in liquid, remove them slowly to avoid spills. Locate drip pans, drain boards, and drying racks to direct drips back into a sink or fluid holding tank for reuse.

Q. Have you looked into using nontoxic or less toxic cleaners or solvents?

If possible, eliminate or reduce the number or amount of hazardous materials and waste by substituting nonhazardous or less hazardous materials. For example:

- Use noncaustic detergents instead of caustic cleaning agents for parts cleaning (ask your supplier about alternative cleaning agents).

ACTIVITIES THAT CAN CONTAMINATE STORM WATER:
Engine repair and service:
• Parts cleaning
• Shop cleanup
• Spilled fuel, oil, or other materials
• Replacement of fluids (oil, oil filters, hydraulic fluids, transmission fluid, and radiator fluids)
Outdoor vehicle and equipment storage and parking:
• Dripping engine and automotive fluids from parked vehicles and equipment
Disposal of materials or process wastes:
• Greasy rags
• Oil filters
• Air filters
• Batteries
• Spent coolant, degreasers, etc.

- Use detergent-based or water-based cleaning systems in place of organic solvent degreasers. Wash water may require treatment before it can be discharged to the sanitary sewer. Contact your local sewer authority for more information.

- Replace chlorinated organic solvents (1,1,1-trichloroethane, methylene chloride, etc.) with nonchlorinated solvents. Nonchlorinated solvents like kerosene or mineral spirits are less toxic and less expensive to dispose of but are by no means harmless themselves. Check the list of active ingredients to see whether it contains chlorinated solvents.

- Choose cleaning agents that can be recycled.

Contact your supplier or trade journal for more waste minimization ideas.

Q. Are work areas and spills washed or hosed down with water?

Clean up leaks, drips, and other spills without large amounts of water. Use rags for small spills, a damp mop for general cleanup, and dry absorbent material for larger spills. Consider the following BMPs:

- Avoid hosing down your work areas.

- Collect leaking or dripping fluids in drip pans or containers. If different liquids are kept separate, the fluids are easier to recycle.

- Keep a drip pan under the vehicle while you unclip hoses, unscrew filters, or remove other parts. Use a drip pan under any vehicle that might leak while you work on it to keep splatters or drips off the shop floor.

- Promptly transfer used fluids to the proper waste or recycling drums. Don't leave full drip pans or other open containers lying around

- Locate waste and recycling drums in properly controlled areas of the yard, preferably areas with a concrete slab and secondary containment.

Q. Are spills or materials washed or poured down the drain?

Do not pour liquid waste to floor drains, sinks, outdoor storm drain inlets, or other storm drains or sewer connections. Used or leftover cleaning solutions, solvents, and automotive fluids and oil are often toxic and should not be put into the sanitary sewer. Be sure to dispose of these materials prope   or find opportunities for reuse and recycling. If you are unsure of how to dispose of chemical wastes, contact your State hazardous waste management agency or the RCRA hotline at 1-800- 424-9346. Post signs at sinks to remind employees, and paint stencils at outdoor drains to tell customers and others not to pour wastes down drains.

Q. Are oil filters completely drained before recycling or disposal?

Oil filters disposed of in trash cans or dumpsters can leak oil and contaminate storm water. Place the oil filter in a funnel over the waste oil recycling or disposal collection tank to drain excess oil before disposal. Oil filters can be crushed and recycled. Ask your oil supplier or recycler about recycling oil filters.

Q. Are incoming vehicles and equipment checked for leaking oil and fluids?

If possible, park vehicles indoors or under a roof so storm water does not contact the area. If you park vehicles outdoors while they await repair, watch them closely for leaks.

Put pans under leaks to collect fluids for proper recycling or disposal. Keeping leaks off the ground reduces the potential for storm water contamination and reduces cleanup time and costs. If the vehicle or equipment is to be stored outdoors, oil and other fluids should be drained first.

Designate a special area to drain and replace motor oil, coolant, and other fluids, where there are no connections to the storm drain or the sanitary sewer and drips and spills can be easily cleaned up.

Q. Are wrecked vehicles or damaged equipment stored onsite?

Be especially careful with *wrecked vehicles*, whether you keep them indoors or out, as well as with vehicles kept onsite for scrap or salvage. Wrecked or damaged vehicles often drip oil and other fluids for several days.

- As the vehicles arrive, place drip pans under them immediately, even if you believe that all fluids have leaked out before the car reaches your shop.

- Build a shed or temporary roof over areas where you park cars awaiting repairs or salvage, especially if you handle wrecked vehicles. Build a roof over vehicles you keep for parts.

- Drain all fluids, including air conditioner coolant, from wrecked vehicles and "parts" cars. Also drain engines, transmissions, and other used parts.

- Store cracked batteries in a nonleaking secondary container. Do this with all cracked batteries, even if you think all the acid has drained out. If you drop a battery, treat it as if it is cracked. Put it into the containment area until you are sure it is not leaking.

> **BATTERY ACID SPILLS:**
>
> Handle spilled acid from broken batteries with care. If you use baking soda to neutralize spilled acid during cleanup, remember that the residue is still dangerous to handle and must be disposed of as a hazardous waste because it may contain lead and other contaminants.

Q. Do you recycle any of these materials?

- Degreasers

- Used oil or oil filters

- Antifreeze

- Cleaning solutions

- Automotive batteries

- Hydraulic fluid.

All of these materials can be either recycled at your facility or sent offsite for recycling. Some recycling options, ranked by level of effort required, follow.

Least Effort:
• Arrange for collection and transportation of car batteries, used oil and other fluids, cleaning solutions, and degreasers to a commercial recycling facility. This requires that you separate wastes and store them until they are picked up by the recycling company. • "Dirty" solvent can be reused. Presoak dirty parts in used solvent before cleaning the parts in fresh solvent.
Moderate Effort:
• Used oil, antifreeze, and cleaning solutions can be recycled onsite using a filtration system that removes impurities and allows the fluid to be reused. Filtration systems are commercially available.
Most Effort:
• Install an onsite solvent recovery unit. If your facility creates large volumes of used solvents, you may consider purchasing or leasing an onsite still to recover the solvent for reuse. Contact your State hazardous waste management agency for more information about onsite recycling of used solvents.

Q. Can you reduce the number of different solvents used?

Reducing the number of solvents makes recycling easier and reduces hazardous waste management costs. Often, one solvent can perform a job as well as two different solvents.

Q. Are wastes separated?

Separating wastes allows for easier recycling and may reduce treatment costs. Keep hazardous and non-hazardous wastes separate, do not mix used oil and solvents, and keep chlorinated solvents (like 1,1,1-trichloroethane) separate from nonchlorinated solvents (like kerosene and mineral spirits). Proper labeling of all wastes and materials will help accomplish this goal (see Signs and Labels BMP).

> **EMPLOYEE INVOLVEMENT IS THE KEY:**
>
> Getting employees interested in reducing waste generation is the key to a successful storm water pollution prevention plan. Discuss pollution prevention with your employees. They are most familiar with the operations that generate wastes and may have helpful waste reduction suggestions. Consider setting up an employee reward program to promote pollution prevention.

Q. Do you use recycled products?

Many products made of recycled (i.e., refined or purified) materials are available. Engine oil, transmission fluid, antifreeze, and hydraulic fluid are available in recycled form. Buying recycled products supports the market for recycled materials.

SUMMARY OF VEHICLE MAINTENANCE AND REPAIR BMPs

- Check for leaking oil and fluids.

- Use nontoxic or low-toxicity materials.

- Drain oil filters before disposal or recycling.

- Don't pour liquid waste down drains.

- Recycle engine fluids and batteries.

- Segregate and label wastes.

- Buy recycled products.

3.3 BMPs FOR PAINTING OPERATIONS

Many painting operations use materials or create wastes that are harmful to humans and the environment. Storm water runoff from areas where these activities occur can become polluted by a variety of contaminants such as solvents and dusts from sanding and grinding that contain toxic metals like cadmium and mercury. These and other potentially harmful substances in storm water can enter water bodies directly through storm drains where they can harm fish and wildlife.

The following questions will help you identify potential sources of storm water contamination from painting operations on your site and BMPs that can reduce or eliminate these sources. Reading this section can help you eliminate, reduce, or recycle pollutants that may otherwise contaminate storm water.

Q. Is care taken to prevent paint wastes from contaminating storm water runoff?

Use tarps and vacuums to collect solid wastes produced by sanding or painting. Tarps, drip pans, or other spill collection devices should be used to collect spills of paints, solvents, or other liquid materials. These wastes should be disposed of properly to keep them from contaminating storm water.

PAINTING ACTIVITIES THAT CAN CONTAMINATE STORM WATER:

- Painting and paint removal
- Sanding or paint stripping
- Spilled paint or paint thinner

Q. Are wastes from sanding contained?

Prevent paint chips from coming into contact with storm water. Paint chips may contain hazardous metallic pigments or biocides. You can reduce contamination of storm water with paint dust and chips from sanding by the following practices:

- Avoid sanding in windy weather when possible.

- Enclose outdoor sanding areas with tarps or plastic sheeting. Be sure to provide adequate ventilation and personal safety equipment. After sanding is complete, collect the waste and dispose of it properly.

- Keep workshops clean of debris and grit so that the wind will not carry any waste into areas where it can contaminate storm water.

- Move the activity indoors if you can do so safely.

Q. Are parts inspected before painting?

Inspect the part or vehicle to be painted to ensure that it is dry, clean, and rust free. Paint sticks to dry, clean surfaces, which in turn means a better, longer-lasting paint job.

Q. Are you using painting equipment that creates little waste?

As little as 30 percent of the paint may reach the target from conventional airless spray guns; the rest is lost as overspray. Paint solids from overspray are deposited on the ground where they can contaminate storm water. Other spray equipment that delivers more paint to the target and less overspray should be used:

- Electrostatic spray equipment

- Air-atomized spray guns

- High-volume/low-pressure spray guns

- Gravity-feed guns.

Q. Are employees trained to use spray equipment correctly?

Operator training can reduce overspray and minimize the amount of paint solids that can contaminate storm water. Correct spraying techniques also reduce the amount of paint needed per job. If possible, avoid spraying on windy days. When spraying outdoors, use a drop cloth or ground cloth to collect and dispose of overspray.

Q. Do you recycle paint, paint thinner, or solvents?

These materials can either be recycled at the facility or sent offsite for recycling. Some recycling options ranked by the level of effort required follow.

Least Effort:
• Dirty solvent can be reused for cleaning dirty spray equipment and parts before equipment is cleaned in fresh solvent.
• Give small amounts of left-over paint to the customer for touchup.
Moderate Effort:
• Arrange for collection and transportation of paints, paint thinner, or spent solvents to a commercial recycling facility.
Most Effort:
• Install an onsite solvent recovery unit. If your facility creates large volumes of used solvents, paint, or paint thinner, you may consider buying or leasing an onsite still to recover used solvent for reuse. Contact your State hazardous waste management agency for more information about onsite recycling of used solvents.

Q. Are wastes separated?

Separating wastes makes recycling easier and may reduce treatment costs. Keep hazardous and nonhazardous wastes separate, and keep chlorinated solvents (like 1,1,1-trichloroethane) separate from nonchlorinated solvents (like petroleum distillate and mineral spirits). Check the materials data sheet for ingredients, or talk with your waste hauler or recycling company to learn which waste types can be stored together and which should be separated.

Q. Can you reduce the number of solvents you use?

Reducing the number of solvents makes recycling easier and reduces hazardous waste management costs. Often, one solvent can do a job as well as two different solvents.

Q. Do you use recycled products?

Many products made of recycled (i.e., refined or purified) materials are available. Buying recycled paints, paint thinner, or solvent products helps build the market for recycled materials.

SUMMARY OF PAINTING OPERATION BMPs

- Inspect parts prior to painting.
- Contain sanding wastes.
- Prevent paint waste from contacting storm water.
- Proper interim storage of waste paint, solvents, etc.
- Evaluate efficiency of equipment.
- Recycle paint, paint thinner, and solvents.
- Segregate wastes.
- Buy recycled products.

3.4 BMPs FOR VEHICLE AND EQUIPMENT WASHING

Washing vehicles and equipment outdoors or in areas where wash water flows onto the ground can pollute storm water. Wash water can contain high concentrations of oil and grease, phosphates, and high suspended solid loads (these and other potentially harmful substances can pollute storm water when deposited on the ground where they can be picked up by rainfall runoff). Vehicle wash water is considered to be a process wastewater and needs to be covered by an NPDES permit. Contact your permitting authority for information about how vehicle wash water is being regulated in your area.

The following questions are designed to help you find sources of storm water contamination from vehicle and equipment washing and to select BMPs to reduce those sources. Reading this section can help you eliminate, reduce, or recycle pollutants that otherwise may contaminate storm water. Also refer to Vehicle Washing BMP in Section 4.4.

Q. Have you considered using phosphate-free biodegradable detergents?

Phosphates, which are plant nutrients, can cause excessive growth of nuisance plants in water when they enter lakes or streams in wash water. Some States ban the use of detergents containing high amounts of phosphates. Contact your supplier about phosphate-free biodegradable detergents that are available on the market.

VEHICLE AND EQUIPMENT WASHING ACTIVITIES THAT CAN CONTAMINATE STORM WATER:

- Outside equipment or vehicle cleaning (washing or steam cleaning)

- Wash water discharged directly to the ground or storm water drain

---

Q. Are vehicles, equipment, or parts washed over the open ground?

Used wash water contains high concentrations of solvents, oil and grease, detergents, and metals. Try not to wash parts or equipment outside. Washing over impervious surfaces like concrete, blacktop, or hardpacked dirt allows wash water to enter storm drains directly or deposits contaminants on the ground, where they are washed into storm drains when it rains. Washing over pervious ground such as sandy soils potentially can pollute ground water. Therefore, small parts and equipment washing should be done over a parts washing container where the wash water can be collected and recycled or disposed of properly.

EMPLOYEE INVOLVEMENT IS THE KEY:

Getting employees interested in reducing waste is the key to a successful storm water pollution prevention plan. Discuss pollution prevention with your employees. They are most familiar with the operations that generate wastes and may have helpful waste reduction suggestions. Consider setting up an employee award program to promote pollution prevention.

If you are washing large equipment or vehicles, and have to wash outside, designate a specific area for washing. This area should be bermed to collect the wastewater and graded to direct the wash water to a treatment facility. Consider filtering and recycling vehicle wash water. If recycling is not practical, the wastewater can be discharged to the sanitary sewer. Contact your local sewer authority to find out whether treatment is required before wash water is discharged to the sewer (pretreatment).

SUMMARY OF VEHICLE AND EQUIPMENT WASHING BMPs

- Consider use of phosphate-free detergents.

- Use designated cleaning areas.

- Consider recycling wash water.

3.5 BMPs FOR LOADING AND UNLOADING MATERIALS

Loading/unloading operations usually take place outside on docks or terminals. Materials spilled, leaked, or lost during loading/unloading may collect in the soil or on other surfaces and be carried away by rainfall runoff or when the area is cleaned. Rainfall may wash off pollutants from machinery used to unload or load materials. The following questions are designed to help you find sources of storm water contamination from loading and unloading materials and choose BMPs to reduce or eliminate those sources. Reading this section can start you on the road to eliminating, reducing, or recycling pollutants that otherwise may contaminate storm water. Also refer to the BMP on Loading and Unloading by Air Pressure or Vacuum in Section 4.2.

Q. Are tank trucks and material delivery vehicles located where spills or leaks can be contained?

Loading/unloading equipment and vehicles should be located so that leaks can be contained in existing containment and flow diversion systems.

Q. Is loading/unloading equipment checked regularly for leaks?

Check vehicles and equipment regularly for leaks, and fix any leaks promptly. Common areas for leaks are valves, pumps, flanges, and connections. Look for dust or fumes. These are signs that material is being lost during unloading/loading operations.

LOADING AND UNLOADING ACTIVITIES THAT CAN CONTAMINATE STORM WATER:

- Pumping of liquids or gases from barge, truck or rail car to a storage facility or vice versa

- Pneumatic transfer of dry chemicals to or from the loading and unloading vehicles

- Transfer by mechanical conveyor systems

- Transfer of bags, boxes, drums, or other containers by forklift, trucks, or other material handling equipment

---

Q. Are loading/unloading docks or areas covered to prevent exposure to rainfall?

Covering loading and unloading areas, such as building overhangs at loading docks, can reduce exposure of materials, vehicles, and equipment to rain.

Q. Are loading/unloading areas designed to prevent storm water runon?

Runon is storm water created from other areas that flows or "runs on" to your property or site. Runon flowing across loading/unloading areas can wash contaminants into storm drains. Runon can be minimized by:

- Grading, berming, or curbing the area around the loading area to direct runon away from the area

- Positioning roof down spouts so storm water is directed away from loading sites and equipment and preferably to a grassy or vegetated area where the storm water can soak into the ground.

SUMMARY OF LOADING/UNLOADING OPERATIONS BMPs

- Contain leaks during transfer.

- Check equipment regularly for leaks.

- Limit exposure of material to rainfall.

- Prevent storm water runon.

3.6 BMPs FOR LIQUID STORAGE IN ABOVE-GROUND TANKS

Accidental releases of chemicals from above-ground liquid storage tanks can contaminate storm water with many different pollutants. Materials spilled, leaked, or lost from storage tanks may accumulate in soils or on other surfaces and be carried away by rainfall runoff. The following questions can help you find sources of storm water contamination from above-ground storage tanks and select BMPs to reduce or eliminate those sources. Also refer of the BMPs listed in Section 4.2 on exposure minimization and Section 4.3 on exposure mitigation for more information.

> **Q. Do storage tanks contain liquid hazardous materials, hazardous wastes, or oil?**

THE MOST COMMON CAUSES OF UNINTENTIONAL RELEASES FROM TANKS:
• External corrosion and structural failure
• Installation problems
• Spills and overfills due to operator error
• Failure of piping systems (pipes, pumps, flanges, couplings, hoses, and valves)
• Leaks or spills during pumping of liquids or gases from barges, trucks, or rail cars to a storage facility or vice versa

Storage of oil and hazardous materials must meet specific standards set by Federal and State laws. These standards include SPCC plans, secondary containment, installation, integrity and leak detection monitoring, and emergency preparedness plans. Federal regulations set specific standards for preventing runon and collecting runoff from hazardous waste storage, disposal, or treatment areas. These standards apply to container storage areas and other areas used to store, treat, or dispose of hazardous waste. If the collected storm water is a hazardous waste, it must be managed as a hazardous waste in accordance with all applicable State and Federal environmental regulations. States may also have standards about controlling runon and runoff from hazardous waste treatment, storage, and disposal areas. To find out more about storage requirements, call the toll-free EPA RCRA hotline at 1-800-424-9346 or contact your State hazardous waste management agency.

> **Q. Are operators trained in correct operating procedures and safety activities?**

Well-trained employees can reduce human errors that lead to accidental releases or spills.

> **Q. Do you have safeguards against accidental releases?**

Engineered safeguards can help prevent operator errors that may cause the accidental release of pollutants. Safeguards include:

• Overflow protection devices on tank systems to warn the operator or to automatically shut down transfer pumps when the tank reaches full capacity

• Protective guards around tanks and piping to prevent vehicle or forklift damage

• Clearly tagging or labeling of valves to reduce human error.

Q. Are the tank systems inspected and is tank integrity tested regularly?

Visually inspect the tank system to identify problem areas before they lead to a release. Correct any problems or potential problems as soon as possible. An audit of a newly installed tank system by a registered and specially trained professional engineer can identify and correct potential problems such as loose fittings, poor welding, and improper or poorly fitted gaskets. After installation, have operators visually inspect the tank system on a routine basis. Areas to inspect include tank foundations, connections, coatings, tank walls, and the piping system. Look for corrosion, leaks, straining of tank support structures from leaks, cracks, scratches in protective coatings, or other physical damage that may weaken the tank system. Integrity testing should be done periodically by a qualified professional.

Q. Are tanks bermed or surrounded by a secondary containment system?

A secondary containment system around both permanent and temporary tanks allows leaks to be more easily detected and contains spills or leaks. Methods include berms, dikes, liners, vaults, and double-walled tanks. See Chapter 4 for additional information on containment and spill control.

SUMMARY OF BMPs FOR LIQUID STORAGE IN ABOVE-GROUND TANKS

- Comply with applicable State and Federal laws.

- Properly train employees.

- Install safeguards against accidental releases.

- Routinely inspect tanks and equipment.

- Consider installing secondary containment.

3.7 BMPs FOR INDUSTRIAL WASTE MANAGEMENT AREAS AND OUTSIDE MANUFACTURING

Storm water runoff from areas where industrial waste is stored, treated, or disposed of can be polluted. Outside manufacturing activities can also contaminate storm water runoff. Activities such as rock grinding or crushing, painting or coating, grinding or sanding, degreasing or parts cleaning, or operations that use hazardous materials are particularly dangerous. Wastes spilled, leaked, or lost from waste management areas or outside manufacturing activities may build-up in soils or on other surfaces and be carried away by rainfall runoff. There is also a potential for liquid wastes from lagoons or surface impoundments to overflow to surface waters or soak the soil where they can be picked up by storm water runoff. Possible storm water contaminants include toxic compounds, oil and grease, paints or solvents, heavy metals, and high levels of suspended solids.

The best way to reduce the potential for storm water contamination from both waste management areas and outside manufacturing activities is to reduce the amount of waste that is created and, consequently, the amount that must be stored or treated. The following questions are designed to help you find BMPs that can eliminate or reduce the amount or toxicity of industrial wastes as well as minimize contamination of storm water from existing waste management areas. Waste reduction BMPs are appropriate for a wide range of industries and are designed to provide ideas on ways to reduce wastes. Turn to Appendix D for a list of State and Federal pollution prevention resources that can provide more information and assistance in choosing industrial waste reduction BMPs.

Q. Have you looked for ways to reduce waste at your facility?

The first step to reducing wastes is to assess activities at your facility. The assessment is designed to find situations at your facility where you can eliminate or reduce waste generation, emissions, and environmental damage. The assessment involves steps very similar to those used to develop your Storm Water Pollution Prevention Plan, such as collecting process-specific information; setting pollution prevention targets; and developing, screening, and selecting waste reduction options for further study. Starting a waste reduction program at your facility has many potential benefits. Some of these benefits are direct (e.g., cost savings from reduced raw material use), while others are indirect (e.g., avoided waste disposal fees).

EPA has developed a series of industry-specific pollution prevention waste minimization guidance manuals. The manuals contain steps for assessing your facility's opportunity for reducing waste and describe source reduction and recycling choices. The manuals currently available are listed in Appendix D.

INDUSTRIAL WASTE MANAGEMENT ACTIVITIES OR AREAS THAT CAN CONTAMINATE STORM WATER:

- Landfills

- Waste piles

- Wastewater and solid waste treatment and disposal:

   - Waste pumping
   - Additions of treatment chemicals
   - Mixing
   - Aeration
   - Clarification
   - Solids dewatering

- Land application

Q. Have you considered waste reduction BMPs?

There are many different types of BMPs that can help eliminate or reduce the amount of industrial waste generated at your facility. Some of these BMPs are listed below and referenced in Appendix D.

- Production planning and sequencing

- Process or equipment modification

- Raw material substitution or elimination

- Loss prevention and housekeeping

- Waste segregation and separation

- Closed-loop recycling

- Training and supervision

- Reuse and recycling.

OUTSIDE MANUFACTURING ACTIVITIES OR SITUATIONS THAT CAN CONTAMINATE STORM WATER:

- Processes or equipment that generate dusts, vapors, or emissions

- Outside storage of hazardous materials or raw materials

- Dripping or leaking fluids from equipment or processes

- Liquid wastes discharged directly onto the ground or into the storm sewer

Q. Are industrial waste management and outside manufacturing areas checked often for spills and leaks?

By checking waste management areas for leaking containers or spills, you can prevent wastes from contaminating storm water. Look for containers that are rusty, corroded, or damaged. Transfer wastes from these damaged containers into safe containers. Close the lids on dumpsters to prevent rain from washing wastes out of holes or cracks in the bottom of the dumpster. In outside manufacturing areas, look for leaking equipment (e.g., valves, lines, seals, or pumps) and fix leaks promptly. Inspect rooftop and other outdoor equipment (e.g., HVAC devices, air pollution control devices, transformers, piping, etc.) for leaks or dust concentrations.

Q. Are industrial waste management areas or manufacturing activities covered, enclosed, or bermed?

The best way to avoid contaminating storm water from existing waste management and manufacturing areas is to prevent storm water runon or rain from entering or contacting these areas. This can be done by:

- Preventing direct contact with rain

- Moving the activity indoors after ensuring that all safety concerns such as fire hazard and ventilation are addressed

- Covering the area with a permanent roof

- Covering waste piles with a temporary covering material such as a reinforced tarpaulin, polyethylene, polyurethane, polypropylene, or Hypalon

- Minimizing storm water runon by enclosing the area or building a berm around the area.

Q. Are vehicles used to transport wastes to the land disposal or treatment site equipped with anti-spill equipment?

Transport vehicles equipped with spill prevention equipment can prevent spills of wastes during transport. Examples include:

- Vehicles equipped with baffles for liquid wastes

- Trucks with sealed gates and spill guards for solid wastes

- Trucks with tarps.

Q. Do you use loading systems that minimize spills and fugitive losses such as dust or mists?

Wastes lost during loading or unloading can contaminate storm water. Vacuum transfer systems minimize waste loss.

Q. Are sediments or wastes prevented from being tracked offsite?

Wastes and sediments tracked offsite can end up on streets where they are picked up by storm water runoff. This can be avoided by using vehicles with specially designed tires, washing vehicles in a designated area before they leave the site, and controlling the wash water.

Q. Is storm water runoff minimized from the land disposal site?

You can take certain precautions to minimize the runoff of polluted storm water from land application sites. Some precautions are detailed below.

DO YOU OWN OR OPERATE A HAZARDOUS WASTE TREATMENT, STORAGE, AND DISPOSAL FACILITY?

Federal and State laws establish strict standards for managing solid and hazardous wastes. If you are not sure whether you own or operate a hazardous waste treatment, storage, or disposal facility, call the toll-free EPA RCRA hotline at 1-800-424-9346 or contact your State hazardous waste management agency. Federal regulations contain specific standards about preventing runon and collecting runoff from hazardous waste storage, disposal, or treatment areas. These standards apply to land treatment units, landfills, waste piles, container storage areas, and other areas used to store, treat, or dispose of hazardous waste. If the collected storm water is a hazardous waste, it must be managed in accordance with all applicable State and Federal environmental regulations. States may also have standards about controlling runon and runoff from hazardous waste treatment, storage, and disposal areas.

- Choose the land application site carefully. Characteristics that help prevent runoff include slopes under 6 percent, permeable soils, a low water table, locations away from wetlands or marshes, and closed drainage systems.

- Avoid applying waste to the site when it is raining or when the ground is frozen or saturated with water. Grow vegetation on areas dedicated to land disposal to stabilize the soils and reduce the volume of surface water runoff from the site.

- Maintain adequate barriers between the land application site and receiving waters.

- Erosion control techniques might include mulching and matting, filter fences, straw bales, diversion terracing, or sediment basins. For a detailed description of erosion control techniques, see Chapter 4.

- Perform routine maintenance to ensure that erosion control or site stabilization measures are working.

SUMMARY OF INDUSTRIAL WASTE MANAGEMENT AND OUTSIDE MANUFACTURING BMPs

- Conduct a waste reduction assessment.

- Institute industrial waste source reduction and recycling BMPs.

- Prevent runoff and runon from contacting the waste management area.

- Minimize runoff from land application sites.

3.8 BMPs FOR OUTSIDE STORAGE OF RAW MATERIALS, BY-PRODUCTS, OR FINISHED PRODUCTS

Raw materials, by-products, finished products, containers, and material storage areas exposed to rain and/or runoff can pollute storm water. Storm water can become contaminated by a wide range of contaminants (e.g., metals, oil, and grease) when solid materials wash off or dissolve into water, or by spills or leaks. The following questions are designed to help you identify potential sources of storm water contamination and select BMPs that can reduce or eliminate those sources. Reading this section can help you eliminate or reduce pollutants that otherwise may contaminate storm water.

Q. Are materials protected from rainfall, runon, and runoff?

The best way to avoid contaminating storm water from outside material storage areas is to prevent storm water runon or rain from coming in contact with the materials. This can be done by:

- Storing the material indoors

- Covering the area with a roof

- Covering the material with a temporary covering made of polyethylene, polyurethane, polypropylene, or Hypalon.

- Minimizing storm water runon by enclosing the area or building a berm around the area.

> **ARE ANY OF THESE MATERIALS STORED OUTSIDE OR IN AREAS WHERE THEY CAN CONTAMINATE STORM WATER?**
>
> - Fuels
> - Raw materials
> - By-products
> - Intermediates
> - Final products
> - Process residuals

> **SUMMARY OF BMPs FOR OUTSIDE STORAGE OF RAW MATERIALS, BY-PRODUCTS, OR FINISHED PRODUCTS**
>
> - Cover or enclose materials.

3.9 BMPs FOR SALT STORAGE FACILITIES

Salt left exposed to rain or snow can be lost. Salt spilled or blown onto the ground during loading and unloading will dissolve in storm water runoff. Storm water contaminated with salt can be harmful to vegetation and aquatic life. Salty storm water runoff soaking into the ground may contaminate ground water and make it unsuitable as a drinking water supply. The following BMPs will help reduce storm water contamination from salt storage and transfer activities. See Chapter 4 for more detailed information on covering storage areas.

Q. Are salt piles protected from rain?

The best way to prevent salt from contaminating storm water is to eliminate or limit the exposure of salt to rain. Preventing contact with rain also protects against salt loss and prevents salt from absorbing moisture and becoming caked or lumpy and making it difficult to handle and use.

- Store salt under a roof. This is the best way to stop direct contact with rain.

If salt must be stored outside:

- Build the storage pile on asphalt to reduce the potential for ground water contamination

- Cover the pile with a temporary covering material such as polyethylene, polyurethane, polypropylene, or Hypalon.

SALT STORAGE ACTIVITIES THAT CAN CONTAMINATE STORM WATER:

- Salt stored outside in piles or bags that are exposed to rain or snow

- Salt loading and unloading areas located outside or in areas where spilled salt can contaminate storm water.

Q. Is storm water runon prevented from contacting storage piles and loading and unloading areas?

Storm water runon can be minimized by enclosing the area or building berms around storage, loading, and unloading areas.

SUMMARY OF SALT STORAGE FACILITIES BMPs

- Put it under a roof.

- Use temporary covers.

- Enclose or berm transfer areas.

CHAPTER

4

SITE-SPECIFIC INDUSTRIAL STORM WATER BMPs

This chapter describes some of the possible Best Management Practices (BMPs) that you might include in your Storm Water Pollution Prevention Plan so that pollutants from your site do not mix with storm water.

Table 4.1 provides an easy index of the BMP descriptions that follow. The BMPs are grouped by section into six categories: Flow Diversion Practices; Exposure Minimization Practices; Mitigative Practices; Other Preventive Practices; Sediment and Erosion Prevention Practices; and Infiltration Practices.

The following information is provided for each BMP: (1) description of the BMP; (2) when and where the BMP can be used; (3) factors that should be considered when using the BMP; and (4) advantages and disadvantages of the BMP. More detailed fact sheets for a limited number of the Sediment and Erosion Prevention Practices are included as Appendix E. When designing these structural controls, EPA recommends that you refer to any State or local storm water management design standards.

TABLE 4.1 INDEX OF SITE-SPECIFIC INDUSTRIAL STORM WATER BMPs (Continued)

4.1 FLOW DIVERSION PRACTICES

Structures that divert stream flow (such as gutters, drains, sewers, dikes, and graded pavement) are used as BMPs in two ways. First, flow diversion structures, called storm water conveyances, may be used to channel storm water away from industrial areas so that pollutants do not mix with the storm water. Second, they also may be used to carry pollutants directly to a treatment facility. This section briefly describes flow diversion as a BMP for industrial storm water.

Storm Water Conveyances
(Channels/Gutters/Drains/Sewers)

What Are They

Storm water conveyances such as channels, gutters, drains, and sewers, collect storm water runoff and direct its flow. A group of connecting conveyances is sometimes installed at an industrial facility to create a storm water collection system. Storm water conveyances can be used for two different purposes. The first purpose is to keep uncontaminated storm water from coming in contact with areas of an industrial site where it may become contaminated with pollutants. This can be accomplished by collecting the storm water in a conveyance and by changing the direction of flow away from those areas. The second purpose is to collect and carry the storm water that has already come into contact with industrial areas and become contaminated to a treatment facility.

Storm water conveyances can be constructed or lined with many different materials, including concrete, clay tiles, asphalt, plastics, metals, riprap, compacted soils, and vegetation. The type of material used depends on the use of the conveyance. These conveyances can be temporary or permanent.

When and Where to Use Them

Storm water conveyances work well at most industrial sites. Storm water can be directed away from industrial areas by collecting it in channels or drains before it reaches the areas. In addition, conveyances can be used to collect storm water downhill from industrial areas and keep it separate from runoff that has not been in contact with those areas. When potentially contaminated storm water is collected in a conveyance like this, it can be directed to a treatment facility on the site if necessary. (If a pollutant is spilled, it should not be allowed to enter a storm water conveyance or drain system.)

What to Consider

In planning for storm water conveyances, consider the amount and speed of the typical storm water runoff. Also, consider the patterns in which the storm water drains so that the channels may be located to collect the most flow and can be built to handle the amount of water they will receive. When deciding on the type of material for the conveyance, consider the resistance of the material, its durability, and compatibility with any pollutants it may carry.

Conveyance systems are most easily installed when a facility is first being constructed. Use of existing grades will decrease costs. Grades should be positive to allow for the continued movement of the runoff through the conveyance system; however, grades should not create an increase in velocity that causes an increase in erosion (this will also depend upon what materials the conveyance is lined with and the types of outlet controls that are provided).

Ideally, storm water conveyances should be inspected to remove debris within 24 hours of rainfall, or daily during periods of prolonged rainfall, since heavy storms may clog or damage them. It is important to repair damages to these structures as soon as possible.

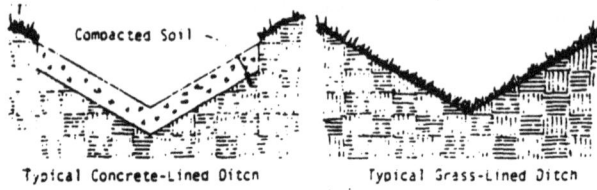

Typical Concrete-Lined Ditch   Typical Grass-Lined Ditch

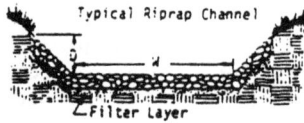

Vegetated V-shaped Waterway with Stone Center Drain

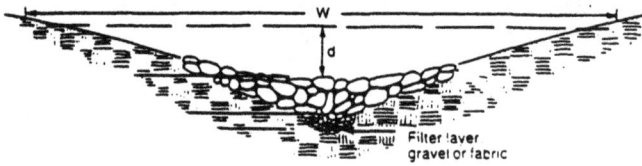

Trapezoidal Riprap Channel

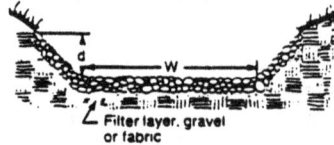

Vegetated Parabolic-shaped Waterway with Stone Center Drain

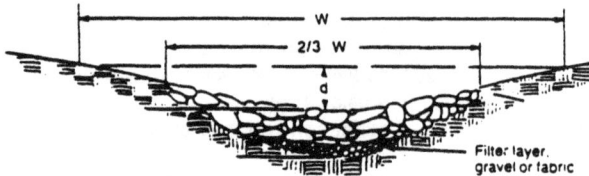

FIGURE 4.1 TYPICAL STORM WATER CONVEYANCE CROSS SECTIONS
(Modified from Commonwealth of Virginia, 1980)

Advantages of Storm Water Conveyances (Channels/Gutters/Drains/Sewers)
• Direct storm-water flows around industrial areas
• Prevent temporary flooding of industrial site
• Require low maintenance
• Provide erosion resistant conveyance of storm water runoff
• Provide long-term control of storm water flows
Disadvantages of Storm Water Conveyances (Channels/Gutters/Drains/Sewers)
• Once flows are concentrated in storm water conveyances, they must be routed through stabilized structures all the way to their discharge to the receiving water or treatment plant to minimize erosion
• May increase flow rates
• May be impractical if there are space limitations
• May not be economical, especially for small facilities or after a site has already been constructed

Diversion Dikes

What Are They

Diversion dikes or berms are structures used to block runoff from passing beyond a certain point. Temporary dikes are usually made with compacted soil. More permanent ridges are constructed out of concrete, asphalt, or similar materials.

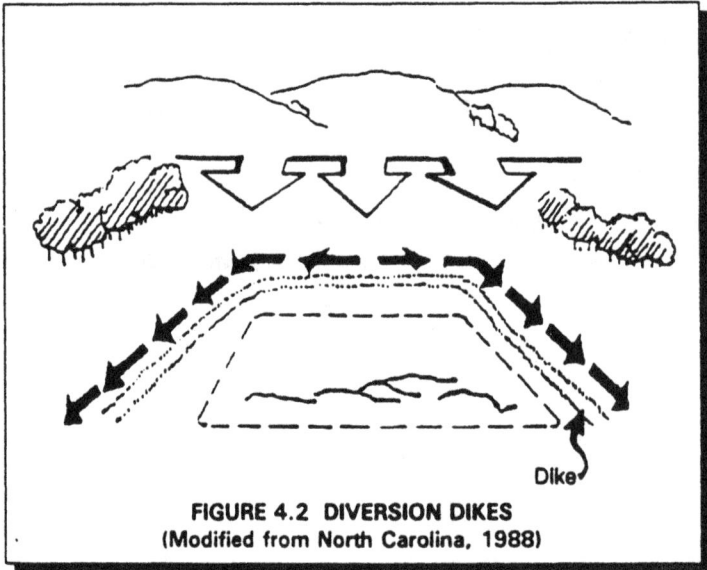

FIGURE 4.2 DIVERSION DIKES
(Modified from North Carolina, 1988)

When and Where to Use Them

Diversion dikes are used to prevent the flow of storm water runoff onto industrial areas. Limiting the volume of flow across industrial areas reduces the volume of storm water that may carry pollutants from the area, requiring treatment for pollutant removal. This BMP is suitable for industrial sites where significant volumes of storm water runoff tend to flow onto active industrial areas. Typically, dikes are built on slopes just uphill from an industrial area together with some sort of a conveyance such as a swale. The storm water conveyance is necessary to direct the water away from the dike so that the water will not pool and seep through the dike.

What to Consider

In planning for the installation of dikes, consider the slope of the drainage area, the height of the dike, the size of rainfall event it will need to divert, and the type of conveyance that will be used with the dike. Steeper slopes result in higher volumes of runoff and higher velocities; therefore, the dike must be constructed to handle this situation. Remember that dikes are limited in their ability to manage large volumes of runoff.

Ideally, dikes are installed before industrial activity begins. However, dikes can be easily constructed at any time. Temporary dikes (usually made of dirt) generally only last for 18 months or less, but they can be made into permanent structures by stabilizing them with vegetation. Vegetation is crucial for preventing the erosion of the dike.

Dikes should be inspected regularly for damage. This is especially important after storm events since a heavy rain may wash parts of a temporary dike away. Any necessary repairs should be made immediately to make sure the structure continues to do its job.

Advantages of Diversion Dikes
• Effectively limit storm water flows over industrial site areas
• Can be installed at any time
• Are economical temporary structures, when built from soil onsite
• Can be converted from temporary to permanent at any time
Disadvantages of Diversion Dikes
• Are not suitable for large drainage areas unless there is a gentle slope
• May require maintenance after heavy rains

Graded Areas and Pavement

What Is It

Land surfaces can be graded or graded and paved so that storm water runoff is directed away from industrial activity areas. The slope of the grade allows the runoff to flow, but limits the runoff from washing over areas that may be contaminated with pollutants. Like conveyances and dikes, graded areas can prevent runoff from contacting industrial areas and becoming contaminated with pollutants from these areas. Grading can be a permanent or temporary control measure.

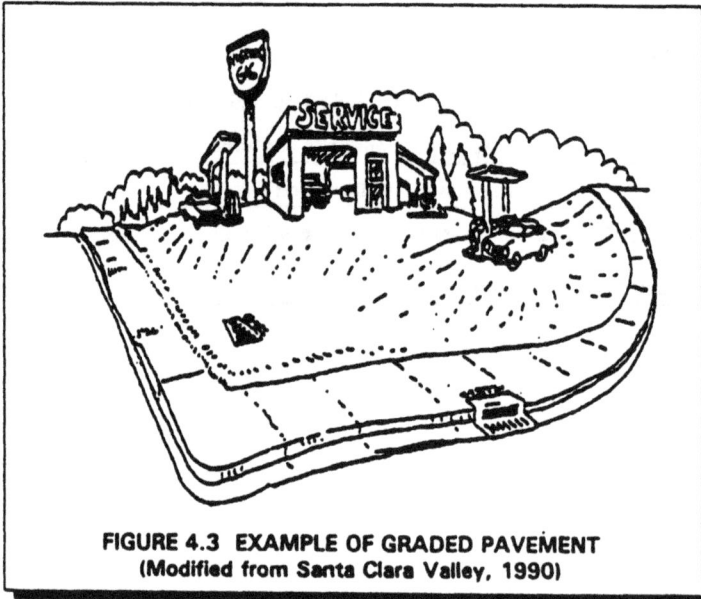

FIGURE 4.3 EXAMPLE OF GRADED PAVEMENT
(Modified from Santa Clara Valley, 1990)

When and Where to Use It

Grading land surfaces is appropriate for any industrial site that has outdoor activities that may contaminate storm water runoff, such as parking lots or outdoor storage areas. Figure 4.3 illustrates the use of graded pavement in preventing runoff from washing over a service station site. Grading is often used with other practices, such as coverings, buffer zones, and other practices to reduce the runoff velocity and provide infiltration of the uncontaminated runoff, or to direct pollutant runoff to storm water treatment facilities.

What to Consider

When designing graded areas and pavement, both control and containment of runoff flows should be considered. The grading should control the uncontaminated flow by diverting it around areas that may have pollutants. The grading should also contain the contaminated flows or divert them to treatment facilities.

When regrading and paving an industrial area, the use of concrete paving instead of asphalt should be considered. This is especially important in potential spill sites or hazardous material storage areas. Asphalt absorbs organic pollutants and can be slowly dissolved by some fluids, thus becoming a possible source of contaminants itself. This control measure should be used with a cover, such as a roof, in areas where contaminants are of concern (see Covering BMP) so that rain or snow does not fall on the area and wash the contaminants down slope.

Inspect paving regularly for cracks that may allow contaminants to seep into the ground. Also, check to make sure that the drains receiving the storm water flow from the paved area remain unclogged with sediment or other debris so that low areas do not flood and wash over the areas where the contaminants may be.

Advantages of Graded Areas and Pavement
• Is effective in limiting storm water contact with contaminants
• Is relatively inexpensive and easily implemented
Disadvantages of Graded Areas and Pavement
• May be uneconomical to regrade and resurface large areas
• May not be effective during heavy precipitation

4.2 EXPOSURE MINIMIZATION PRACTICES

By eliminating or minimizing the possibility of storm water coming into contact with pollutants, facilities can eliminate or minimize the contamination of storm water discharges associated with their industrial activity. As a result, fewer materials will be carried away by storm water runoff, the costs of collecting and treating contaminated storm water will be decreased, and safety and environmental liabilities that result from spills and leaks will be reduced.

Completely eliminating the exposure of materials to storm water is not always possible, however. For many industrial facilities, enclosure of facility grounds is not technologically or economically possible. Therefore, this section describes several simple and inexpensive structural and nonstructural BMPs that a facility can use to minimize the exposure of materials to storm water.

Containing spills is one of the primary methods of minimizing exposure of contaminants to storm water runoff. Spill containment is used for enclosing any drips, overflows, leaks, or other liquid material releases, as well as for isolating and keeping pollutant spills away from storm water runoff.

There are numerous spill containment methods, ranging from large structural barriers to simple, small drip pans. The benefits of each of these practices vary based on cost, need for maintenance, and size of the spill they are designed to control. This section describes several containment methods, including:

- Containment Diking

- Curbing

- Drip Pans

- Catch Basins

- Sumps.

Other practices commonly used to minimize exposure of contaminants are also discussed, including the following:

- Covering

- Vehicle Positioning

- Loading and Unloading by Air Pressure or Vacuum.

Containment Diking

What Is It

Containment dikes are temporary or permanent earth or concrete berms or retaining walls that are designed to hold spills. Diking, one of the most common types of containment, is an effective method of pollution prevention for above-ground liquid storage tanks and rail car or tank truck loading and unloading areas. Diking can provide one of the best protective measures against the contamination of storm water because it surrounds the area of concern and holds the spill, keeping spill materials separated from the storm water outside of the diked area.

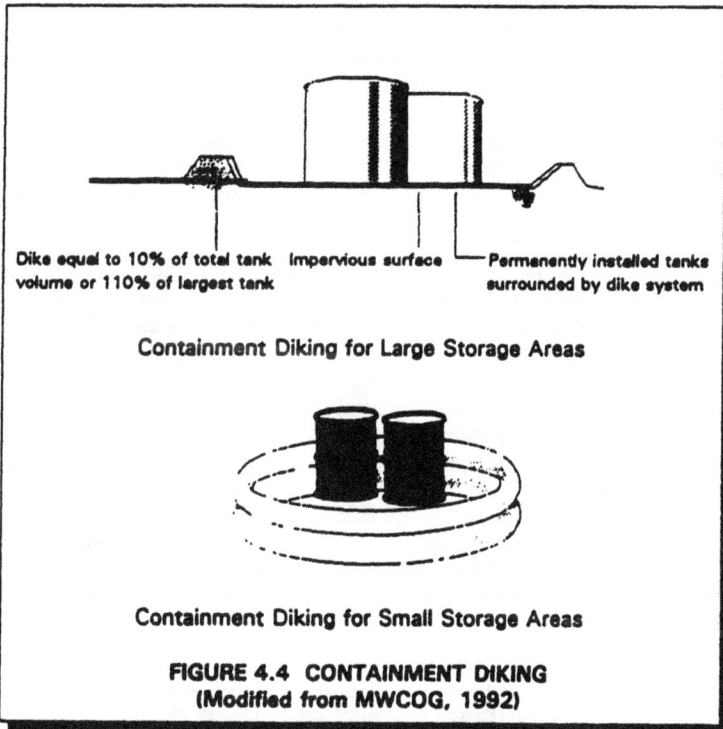

Dike equal to 10% of total tank   Impervious surface   Permanently installed tanks
volume or 110% of largest tank   surrounded by dike system

Containment Diking for Large Storage Areas

Containment Diking for Small Storage Areas

FIGURE 4.4 CONTAINMENT DIKING
(Modified from MWCOG, 1992)

When and Where to Use It

Diking can be used at any industrial facility but is most commonly used for controlling large spills or releases from liquid storage areas and liquid transfer areas.

What to Consider

Containment dikes should be large enough to hold an amount equal to the largest single storage tank at the particular facility plus the volume of rainfall. For rail car and tank truck loading and unloading operations, the diked area should be capable of holding an amount equal to any single tank truck compartment. Materials used to construct the dike should be strong enough to safely hold spilled materials. The materials used usually depend on what is available onsite and the substance to be contained, and may consist of earth (i.e., soil or clay), concrete, synthetic materials (liners), metal, or other impervious materials. In general, strong acids and bases may react with metal containers, concrete, and some plastics, so where spills may consist of these substances, other alternatives should be considered. Some of the more reactive organic chemicals may also need to be contained with special liners. If there are any questions about storing chemicals in certain dikes because of their construction materials, refer to the Material Safety Data Sheets (MSDSs).

Containment dikes may need to be designed with impervious materials to prevent leaking or contamination of storm water, surface, and ground water supplies.

Similarly, uncontrolled overflows from diked areas containing spilled materials or contaminated storm water should be prevented to protect nearby surface waters or ground waters. Therefore, dikes should have either pumping systems (see Sumps BMP) or vacuum trucks available to remove the spilled materials. When evaluating the performance of the containment system, you should pay special attention to the overflow system, since it is often the source of uncontrolled leaks. If overflow systems do not exist, accumulated storm water should be released periodically. Contaminated storm water should be treated prior to release. Mechanical parts, such as pumps or even manual systems (e.g., slide gates, stopcock valves), may require regular cleaning and maintenance.

When considering containment diking as a BMP, you should consult local authorities about any regulations governing construction of such structures to comply with local and State requirements. Facilities located in a flood plain should contact their local flood control authority to ensure that construction of the dikes is permitted.

Inspections of containment dikes should be conducted during or after significant storms or spills to check for washouts or overflows. In addition, regular checks of containment dikes (i.e., testing to ensure that dikes are capable of holding spills) is recommended. Soil dikes may need to be inspected on a more frequent basis.

Changes in vegetation, inability of the structure to retain storm water dike erosion, or soggy areas indicate problems with the dike's structure. Damaged areas should be patched and stabilized immediately, where necessary. Earthen dikes may require special maintenance of vegetation, such as mowing and irrigation.

Advantages of Containment Diking
• Contains spills, leaks, and other releases and prevent them from flowing into runoff conveyances, nearby streams, or underground water supplies
• Permits materials collected in dikes to be recycled
• Is a common industry practice for storage tanks and already required for certain chemicals
Disadvantages of Containment Diking
• May be too expensive for some smaller facilities
• Requires maintenance
• Could collect contaminated storm water, possibly resulting in infiltration of storm water to ground water

Curbing

What Is It

Like containment diking, curbing is a barrier that surrounds an area of concern. Curbing functions in a similar way to prevent spills, leaks, etc. from being released to the environment by routing runoff to treatment or control areas. The terms curbing and diking are sometimes used interchangeably.

Because curbing is usually small-scale, it cannot contain large spills like diking can, however, curbing is common at many facilities in small areas where handling and transferring liquid materials occur.

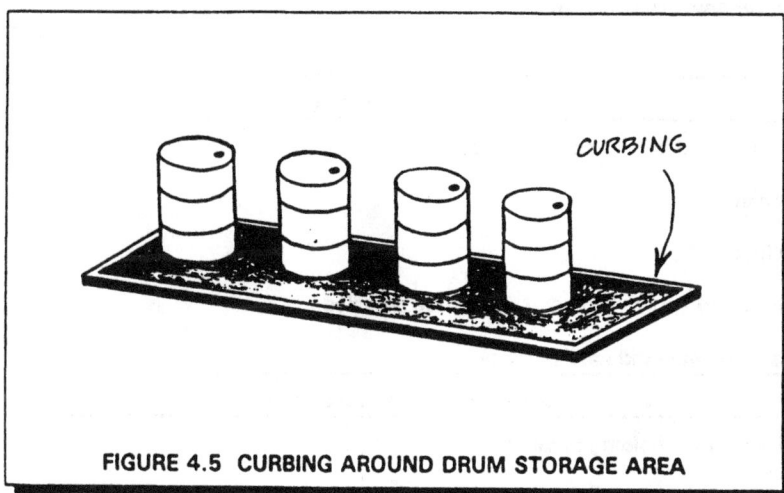

CURBING

FIGURE 4.5 CURBING AROUND DRUM STORAGE AREA

When and Where to Use It

Curbing can be used at all industrial facilities. It is particularly useful in areas where liquid materials are transferred and as a storm water runoff control.

As with diking, common materials for curbing include earth, concrete, synthetic materials, metal, or other impenetrable materials. Asphalt is also a common material used in curbing.

What to Consider

For maximum efficiency of curbing, spilled materials should be removed immediately, to allow space for future spills. Curbs should have pumping systems, rather than drainage systems, for collecting spilled materials. Manual or mechanical methods, such as those provided by sump systems (see Sump BMP), can be used to remove the material. Curbing systems should be maintained through curb repair (patching and replacement).

When using curbing for runoff control, facilities should protect the berm by limiting traffic and installing reinforced berms in areas of concern.

Spills of materials that are stored within a curbed area can be tracked outside of that area when personnel and equipment leave the area. This tracking can be minimized by grading within the curbing to direct the spilled materials to a down-slope side of the curbing. This will keep the materials away from personnel and equipment that pass through the area. It will also allow the materials to accumulate in one area making cleanup much easier.

Inspections should also be conducted <u>before</u> forecasted rainfall events and immediately after storm events. If spilled or leaked materials are observed, cleanup should start immediately. This will prevent overflows and/or contamination of storm water runoff. In addition, prompt cleanup of materials will prevent dilution by rainwater, which can adversely affect recycling opportunities. Inspection of curbed areas should be conducted regularly, to clear clogging debris. Because curbing is sized to contain small spill volumes, maintenance should also be conducted frequently to prevent overflow of any spilled materials.

Advantages of Curbing
• Is an excellent method to control <u>runon</u>
• Is inexpensive
• Is easily installed
• Materials spilled within curbed areas can be recycled
• Exists as a common industry practice
Disadvantages of Curbing
• Is not effective for holding large spills
• May require more maintenance than diking

Drip Pans

What Are They

Drip pans are small depressions or pans used to contain very small volumes of leaks, drips, and spills that occur at a facility. Drip pans can be depressions in concrete, asphalt, or other impenetrable materials or they can be made of metals, plastic, or any material that does not react with the dripped chemicals. Drip pans can be temporary or permanent.

Drip pans are used to catch drips from valves, pipes, etc. so that the materials or chemicals can be cleaned up easily or recycled before they can contaminate storm water. Although leaks and drips should be repaired and eliminated as part of a preventive maintenance program, drip pans can provide a temporary solution where repair or replacement must be delayed. In addition, drip pans can be an added safeguard when they are positioned beneath areas where leaks and drips may occur.

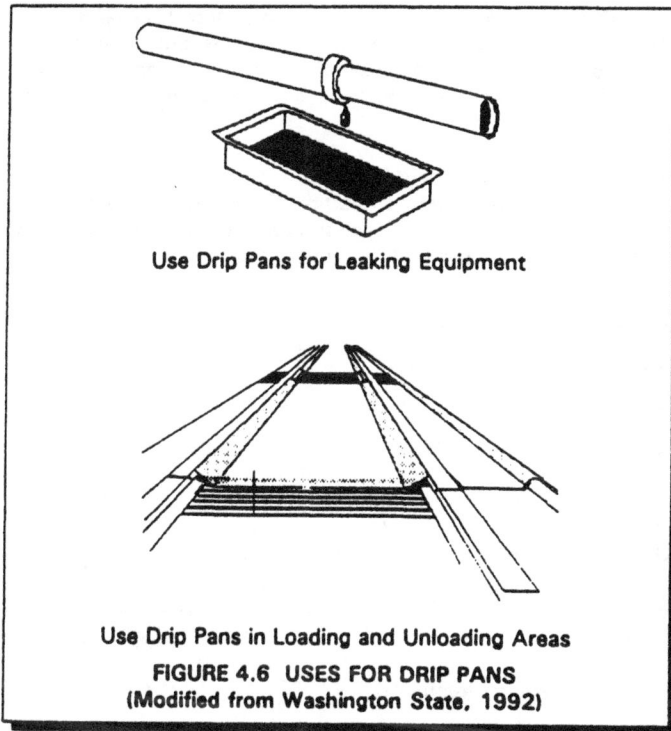

Use Drip Pans for Leaking Equipment

Use Drip Pans in Loading and Unloading Areas

FIGURE 4.6 USES FOR DRIP PANS
(Modified from Washington State, 1992)

When and Where to Use Them

Drip pans can be used at any industry where valves and piping are present and the potential for small volume leakage and dripping exist.

What to Consider

When using drip pans, consider the location of the drip pan, weather conditions, the type of material to be used for the drip pan, and how it will be cleaned.

The location of the drip pan is important. Because drip pans must be inspected and cleaned frequently, they must be easy to reach and remove. In addition, take special care to avoid placing drip pans in precarious positions such as next to walkways, on uneven pavement/ground, or sitting on pipelines. Drip pans in these locations are easily overturned and may present a safety hazard, as well as an environmental hazard.

Weather conditions are also important factors. Heavy winds and rainfall move or damage drip pans because of their small size and their light weight (if not built-in). To prevent this, secure the pans by installing or anchoring them. Drip pans may be placed on platforms or behind wind blocks or tied down.

For drip pans to be effective, employees must pay attention to the pans and empty them when they are nearly full. Because of their small holding capacities, drip pans will easily overflow if not emptied. Also, recycling efforts can be affected if storm water accumulates in drip pans and dilutes the spilled material. It is important to have clearly specified and easily followed practices of reuse/recycle and/or disposal, especially the disposal of hazardous materials. Many facilities dump the drip pan contents into a nearby larger volume storage container and periodically recycle the contents of the storage container.

In addition, frequent inspection of the drip pans is necessary due to the possibility of leaks in the pan itself or in piping or valves that may occur randomly or irregular slow drips that may increase in volume. Conduct inspections before forecasted rainfall events to remove accumulated materials and immediately after storm events to empty storm water accumulations.

Advantages of Drip Pans
• Are inexpensive
• Are easily installed and simple to operate
• Allow for reuse/recycle of collected material
• Empty or discarded containers may be reused as drip pans
Disadvantages of Drip Pans
• Contain small volumes only
• Must be inspected and cleaned frequently
• Must be secured during poor weather conditions
• Contents may be disposed of improperly unless facility personnel are trained in proper disposal methods

Collection Basins

What Are They

Collection basins, or storage basins, are permanent structures where large spills or contaminated storm water are contained and stored before cleanup or treatment. Collection basins are designed to receive spills, leaks, etc. that may occur and prevent these materials from being released to the environment. Unlike containment dikes, collection basins can receive and contain materials from many locations across a facility.

Collection basins are commonly confused with treatment units such as ponds, lagoons, and other containment structures. Collection basins differ from these structures because they are designed to temporarily store storm water rather than treat it.

When and Where to Use Them

Collection basins are appropriate for all industrial sites where space allows. Collection basins are particularly useful for areas that have a high spill potential.

What to Consider

The design and installation considerations for collection basins include sizing the basin either to hold a certain amount of spill or a certain size storm, or both. In designing the collection system, the type of material for the conveyances, compatibility of various materials to be carried through the system, and requirements for compliance with State and local regulations should be considered. Ideally, the system should function to route the materials quickly and easily to the collection basin.

When spills occur, the collection system must route the spill or storm water immediately to the collection basin. After a spill is contained, the collection system and basin may require cleaning. Remove the collection basin contents immediately to prevent an unintentional release and recycle the spilled material as much as possible. Inspect the structure on a regular basis and after storm events or spills. Depending upon the types of pollutants that may be in the storm water, or are collected as spills, design of the basin may require a liner to prevent infiltration into the ground water. Make sure that the installation of this BMP does not violate State ground water regulations.

If it is possible that the collection basin may handle combustible or flammable spilled materials, explosion-proof pumping equipment and controls or other appropriate precautions should be taken to prevent explosions or fires. Consult OSHA and local safety codes and standards for specific requirements and guidance.

Advantages of Collection Basins
• Can store contaminated storm water until directed to a treatment facility
• Can collect spills for recycling where materials are separated
Disadvantages of Collection Basins
• May need a conveyance system for increased effectiveness
• May collect materials that are not compatible
• May reduce the potential for recycling materials by collecting storm water, which dilutes the materials
• May create ground water problems if pollutants infiltrate into ground

Sumps

What Are They

Sumps are holes or low areas that are structured so that liquid spills or leaks will flow down toward a particular part of a containment area. Frequently, pumps are placed in a depressed area and are turned on automatically to transfer liquids away from the sump when the level of liquids gets too high. Sumps can be temporary or permanent.

When and Where to Use Them

Sumps can be used at all facilities. Sumps are used with other spill containment and treatment measures and can be located almost anywhere onsite. Sumps are frequently located in low lying areas within material handling or storage areas.

What to Consider

When designing and installing a sump system, consider the pump location, function, and system alarms. Design and install the sump in the lowest lying area of a containment structure, allowing for materials to gather in the area of the sump. Construct the sump of impenetrable materials and provide a smooth surface so that liquids are funneled toward the pump. It may be appropriate to house the pumps in a shed or other structure for protection and stabilization.

There are numerous factors that should be considered when purchasing a pump. Base the size of the pump on the maximum expected volume to be collected in the containment structure. In some cases, more than one pump may be appropriate. Typically, pumps that can be submerged under the spill are the most appropriate for areas where large spills may occur and that may submerge the sump area. The viscosity (thickness) of the material and the distance that the material must be pumped are also important considerations. Install pumps according to the manufacturer's recommendations.

An alarm system can be installed for pumps that are used to remove collected materials. An alarm system can indicate that a pump should be operated by hand or that an automatically operated pump has failed to function. Ultimately, facility personnel should have some mechanism to take action to prevent spills from by-passing and overflowing containment structures.

The pumps and the alarm system used in the sump generally require regular inspections for service and maintenance of parts based on manufacturers' recommendations.

If it is possible that the sump may handle combustible or flammable spilled materials, explosion-proof pumping equipment and controls or other appropriate precautions should be taken to prevent explosions or fires. Consult OSHA and local safety codes and standards for specific requirements and guidance.

Advantages of Sumps
• Provide a simple and quick collection method for recycling, reusing, or treating materials in a containment structure
• Are commonly used at industrial facilities
Disadvantages of Sumps
• Pumps may clog easily if not designed correctly
• May require maintenance/servicing agreements with pump dealers
• Costs for purchasing and/or replacing pumps may be high

Covering

What Is It

Covering is the partial or total physical enclosure of materials, equipment, process operations, or activities. Covering certain areas or activities prevents storm water from coming into contact with potential pollutants and reduces material loss from wind blowing. Tarpaulins, plastic sheeting, roofs, buildings, and other enclosures are examples of covering that are effective in preventing storm water contamination. Covering can be temporary or permanent.

When and Where to Use It

Covering is appropriate for outdoor material storage piles (e.g., stockpiles of dry materials, gravel, sand, compost, sawdust, wood chips, de-icing salt, and building materials) and areas where liquids and solids in containers are stored or transferred. Although it may be too expensive to cover or enclose all industrial activities, cover high-risk areas (identified during the storm water pollutant source identification). For example, cover chemical preparation areas, vehicle maintenance areas, areas where chemically treated products are stored, and areas where salts are stored.

If covering or enclosing the entire activity is not possible, the high-risk part of the activity can often be separated from other processes and covered. Another option that reduces the cost of building a complete enclosure is to build a roof over the activity. A roof may also eliminate the need for ventilation and lighting systems (Washington State, 1992).

What to Consider

Evaluate the strength and longevity of the covering, as well as its compatibility with the material or activity being enclosed. When designing an enclosure, consider access to materials, their handling, and transfer. Materials that pose environmental and safety dangers because they are radioactive, biological, flammable, explosive, or reactive require special ventilation and temperature considerations.

Covering alone may not protect exposed materials from storm water contact. Place the material on an elevated, impermeable surface or build curbing around the outside of the materials to prevent problems from runon of uncontaminated storm water from adjacent areas.

Frequently inspect covering, such as tarpaulins, for rips, holes, and general wear. Anchor the covering with stakes, tie-down ropes, large rocks, tires, or other easily available heavy objects.

Practicing proper materials management within an enclosure or underneath a covered area is essential. For example, floor drainage within an enclosure should be properly designed and connected to the wastewater sewer where appropriate and allowed. If connection to an offsite wastewater sewer is considered, the local Publicly Owned Treatment Works (POTW) should be consulted to find out if there are any pretreatment requirements or restrictions that must be followed.

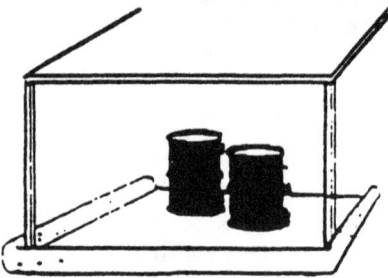

Small Chemical Storage Area
with Curbing and Cover

Raw Material Storage Covered with Tarpaulin

Covered Area for Raw Materials

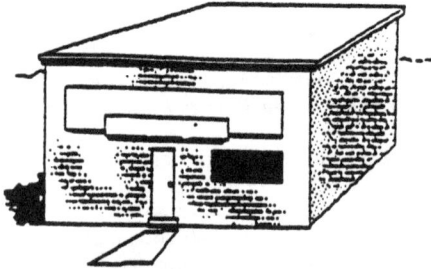

Enclosed Area for Storage of
Raw Materials or Chemicals

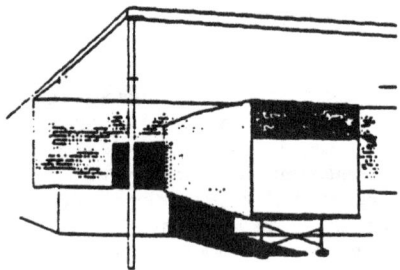

Covered Area for Loading and Unloading

FIGURE 4.7 EXAMPLE COVERING FOR INDUSTRIAL ACTIVITIES
(Modified from Washington State, 1992; Salt Institute, 1987)

Advantages of Covering
• Is simple and effective
• Is commonly inexpensive

Disadvantages of Covering
• Requires frequent inspection
• May pose health or safety problems if enclosure is built over certain activities

Vehicle Positioning

What Is It

Vehicle positioning is the practice of locating trucks or rail cars while transferring materials to prevent spills of materials onto the ground surface, which may then contaminate storm water runoff. Vehicle positioning is a simple and effective method of material spill prevention and yet it is commonly overlooked.

When and Where to Use It

Vehicle positioning can be used at all types of industrial facilities. This practice is appropriate for any area where materials are transferred from or to vehicles, such as loading and unloading areas, storage areas, and material transfer areas. Use vehicle positioning in conjunction with other practices such as covering, sumps, drip pans, or loading and unloading by air pressure or vacuum where chemical spills are of concern.

What to Consider

The purpose of vehicle positioning is to locate vehicles in a stable and appropriate position to prevent problems, such as spills resulting from broken material storage containers, spills caused by vehicle movement during materials transfer activities, and spills caused by improperly located vehicles. Vehicles should also be positioned near containment or flow diversion systems to collect unexpected spills from leaks in transfer lines or connections. The following activities are included in this practice:

- Constructing walls that help in positioning the vehicles

- Positioning vehicle either over a drain or on a sloped surface that drains to a containment structure

- Outlining required vehicle positions on the pavement

- Using wheel guards or wheel blocks

- Posting signs requiring the use of emergency brakes

- Requiring vehicles to shut off engines during materials transfer activities.

Advantages of Vehicle Positioning
• Is inexpensive
• Is easy and effective
Disadvantages of Vehicle Positioning
• May require redesign of loading and unloading areas

Loading and Unloading by Air Pressure or Vacuum

What Is It

Air pressure and vacuum systems are commonly used for transporting and loading and unloading materials. These systems are simple to use and effective in transferring dry chemicals or solids from one area to another, but are less effective as the particles of material become more dense.

In an air pressure system, a safety-relief valve and a dust collector are used to separate the dry materials from the air and then release the air accumulated during transfer operations. In a vacuum system, a dust collection device and an air lock, such as a rotary gate or trap door feeder, are typically used.

The use of mechanical equipment that involves enclosed lines, such as those provided by air pressure (also referred to as pneumatic) and vacuum loading systems, are effective methods for minimizing releases of pollutants into the environment. Because of the enclosed nature of the system, pollutants are not exposed to wind or precipitation and therefore have less potential to contaminate storm water discharges.

When and Where to Use It

Air pressure and vacuum systems can be used at all types of industrial facilities. This equipment is located in material handling areas to use for storing, loading and unloading, transporting, or conveying materials.

What to Consider

Unlike many of the other BMPs discussed in this manual, air pressure and vacuum systems may be expensive because of the costs of purchasing the system and retrofitting the system to existing materials handling procedures. In many cases, these systems can be shipped to a facility and be installed onsite without contractor help. Manufacturer's recommendations should be followed closely to ensure proper installation. In other cases, systems may have to be designed specifically for a site. Proper design and installation are very important for air pressure and vacuum systems to be as effective as possible. The equipment may be weatherproof or, if not, consider enclosing or covering the equipment.

Conduct routine inspections of air pressure and vacuum systems. Regular maintenance is required of these systems, especially the dust collectors. Conduct maintenance activities based on manufacturers' recommendations. Inspect air pressure systems more frequently due to the greater potential for leaks to the environment.

Advantages of Loading and Unloading by Air Pressure or Vacuum
• Is quick and simple
• May be economical if materials can be recovered
• Will minimize exposure of pollutants to storm water
Disadvantages of Loading and Unloading by Air Pressure or Vacuum
• May be costly to install and maintain
• May not be appropriate for some denser materials
• May require site-specific design
• Dust collectors may need a permit under the Clean Air Act to install

4.3 MITIGATIVE PRACTICES

Mitigation involves cleaning up or recovering a substance after it has been released or spilled to reduce the potential impact of a spill before it reaches the environment. Therefore, pollution mitigation is a second line of defense where pollution prevention practices have failed or are impractical. Because spills cannot always be avoided at industrial sites, it is necessary to plan for these events and to design proper response procedures. This section discusses mitigative BMPs to avoid contamination of storm water. Most of the mitigative practices discussed are simple and should be incorporated in your facility's good housekeeping and spill response plans. The mitigation practices discussed include manual cleanup methods, such as sweeping and shoveling, mechanical cleanup by excavation or vacuuming, and cleanup with sorbents and gels.

Facilities are cautioned that spills of certain toxic and hazardous substances and their cleanup may be covered under regulations, including those imposed under the Superfund Amendments and Reauthorization Act (SARA), the Comprehensive Environmental Responsibility, Compensation, and Liability Act (CERCLA), and the Resource Conservation and Recovery Act (RCRA).

Sweeping

What Is It

Sweeping with brooms, squeegees, or other mechanical devices is used to remove small quantities of dry chemicals and dry solids from areas that are exposed to precipitation or storm water runoff. These areas may include dust or contaminant covered bags, drums containing remaining materials on their lids, areas housing enclosed or covered materials, and spills of dry chemicals and dry solids in locations on the industrial site. Cleaning by sweeping with brooms is a low cost practice that can be performed by all employees and requires no special equipment or training.

When and Where to Use It

Sweeping can be used at many material handling areas and process areas in all types of industrial facilities. Timing is an important consideration for all mitigative practices. To be effective as a storm water control, cleanup must take place before rainfall or contact with storm water runoff or before an outside area is hosed down.

Do not limit your cleanup activities to those outside activities that are exposed to rainfall. In many cases, tracking of materials to the outside from areas that are enclosed or covered (e.g., on shoes) may also occur.

What to Consider

Store brooms appropriately and do not expose them to precipitation. In addition, rules of compatibility also apply. Do not use the same broom to clean up two chemicals that are incompatible. Determine the compatibility between the brooms themselves and the chemical of concern before using this practice. In some instances, chemicals should be vacuumed instead of swept. Be sure that swept material is disposed of properly.

Advantages of Sweeping
• Is inexpensive
• Requires no special training
• Provides recycling opportunities
Disadvantages of Sweeping
• Is a labor-intensive practice
• Is limited to small releases of dry materials

Shoveling

What Is It

Shoveling is another manual cleanup method that is simple and low in cost. Generally, shoveling can be used to remove larger quantities of dry chemicals and dry solids, as well as to remove wetter solids and sludge. Shoveling is also useful in removing accumulated materials from sites not accessible by mechanical cleanup methods.

When and Where to Use It

Shoveling can be used at any facility. Shoveling provides an added advantage over sweeping because cleanup methods are not limited to dry materials. In many cases, accumulated solids and sludges that are in ditches, sumps, or other facility locations can be effectively and quickly removed by shoveling.

Shovels can also be used to clean up contaminated snows. Timing is an important consideration in any mitigative practice. Materials that could contaminate storm water runoff should be removed before any storm event.

What to Consider

As with brooms, clean and store shovels properly. Also, consider planning for the transport and disposal or reuse of the shoveled materials.

Advantages of Shoveling
• Is inexpensive
• Provides recycling opportunities
• Can remediate larger releases and is effective for dry and wet materials
Disadvantages of Shoveling
• Is labor-intensive
• Is not an appropriate practice for large spills

Excavation Practices

What Are They

Excavation (i.e., removal of contaminated material) of released materials is typically conducted by mechanical equipment, such as plows and backhoes. Generally, plowing and backhoeing can be done using a specifically designed vehicle, tractor, or truck.

Excavation removes the materials of concern and any deposition of contaminants, thereby reducing the potential for storm water contamination. Mechanical cleanup methods are typically less precise than manual cleanup methods, resulting in reduced opportunities for recycle and reuse.

When and Where to Use Them

Excavation practices are most useful for large releases of dry materials and for areas contaminated by liquid material releases. In excavation, you want to be sure that all of the contaminated material is removed.

Timing is an important consideration for all mitigative practices. To be effective as a storm water control, cleanup must take place before a rainfall event.

What to Consider

Conduct inspections and operations and maintenance in accordance with a manufacturer's recommendations, which may include the following:

- A specified frequency for inspection, maintenance, and servicing of the equipment

- Parts replacement, rotation, and lubrication specifications

- Procedures for evaluating all parts.

As with any equipment used during cleanup, other considerations apply, including the following:

- Plows, backhoes, etc. should be stored appropriately with no exposure to precipitation

- Excavated materials should be properly handled or disposed of.

Advantages of Excavation Practices
• Are a cost effective method for cleaning up dry materials release
• Are common and simple

Disadvantages of Excavation Practices
• Are less precise, resulting in less recycling and reuse opportunities

Vacuum and Pump Systems

What Are They

Vacuum and pump systems are effective for cleaning up spilled or exposed materials.

The benefits of vacuum and pump cleanup systems include simplicity and speed. With such systems, only the spilled materials need be collected. Also, these systems are often portable and can be used at many locations to clean up releases to the environment. Portable systems can usually be rented.

When and Where to Use Them

Vacuum and pump systems can be used at any industrial facility. Both wet and dry materials can be collected with these systems. Vacuum systems can be used in material handling areas and process areas.

What to Consider

Consider the area of use and the most appropriate size for the system. Since these systems can be portable, size is important, especially if materials will be stored in the unit. In this case, the portable system must have enough suction or positive air pressure to transport materials over long distances. Include plans for proper disposal or reuse of the collected materials.

Advantages of Vacuum and Pump Systems
• Remove materials by air pressure or vacuum quickly and simply
• Collect materials accurately
• Offer good recycling opportunities
Disadvantages of Vacuum and Pump Systems
• May require high initial capital cost
• Require equipment maintenance

Sorbents

What Are They

Sorbents are materials that are capable of cleaning up spills through the chemical processes of adsorption and absorption. Sorbents adsorb (an attraction to the outer surface of a material) or absorb (taken in by the material like a sponge) only when they come in contact with the sorbent materials. The sorbents must be mixed with a spill or the liquid must be passed through the sorbent. Sorbent materials come in many different forms from particles to foams. Often the particles are held together in structures called booms, pads, or socks. Sorbents include, but are not limited to, the following:

- **Common Materials (clays, sawdust, straw, and flyash)**—Generally come in small particles that can be thrown onto a spill that is on a surface. The materials absorb the spill by taking up the liquid.

- **Polymers (polyurethane and polyolefin)**—Come in the form of spheres, beads, or foam tablets. These materials absorb a chemical spill by taking up the liquid into their open-pore structure.

- **Activated Carbon**—Comes in a powdered or granular form and can be mixed with liquids to remove pollutants. This sorbent works by adsorbing the organics to its surface and can be recycled and then reused by a process called regeneration.

- **"Universal Sorbent Material"**—Is a silicate glass foam consisting of rounded particles that can absorb the material.

When and Where to Use Them

Sorbents are useful BMPs for facilities with liquid materials onsite. Timing is important for these practices. To be effective as a storm water BMP, cleanup must take place before a rainfall. Sorbents are often used in conjunction with curbing to provide cleanup of small spills within a containment area.

"Universal Sorbent Materials" are suitable for use on many compounds including acids, alkalis, alcohols, aldehydes, arsenate, ketones, petroleum products, and chlorinated solvents.

Activated carbon is useful for adsorbing many organic compounds. Organics that are diluted in water can be passed through a column that is filled with the activated carbon material to remove the organics, or the activated carbon can be mixed into the water and can then be filtered out.

Polyurethane is good with chemical liquids such as benzene, chlorinated solvents, epicholorhydrin, and phenol. Polyolefin is used to remove organic solvents, such as phenol and various chlorinated solvents. The beads and spheres are usually mixed into a spill by use of a blower and then are skimmed from the top surface by use of an oil boom.

More common materials such as clay, sawdust, straw, and fly-ash can be used for a liquid spill on a surface that is relatively impenetrable, and are usually spread over the spill area with shovels.

Booms, pads, and socks are also useful in areas where there are small liquid spills or drips or where small amounts of solids may mix with small amounts of storm water runoff. They can function both to absorb the pollutants from the storm water and restrict the movement of a spill. Socks are often used together with curbing to clean up small spills.

What to Consider

Because sorbents work by a chemical or physical reaction, some sorbents are better than others for certain types of spills. Therefore, the use of sorbents requires that personnel know the properties of the spilled material(s) to know which sorbent is appropriate. To be effective, sorbents must adsorb the material spilled but must not react with the spilled material to form hazardous or toxic substances. Follow the manufacturers' recommendations.

For sorbents to be effective, they must be applied immediately in the release area. The use of sorbent material is generally very simple: the sorbent is added to the area of release, mixed well, and allowed to adsorb or absorb. Many sorbents are not reusable once they have been used. Proper disposal is required.

Advantages of Sorbents
• Work in water environments (booms and socks)
• Offer recycling opportunities (some types of sorbents)
Disadvantages of Sorbents
• Require a knowledge of the chemical makeup of a spill (to choose the best sorbent)
• Offer no recycling opportunities (some types of sorbents)
• May be expensive practice for large spills
• May create disposal problems and increase disposal costs by creating a solid waste and potentially a hazardous waste.

Gelling Agents

What Are They

Gelling agents are materials that interact with liquids either physically or chemically (i.e., thickening or polymerization). Some of the typical gelling agents are polyelectrolytes, polyacrylamide, butylstyrene copolymers, polyacrylonitrile, polyethylene oxide, and a gelling agent referred to as the universal gelling agent which is a combination of these synthetics.

Gelling interacts with a material by concentrating and congealing it to become semisolid. The semisolid gel later forms a solid material, which can then be cleaned up by manual or mechanical methods. The BMP of using a gelling agent is one of the few ways to effectively control a liquid spill before it reaches a receiving water or infiltrates into the soil and then ground water.

When and Where to Use Them

Gelling agents are useful for facilities with significant amounts of liquid materials stored onsite. Gels cannot be used to clean up spills on surface water unless authorized by the U.S. Coast Guard or EPA Regional Response Team.

What to Consider

Gels can be used to stop the liquid's flow on land, prevent its seeping into the soil, and reduce the surface spreading of a spill. Because of these properties, gels can reduce the need for extensive cleanup methods and reduce the possibility of storm water contamination from an uncontrolled industrial spill. As with sorbents, the use of gels simply involves the addition of the gel to the area of the spill, mixing well, and allowing the mass to congeal. To use gels correctly, however, personnel need to know the properties of the spilled materials so that they can choose the correct gel.

Timing is particularly important for gelling agent use. To prevent the movement of materials, gelling agents must be applied immediately after the spill. The use of gelling agents results in a large bulk of congealed mass that usually cannot be separated. Ultimately, this mass will need to be cleaned up by manual or mechanical methods and disposed of properly.

Advantages of Gelling Agents
• Stop the movement of spilled or released liquid materials
• Require no permanent structure

Disadvantages of Gelling Agents
• May require knowledge of the spilled materials to select correct gelling agents
• Usually offer no recycling opportunities
• May be difficult to clean up
• May create disposal problems and increase disposal costs by creating a solid waste and potentially a hazardous waste

4.4 OTHER PREVENTIVE PRACTICES

A number of preventive measures can be taken at industrial sites to limit or prevent the exposure of storm water runoff to contaminants. This section describes a few of the most easily implemented measures:

- Preventive Monitoring Practices

- Dust Control (Land Disturbance and Demolition Areas)

- Dust Control (Industrial)

- Signs and Labels

- Security

- Area Control Procedures

- Vehicle Washing.

Preventive Monitoring Practices

What Are They

Preventive monitoring practices include the routine observation of a process or piece of equipment to ensure its safe performance. It may also include the chemical analysis of storm water before discharge to the environment.

When and Where to Use Them

Automatic Monitoring System—In areas where overflows, spills, and catastrophic leaks are possible, an automatic monitoring system is recommended. Some Federal, State, and local laws require such systems to be present if threats exist to the health and safety of personnel and the environment. For material management areas, monitoring may include liquid level detectors, pressure and temperature gauges, and pressure-relief devices. In material transfer, process, and material handling areas, automatic monitoring systems can include pressure drop shutoff devices, flow meters, thermal probes, valve position indicators, and operation lights. Loading and unloading operations might use these devices for measuring the volume of tanks before loading, for weighing vehicles or containers, and for determining rates of flow during loading and unloading.

Automatic Chemical Monitoring—Measures the quality of plant runoff to determine whether discharge is appropriate or whether diversion to a treatment system is warranted. Such systems might monitor pH, turbidity, or conductivity. These parameters might be monitored in diked areas, sewers, drainage ditches, or holding ponds. Systems can also be designed to signal automatic diversion of contaminated storm water runoff to a holding pond (e.g., a valve or a gate could be triggered by a certain pollutant in the storm water runoff).

Manned Operations—In material transfer areas and process areas, personnel can be stationed to watch over the operations so that any spills or mismanagement of materials can be corrected immediately. This is particularly useful at loading and unloading areas where vehicles or equipment must be maneuvered into the proper position to unload (see Vehicle Positioning BMP).

Nondestructive Testing—Some situations require that a storage tank or a pipeline system be tested without being physically moved or disassembled. The structural integrity of tanks, valves, pipes, joints, welds, and other equipment can be tested using nondestructive methods. Acoustic emission tests use high frequency sound waves to draw a picture of the structure to reveal cracks, malformations, or other structural damage. Another type of testing is hydrostatic pressure testing. During pressure testing, the tank or pipe is subjected to pressures several times the normal pressure. A loss in pressure during the testing may indicate a leak or some other structural damage. Tanks and containers should be pressure tested as required by Federal, State, or local regulations.

What to Consider

Automated monitoring systems should be placed in an area where plant personnel can easily observe the measurements. Alarms can be used in conjunction with the measurement display to warn personnel. Manned operations should have communication systems available for getting help in case spills or leaks occur. Especially sensitive or spill-prone areas may require back-up instrumentation in case the primary instruments malfunction.

Mechanical and electronic equipment should be operated and maintained according to the manufacturers' recommendations. Equipment should be inspected regularly to ensure proper and accurate operation.

The pollution prevention team, in consultation with a certified safety inspector, should evaluate system monitoring requirements to decide which systems are appropriate based on hazard potential.

Advantages of Preventive Monitoring Practices
• Pressure and vacuum testing can locate potential leaks or damage to vessels early. The primary benefit of such testing is in ensuring the safety of personnel, but it also has secondary benefits including prevention of storm water contamination.
• Automatic system monitors allow for early warnings if a leak, overflow, or catastrophic incident is imminent.
• Manning operations, especially during loading and unloading activities, is effective and generally inexpensive.
• The primary benefit of nondestructive testing is in ensuring the safety of personnel, but it also has secondary benefits including early detection of the potential for contaminating storm water runoff.
Disadvantages of Preventive Monitoring Practices
• Plant personnel often do not have the expertise to maintain automatic equipment.
• Automatic equipment can fail without warning.
• Automated process control and monitoring equipment may be expensive to purchase and operate

Dust Control (Land Disturbance and Demolition Areas)

What Is It

Dust controls for land disturbance and demolition areas are any controls that reduce the potential for particles being carried through air or water. Types of dust control are:

- **Irrigation**—Irrigation is a temporary measure involving a light application of water to moisten the soil surface. The process should be repeated as necessary.

- **Minimization of Denuded Areas**—Minimizing soil exposure reduces the amount of soil available for transport and erosion. Soil exposure can be lessened by temporary or permanent soil stabilization controls, such as seeding, mulching, topsoiling, crushed stone or coarse gravel spreading, or tree planting. Maintaining existing vegetation on a site will also help control dust.

- **Wind Breaks**—Wind breaks are temporary or permanent barriers that reduce airborne particles by slowing wind velocities (slower winds do not suspend particles). Leaving existing trees and large shrubs in place will create effective wind breaks. More temporary types of wind breaks are solid board fences, snow fences, tarp curtains, bales of hay, crate walls, and sediment walls.
- **Tillage**—Deep plowing will roughen the soil surface to bring up to the surface cohesive clods of soil, which in turn rest on top of dusts, protecting them from wind and water erosion. This practice is commonly practiced in arid regions where establishing vegetation may take time.

- **Chemical Soil Treatments (palliatives)**—These are temporary controls that are applied to soil surfaces in the form of spray-on adhesives, such as anionic asphalt emulsion, latex emulsion, resin-water emulsions, or calcium chloride. The palliative is the chemical used. These should be used with caution as they may create pollution if not used correctly.

When and Where to Use It

Dust controls can be used on any site where dust may be generated and where the dust may cause onsite and offsite damage. Dust controls are especially critical in arid areas, where reduced rainfall levels expose soil particles for transport by air and runoff. This control should be used in conjunction with other sedimentation controls such as sediment traps.

What to Consider

To control dust during land disturbance and at demolition areas, exposure of soil should be limited as much as possible. When possible, work that causes soil disturbance or involves demolition should be done in phases and should be accompanied by temporary stabilization measures. These precautions will minimize the amount of soil that is disturbed at any one time and, therefore, control dust.

Oil should not be used to control dust because of its high potential for polluting storm water discharges.

Irrigation will be most effective if site drainage systems are checked to ensure that the right amount of water is used. Too much water can cause runoff problems.

Chemical treatment is only effective on mineral soils, as opposed to muck soils, because the chemicals bond better to mineral soils. Therefore, it should be used only in arid regions. Vehicular traffic should be routed around chemically treated areas to avoid tracking of the chemicals. Certain chemicals may be inappropriate for some types of soils or application areas. For example, spraying chemicals on the soil of an industrial site adjacent to a school may be dangerous. Local governments usually have information about restrictions on the types of palliatives that may be used. Special consideration must be given to preserving ground water quality whenever chemicals are applied to the land.

Since most of these techniques are temporary controls, sites should be inspected often and materials should be reapplied when needed. The frequency for these inspections depends on site-specific conditions, weather conditions, and the type of technique used.

Advantages of Dust Control (Land Disturbance and Demolition Areas)
• Can help prevent wind-and-water based erosion of disturbed areas and will reduce respiratory problems in employees
• Some types can be implemented quickly at low cost and effort (except wind breaks)
• Helps preserve the aesthetics of the site and screens certain activities from view (wind breaks)
• Vegetative wind breaks are permanent and an excellent alternative to chemical use
Disadvantages of Dust Control (Land Disturbance and Demolition Areas)
• Some types are temporary and must be reapplied or replenished regularly
• Some types are expensive (irrigation and chemical treatment) and may be ineffective under certain conditions
• May result in health and/or environmental hazards, e.g., if overapplication of the chemicals leaves large amounts exposed to wind and rain erosion or ground water contamination
• May create excess runoff that the site was not designed to control (irrigation)
• May cause increased offsite tracking of mud (irrigation)
• Is not as effective as chemical treatment or mulching and seeding; requires land space that may not be available at all locations (wind breaks)

Dust Control (Industrial)

What Is It

Dust controls for material handling areas are controls that prevent pollutants from entering storm water discharges by reducing the surface and air transport of dust caused by industrial activities. Consider the following types of controls:

- Water spraying

- Negative pressure systems (vacuum systems)

- Collector systems (bag and cyclone)

- Filter systems

- Street sweeping.

The purpose of industrial dust control is to collect or contain dusts to prevent storm water runoff from carrying the dusts to the sewer collection system or to surface waters.

When and Where to Use It

Dust control is useful in any process area, loading and unloading area, material handling areas, and transfer areas where dust is generated. Street sweeping is limited to areas that are paved.

What to Consider

Mechanical dust collection systems are designed according to the size of dust particles and the amount of air to be processed. Manufacturers' recommendations should be followed for installation (as well as the design of the equipment).

If water sprayers are used, dust-contaminated waters should be collected and taken for treatment. Areas will probably need to be resprayed to keep dust from spreading.

Two kinds of street sweepers are common: brush and vacuum. Vacuum sweepers are more efficient and work best when the area is dry.

Mechanical equipment should be operated according to the manufacturers' recommendations and should be inspected regularly.

Advantages of Dust Control (Industrial)
• May cause a decrease of respiratory problems in employees around the site
• May cause less material to be lost and may therefore save money
• Provides efficient collection of larger dust particles (street sweepers)
Disadvantages of Dust Control (Industrial)
• Is generally more expensive than manual systems
• May be impossible to maintain by plant personnel (the more elaborate equipment)
• Is labor and equipment intensive and may not be effective for all pollutants (street sweepers)

Signs and Labels

What Are They

Signs and labels identify problem areas or hazardous materials at a facility. Warning signs, often found at industrial facilities, are a good way to suggest caution in certain areas. Signs and labels can also provide instructions on the use of materials and equipment. Labelling is a good way to organize large amounts of materials, pipes, and equipment, particularly on large sites.

Labels tell material type and container contents. Accurate labeling can help facilities to quickly identify the type of material released so facility personnel can respond correctly.

Two effective labeling methods include color coding and Department of Transportation (DOT) labeling. Color coding is easily recognized by facility personnel and simply involves painting/coating or applying an adhesive label to the container. Color codes must be consistent throughout the facility to be effective, and signs explaining the color codes should be posted in all areas.

DOT requires that labels be prominently displayed on transported hazardous and toxic materials. Labeling required by DOT could be expanded to piping and containers, making it easy to recognize materials that are corrosive, radioactive, reactive, flammable, explosive, or poisonous.

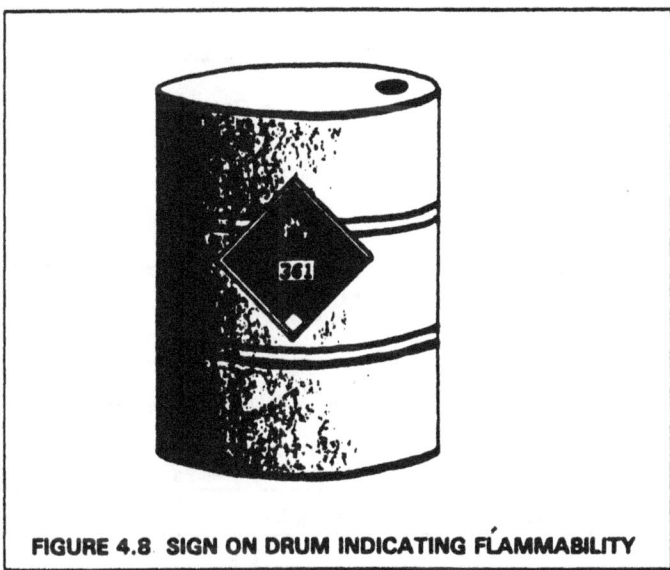

FIGURE 4.8 SIGN ON DRUM INDICATING FLAMMABILITY

When and Where to Use Them

Signs and labels can be used at all types of facilities. Areas where they are particularly useful are material transfer areas, equipment areas, loading and unloading areas, or anywhere information might prevent contaminants from being released to storm water.

What to Consider

Signs and labels should be visible and easy to read. Useful signs and labels might provide the following information:

- Names of facility and regulatory personnel, including emergency phone numbers, to contact in case of an accidental discharge, spill, or other emergency

- Proper uses of equipment that could cause release of storm water contaminants

- Types of chemicals used in high-risk areas

- The direction of drainage lines/ditches and their destination (treatment or discharge)

- Information on a specific material

- Refer to OSHA standards for sizes and numbers of signs required for hazardous material labeling.

Hazardous chemicals might be labeled as follows:

- Danger
- Combustible
- Warning
- Caution
- Flammable

- Poisonous
- Caustic
- Corrosive
- Volatile
- Explosive

Periodic checks can ensure that signs are still in place and labels are properly attached. Signs and labels should be replaced and repaired as often as necessary.

Advantages of Signs and Labels
• Are inexpensive and easily used
Disadvantages of Signs and Labels
• Must be updated and maintained so they are legible

Security

What Is It

Setting up a security system as part of your Plan could help prevent an accidental or intentional release of materials to storm water runoff as a result of vandalism, theft, sabotage, or other improper uses of facility property. If your facility already has a security system, consider improving it by training security personnel about the specifics of the Storm Water Pollution Prevention Plan. Routine patrol, lighting, and access control are discussed below as possible measures to include in your facility's security system.

When and Where to Use It

Routine patrol, lighting, and access control are measures that can be used at any facility.

What to Consider

Security information could be included in the existing training required by the Plan to instruct personnel about where and how to patrol areas within the facility. Instruction might also include what to look for in problem areas and how to respond to problems. During routine patrol, security personnel can actively search the facility site for indications of spills, leaks, or other discharges; respond to any disturbance resulting from intruders or inappropriate facility operations; and generally work as a safeguard to prevent unexpected events. Routine patrols could be an effective part of the Storm Water Pollution Prevention Plan, especially for large facilities with established security measures. To make this practice effective, security personnel can help develop the Storm Water Pollution Prevention Plan, possibly with one person acting as a member of the pollution prevention committee.

Sufficient lighting throughout the facility during daytime and night hours will make it easier to get to equipment during checks and will make it easy to detect spills and leaks that might otherwise be hidden. Routine patrols are also easier with proper lighting.

Controlling access to the industrial site is an important part of plant security and of activity and traffic control. Signs, fencing, guard houses, dog patrols, and visitor clearance requirements are often used to control site access.

- Signs are the simplest, most inexpensive method of access control, but they are limited in their actual control since they provide no physical barriers and require that people obey them voluntarily.

- Fencing provides a physical barrier to the facility site and an added means of security.

- Guard houses used with visitor rules can help to ensure that only authorized personnel enter the facility site and can limit vehicular traffic as well.

- Traffic signs are also useful at facility sites. Restricting vehicles to paved roads and providing direction and warning signs can help prevent accidents. Where restricting vehicles to certain pathways is not possible, it is important to ensure that all above-ground valves and pipelines are well marked.

Advantages of Security
• Provides a preventive safeguard to operational malfunctions or other facility disturbances (routine patrols)
• Allows easier detection of vandals or thieves (lighting)
• Allows easier detection of spills, leaks, or other releases (lighting)
• Prevents spills by providing good visibility (lighting)
• Prevents unauthorized access to facility (access control)
Disadvantages of Security
• May not be feasible for smaller facilities
• May be costly (e.g., installation of lighting systems)
• May increase energy costs as a result of additional lighting
• May not be feasible to have extensive access controls at smaller facilities

Area Control Procedures

What Are They

The activities conducted at an industrial site often result in the materials being deposited on clothes and footwear and the being carried throughout the facility site. As a result, these materials may find their way into the storm water runoff.

Area control procedures involve practicing good housekeeping measures such as maintaining indoor or covered material storage and industrial processing areas. If the area is kept clean, the risk of accumulating materials on footwear and clothing is reduced. In turn, the chance of left over pollutants making contact with storm water and polluting surface water is minimized.

When and Where to Use Them

Area control measures can be used at any facility where materials may be tracked into areas where they can come in contact with storm water runoff. Areas can include material handling areas, storage areas, or process areas.

What to Consider

Materials storage areas and industrial processing areas should be checked regularly to ensure that good housekeeping measures are being implemented. Cover-garments, foot mats, and other devices used to collect residual material near the area should be cleaned regularly.

Other effective practices include the following:

- Brushing off clothing before leaving the area

- Stomping feet to remove material before leaving the area

- Using floor mats at area exits

- Using coveralls, smocks, and other overgarments in areas where exposure to material is of greatest concern (employees should remove the overgarments before leaving the area)

- Posting signs to remind employees about these practices.

Advantages of Area Control Procedures
• Are easy to implement
• Result in a cleaner facility and improved work environment
Disadvantage of Area Control Procedures
• May be seen as tedious by employees and therefore may not be followed

Vehicle Washing

What Is It

Materials that accumulate on vehicles and then scatter across industrial sites represent an important source of storm water contamination. Vehicle washing removes materials such as site-specific dust and spilled materials that have accumulated on the vehicle. If not removed, residual material will be spread by gravity, wind, snow, or rainfall as the vehicles move across the facility site and off the site.

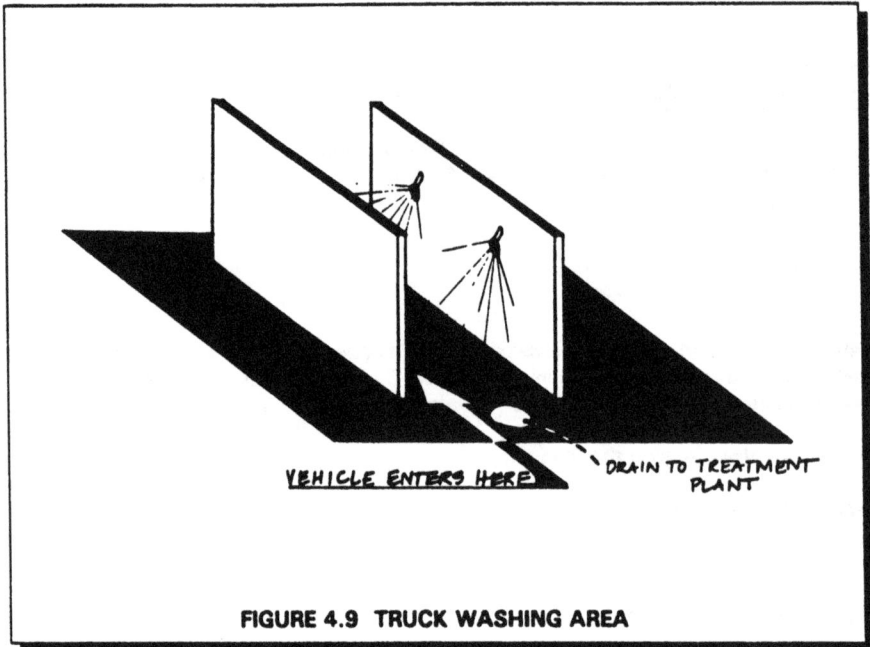

VEHICLE ENTERS HERE   DRAIN TO TREATMENT PLANT

FIGURE 4.9 TRUCK WASHING AREA

When and Where to Use It

This practice is appropriate for any facility where vehicles come into contact with raw materials on a site. If possible, the vehicle washing area should be built near the location where the most vehicle activity occurs. Wastewater from vehicle washing should be directed away from process materials to prevent contact. Those areas include material transfer areas, loading and unloading areas, or areas located just before the site exit.

What to Consider

When considering the method of vehicle washing, the facility should consider using a high-pressure water spray with no detergent additives. In general, water will adequately remove contaminants from the vehicle. If detergents are used, they may cause other environmental impacts. Phosphate- or organic-containing compounds should be avoided.

If this practice is considered, truck wash waters will result in a non-storm water discharge, thus requiring an application for an NPDES permit to cover the discharge.

Blowers or vacuums should be considered where the materials are dry and easily removed by air.

Advantages of Vehicle Washing
• Prevents dispersion of materials across the facility site
• Is necessary only where methods for transferring contained materials and minimizing exposure have not been successfully adopted and implemented
Disadvantages of Vehicle Washing
• May be costly to construct a truck washing facility

4.5 SEDIMENT AND EROSION PREVENTION PRACTICES

Any site where soils are exposed to water, wind or ice can have soil erosion and sedimentation problems. Erosion is a natural process in which soil and rock material is loosened and removed. Sedimentation occurs when soil particles are suspended in surface runoff or wind and are deposited in streams and other water bodies.

Human activities can accelerate erosion by removing vegetation, compacting or disturbing the soil, changing natural drainage patterns, and by covering the ground with impermeable surfaces (pavement, concrete, buildings). When the land surface is developed or "hardened" in this manner, storm water and snowmelt can not seep into or "infiltrate" the ground. This results in larger amounts of water moving more quickly across a site which can carry more sediment and other pollutants to streams and rivers.

EPA's General Permit requires that all industries identify in their Storm Water Pollution Prevention Plans areas that may have a high potential for soil erosion. This includes areas with such heavy activity that plants cannot grow, soil stockpiles, stream banks, steep slopes, construction areas, demolition areas, and any area where the soil is disturbed, denuded (stripped of plants), and subject to wind and water erosion. EPA further requires that you take steps to limit this erosion.

There are seven ways to limit and control sediment and erosion on your site:

- Leave as much vegetation (plants) onsite as possible.

- Minimize the time that soil is exposed.

- Prevent runoff from flowing across disturbed areas (divert the flow to vegetated areas).

- Stabilizing the disturbed soils as soon as possible.

- Slow down the runoff flowing across the site.

- Provide drainage ways for the increased runoff (use grassy swales rather than concrete drains).

- Remove sediment from storm water runoff before it leaves the site.

Using these measures to control erosion and sedimentation is an important part of storm water management. Selecting the best set of sediment and erosion prevention measures for your industry depends upon the nature of the activities on your site (i.e., how much construction or land disturbance there is) and other site-specific conditions (soil type, topography, climate, and season). Section 4.5.1 discusses some temporary and permanent ways to stabilize your site. Section 4.5.2 describes more structural ways to control sediment and erosion.

In some arid regions, growing vegetation to prevent erosion may be difficult. The local Soil Conservation Service Office or County Extension Office can provide information on any special measures necessary to promote the establishment of vegetation.

4.5.1 Vegetative Practices

Preserving existing vegetation or revegetating disturbed soil as soon as possible after construction is the most effective way to control erosion. A vegetation cover reduces erosion potential in four ways: (1) by shielding the soil surface from direct erosive impact of raindrops; (2) by improving the soil's water storage porosity and capacity so more water can infiltrate into the ground; (3) by slowing the runoff and allowing the sediment to drop out or deposit; and (4) by physically holding the soil in place with plant roots.

Vegetative cover can be grass, trees, shrubs, bark, mulch, or straw. Grasses are the most common type of cover used for revegetation because they grow quickly, providing erosion protection within days. Other soil stabilization practices such as straw or mulch may be used during non-growing seasons to prevent erosion. Newly planted shrubs and trees establish root systems more slowly, so keeping existing ones is a more effective practice.

Vegetative and other site stabilization practices can be either temporary or permanent controls. Temporary controls provide a cover for exposed or disturbed areas for short periods of time or until permanent erosion controls are put in place. Permanent vegetative practices are used when activities that disturb the soil are completed or when erosion is occurring on a site that is otherwise stabilized. The remainder of this section describes the common vegetative practices listed below:

- Preservation of Natural Vegetation

- Buffer Zones

- Stream Bank Stabilization

- Mulching, Matting, and Netting

- Temporary Seeding

- Permanent Seeding and Planting

- Sodding

- Chemical Stabilization.

Preservation of Natural Vegetation

What Is It

The preservation of natural vegetation (existing trees, vines, brushes, and grasses) provides natural buffer zones. By preserving stabilized areas, it minimizes erosion potential, protects water quality, and provides aesthetic benefits. This practice is used as a permanent control measure.

When and Where to Use It

This technique is applicable to all types of sites. Areas where preserving vegetation can be particularly beneficial are floodplains, wetlands, stream banks, steep slopes, and other areas where erosion controls would be difficult to establish, install, or maintain.

What to Consider

Preservation of vegetation on a site should be planned before any site disturbance begins. Preservation requires good site management to minimize the impact of construction activities on existing vegetation. Clearly mark the trees to be preserved and protect them from ground disturbances around the base of the tree. Proper maintenance is important to ensure healthy vegetation that can control erosion. Different species, soil types, and climatic conditions will require different maintenance activities such as mowing, fertilizing, liming, irrigation, pruning, and weed and pest control. Some State/local regulations require natural vegetation to be preserved in sensitive areas; consult the appropriate State/local agencies for more information on their regulations. Maintenance should be performed regularly, especially during construction.

Advantages of Preservation of Natural Vegetation
• Can handle higher quantities of storm water runoff than newly seeded areas
• Does not require time to establish (i.e., effective immediately)
• Increases the filtering capacity because the vegetation and root structure are usually denser in preserved natural vegetation than in newly seeded or base areas
• Enhances aesthetics
• Provides areas for infiltration, reducing the quantity and velocity of storm water runoff
• Allows areas where wildlife can remain undisturbed
• Provides noise buffers and screens for onsite operations
• Usually requires less maintenance (e.g., irrigation, fertilizer) than planting new vegetation
Disadvantages of Preservation of Natural Vegetation
• Requires planning to preserve and maintain the existing vegetation
• May not be cost effective with high land costs
• May constrict area available for construction activities

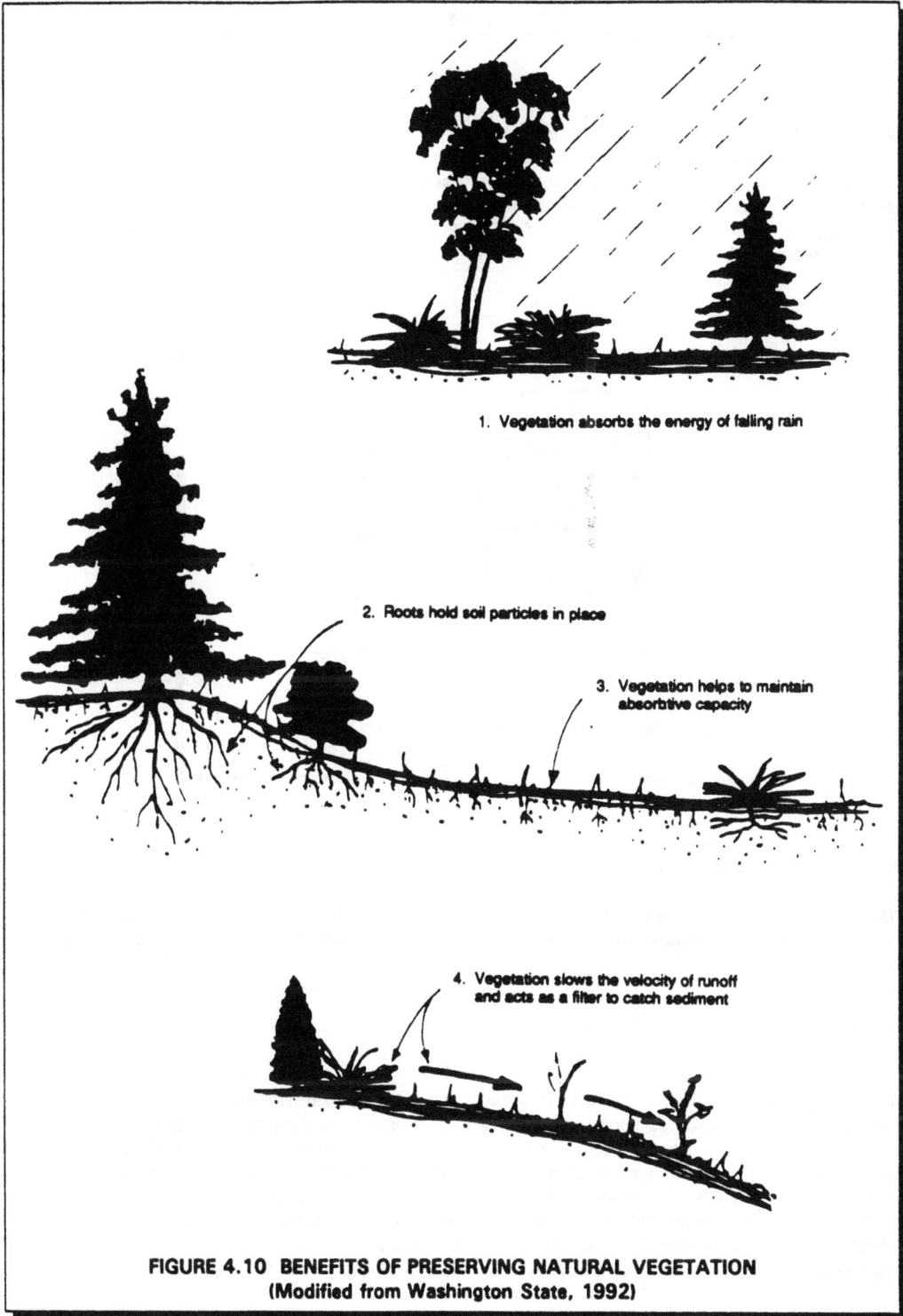

1. Vegetation absorbs the energy of falling rain

2. Roots hold soil particles in place

3. Vegetation helps to maintain absorbtive capacity

4. Vegetation slows the velocity of runoff and acts as a filter to catch sediment

FIGURE 4.10 BENEFITS OF PRESERVING NATURAL VEGETATION
(Modified from Washington State, 1992)

Buffer Zones

What Are They

Buffer zones are vegetated strips of land used for temporary or permanent water quality benefits. Buffer zones are used to decrease the velocity of storm water runoff, which in turn helps to prevent soil erosion. Buffer zones are different from vegetated filter strips (see section on Vegetated Filter Strips) because buffer zone effectiveness is not measured by its ability to improve infiltration (allow water to go into the ground). The buffer zone can be an area of vegetation that is left undisturbed during construction, or it can be newly planted.

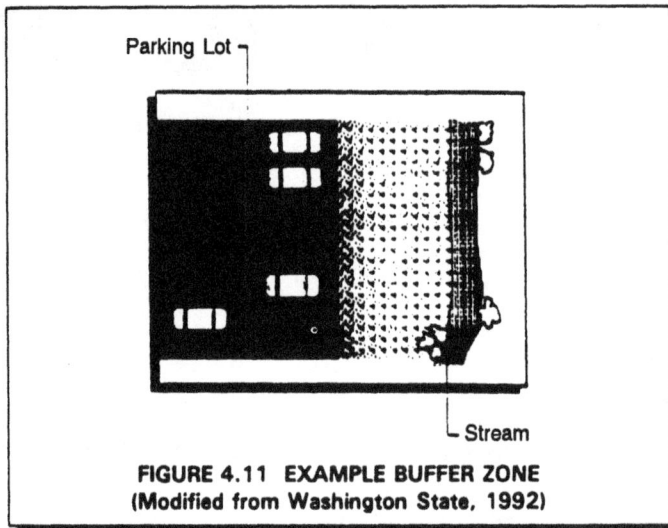

Parking Lot ⌐

⌐ Stream

FIGURE 4.11 EXAMPLE BUFFER ZONE
(Modified from Washington State, 1992)

When and Where to Use Them

Buffer zones technique can be used at any site that can support vegetation. Buffer zones are particularly effective on floodplains, next to wetlands, along stream banks, and on steep, unstable slopes.

What to Consider

If buffer zones are preserved, existing vegetation, good planning, and site management are needed to protect against disturbances such as grade changes, excavation, damage from equipment, and other activities. Establishing new buffer strips requires the establishment of a good dense turf, trees, and shrubs (see Permanent Seeding and Planting). Careful maintenance is important to ensure healthy vegetation. The need for routine maintenance such as mowing, fertilizing, liming, irrigating, pruning, and weed and pest control will depend on the species of plants and trees involved, soil types, and climatic conditions. Maintaining planted areas may require debris removal and protection against unintended uses or traffic. Many State/local storm water program or zoning agencies have regulations which define required or allowable buffer zones especially near sensitive areas such as wetlands. Contact the appropriate State/local agencies for their requirements.

Advantages of Buffer Zones
• Provide aesthetic as well as water quality benefits
• Provide areas for infiltration, which reduces amount and speed of storm water runoff
• Provide areas for wildlife habitat
• Provide areas for recreation
• Provide buffers and screens for onsite noise if trees or large bushes are used
• Low maintenance requirements
• Low cost when using existing vegetation
Disadvantages of Buffer Zones
• May not be cost effective to use if the cost of land is high
• Are not feasible if land is not available
• Require plant growth before they are effective

Stream Bank Stabilization

What is It

Stream bank stabilization is used to prevent stream bank erosion from high velocities and quantities of storm water runoff. Typical methods include the following:

- **Riprap**–Large angular stones placed along the stream bank or lake

- **Gabion**–Rock-filled wire cages that are used to create a new stream bank

- **Reinforced Concrete**–Concrete bulkheads and retaining walls that replace natural stream banks and create a nonerosive surface

- **Log Cribbing**–Retaining walls built of logs to anchor the soils against erosive forces. Usually built on the outside of stream bends

- **Grid Pavers**–Precast or poured-in-place concrete units that are placed along stream banks to stabilize the stream bank and create open spaces where vegetation can be established

- **Asphalt**–Asphalt paving that is placed along the natural stream bank to create a nonerosive surface.

When and Where to Use It

Stream bank stabilization is used where vegetative stabilization practices are not practical and where the stream banks are subject to heavy erosion from increased flows or disturbance during construction. Stabilization should occur before any land development in the watershed area. Stabilization can also be retrofitted when erosion of a stream bank occurs.

What to Consider

Stream bank stabilization structures should be planned and designed by a professional engineer licensed in the State where the site is located. Applicable Federal, State, and local requirements should be followed, including Clean Water Act Section 404 regulations. An important design feature of stream bank stabilization methods is the foundation of the structure; the potential for the stream to erode the sides and bottom of the channel should be considered to make sure the stabilization measure will be supported properly. Structures can be designed to protect and improve natural wildlife habitats; for example, log structures and grid pavers can be designed to keep vegetation. Only pressure-treated wood should be used in log structures. Permanent structures should be designed to handle expected flood conditions. A well-designed layer of stone can be used in many ways and in many locations to control erosion and sedimentation. Riprap protects soil from erosion and is often used on steep slopes built with fill materials that are subject to harsh weather or seepage. Riprap can also be used for flow channel liners, inlet and outlet protection at culverts, stream bank protection, and protection of shore lines subject to wave action. It is used where water is turbulent and fast flowing and where soil may erode under the design flow conditions. It is used to expose the water to air as well as to reduce water energy. Riprap and gabion (wire mesh cages filled with rock) are usually placed over a filter blanket (i.e., a gravel layer or filter cloth). Riprap is either a uniform size or graded (different sizes) and is usually applied in an even layer throughout the stream. Reinforced concrete structures may require positive

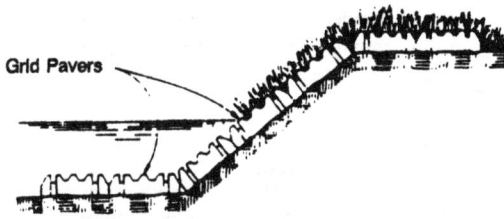

Grid Pavers

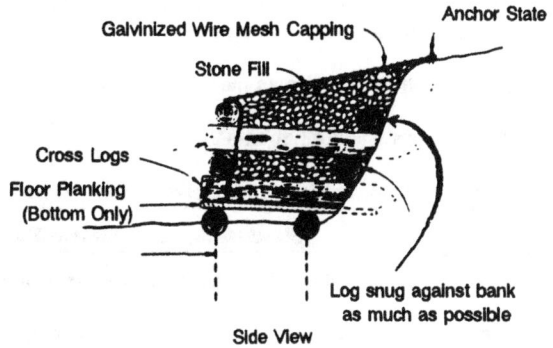

Log Cribbing

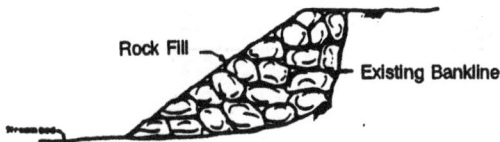

Riprap

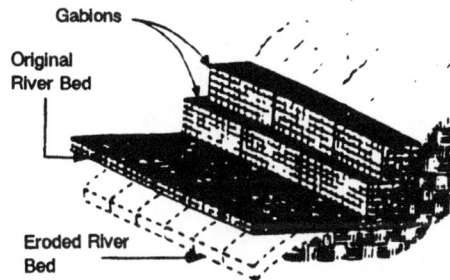

Gabion

FIGURE 4.12 EXAMPLES OF STREAM BANK STABILIZATION PRACTICES
(Modified from Commonwealth of Virginia, 1980, and Commonwealth of Pennsylvania, 1990)

drainage behind the bulkhead or retaining wall to prevent erosion around the structure. Gabion and grid pavers should be installed according to manufacturers' recommendations.

Stream bank stabilization structures should be inspected regularly and after each large storm event. Structures should be maintained as installed. Structural damage should be repaired as soon as possible to prevent further damage or erosion to the stream bank.

Advantages of Stream Bank Stabilization
• Can provide control against erosive forces caused by the increase in storm water flows created during land development
• Usually will not require as much maintenance as vegetative erosion controls
• May provide wildlife habitats
• Forms a dense, flexible, self-healing cover that will adapt well to uneven surfaces (riprap)
Disadvantages of Stream Bank Stabilization
• Does not provide the water quality or aesthetic benefits that vegetative practices could
• Should be designed by qualified professional engineers, which may increase project costs
• May be expensive (materials costs)
• May require additional permits for structure
• May alter stream dynamics which cause changes in the channel downstream
• May cause negative impacts to wildlife habitats

Mulching, Matting, and Netting

What Are They

Mulching is a temporary soil stabilization or erosion control practice where materials such as grass, hay, woodchips, wood fibers, straw, or gravel are placed on the soil surface. In addition to stabilizing soils, mulching can reduce the speed of storm water runoff over an area. When used together with seeding or planting, mulching can aid in plant growth by holding the seeds, fertilizers, and topsoil in place, by preventing birds from eating seeds, helping to retain moisture, and by insulating against extreme temperatures. Mulch mattings are materials (jute or other wood fibers) that have been formed into sheets of mulch that are more stable than normal mulch. Netting is typically made from jute, other wood fiber, plastic, paper, or cotton and can be used to hold the mulching and matting to the ground. Netting can also be used alone to stabilize soils while the plants are growing; however, it does not retain moisture or temperature well. Mulch binders (either asphalt or synthetic) are sometimes used instead of netting to hold loose mulches together.

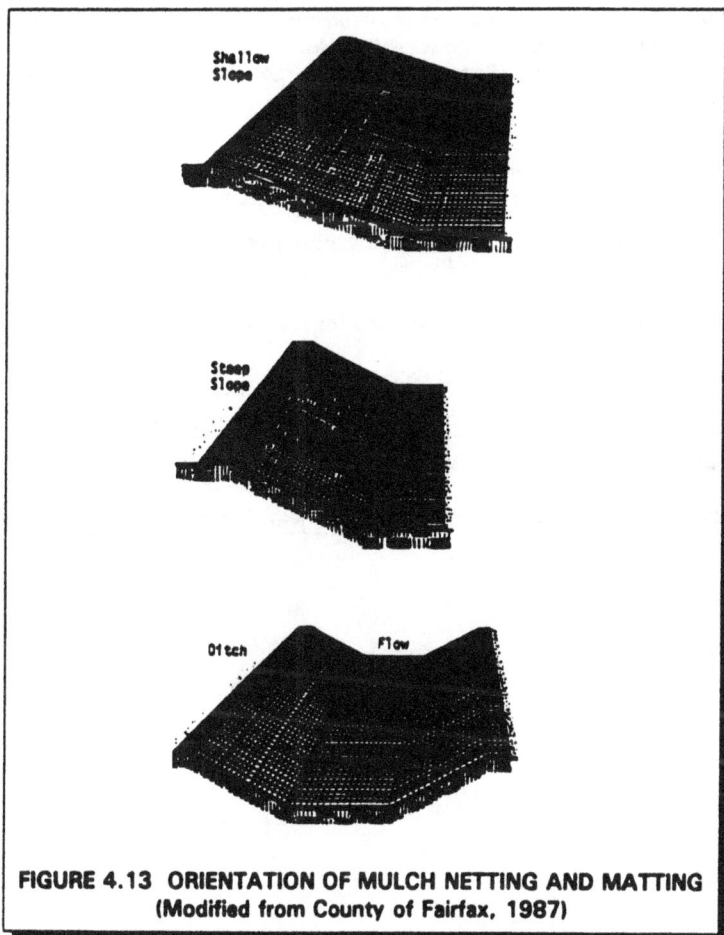

FIGURE 4.13 ORIENTATION OF MULCH NETTING AND MATTING
(Modified from County of Fairfax, 1987)

When and Where to Use Them

Mulching is often used alone in areas where temporary seeding cannot be used because of the season or climate. Mulching can provide immediate, effective, and inexpensive erosion control. On steep slopes and critical areas such as waterways, mulch matting is used with netting or anchoring to hold it in place.

Mulch seeded and planted areas where slopes are steeper than 2:1, where runoff is flowing across the area, or when seedlings need protection from bad weather.

What to Consider

Use of mulch may or may not require a binder, netting, or the tacking of mulch to the ground. Effective netting and matting require firm, continuous contact between the materials and the soil. If there is no contact, the material will not hold the soil and erosion will occur underneath the material. Final grading is not necessary before mulching. Mulched areas should be inspected often to find where mulched material has been loosened or removed. Such areas should be reseeded (if necessary) and the mulch cover replaced immediately. Mulch binders should be applied at rates recommended by the manufacturer or, if asphalt is used, at rates of approximately 480 gallons per acre (Arapahoe County, 1988).

Advantages of Mulching, Matting, and Netting
• Provide immediate protection to soils that are exposed and that are subject to heavy erosion
• Retain moisture, which may minimize the need for watering
• Require no removal because of natural deterioration of mulching and matting
Disadvantages of Mulching, Matting, and Netting
• May delay germination of some seeds because cover reduces the soil surface temperature
• Netting should be removed after usefulness is finished, then landfilled or composted

Temporary Seeding

What Is It

Temporary seeding means growing a short-term vegetative cover (plants) on disturbed site areas that may be in danger of erosion. The purpose of temporary seeding is to reduce erosion and sedimentation by stabilizing disturbed areas that will not be stabilized for long periods of time or where permanent plant growth is not necessary or appropriate. This practice uses fast-growing grasses whose root systems hold down the soils so that they are less apt to be carried offsite by storm water runoff or wind. Temporary seeding also reduces the problems associated with mud and dust from bare soil surfaces during construction.

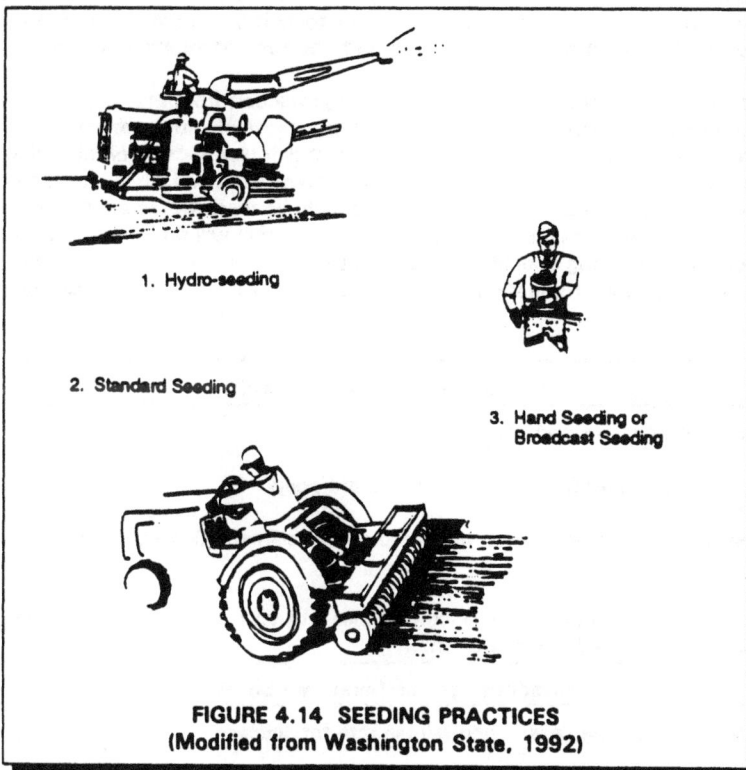

1. Hydro-seeding

2. Standard Seeding

3. Hand Seeding or Broadcast Seeding

FIGURE 4.14 SEEDING PRACTICES
(Modified from Washington State, 1992)

When and Where to Use It

Temporary seeding should be performed on areas which have been disturbed by construction and which are likely to be redisturbed, but not for several weeks or more. Typical areas might include denuded areas, soil stockpiles, dikes, dams, sides of sediment basins, and temporary roadbanks. Temporary seeding should take place as soon as practicable after the last land disturbing activity in an area. Check the requirements of your permit for the maximum amount of time allowed between the last disturbance of an area and temporary stabilization. Temporary seeding may not be an effective practice in arid and semi-arid regions where the climate prevents fast plant growth, particularly during the dry seasons. In those areas, mulching or chemical stabilization may be better for the short-term (see sections on Mulching, Geotextiles, and Chemical Stabilization).

What to Consider

Proper seed bed preparation and the use of high-quality seed are needed to grow plants for effective erosion control. Soil that has been compacted by heavy traffic or machinery may need to be loosened. Successful growth usually requires that the soil be tilled before the seed is applied. Topsoiling is not necessary for temporary seeding; however, it may improve the chances of establishing temporary vegetation in an area. Seed bed preparation may also require applying fertilizer and/or lime to the soil to make conditions more suitable for plant growth. Proper fertilizer, seeding mixtures, and seeding rates vary depending on the location of the site, soil types, slopes, and season. Local suppliers, State and local regulatory agencies, and the USDA Soil Conservation Service will supply information on the best seed mixes and soil conditioning methods.

Seeded areas should be covered with mulch to provide protection from the weather. Seeding on slopes of 2:1 or more, in adverse soil conditions, during excessively hot or dry weather, or where heavy rain is expected should be followed by spreading mulch (see section on Mulching). Frequent inspections are necessary to check that conditions for growth are good. If the plants do not grow quickly or thick enough to prevent erosion, the area should be reseeded as soon as possible. Seeded areas should be kept adequately moist. If normal rainfall will not be enough, mulching, matting, and controlled watering should be done. If seeded areas are watered, watering rates should be watched so that over-irrigation (which can cause erosion itself) does not occur.

Advantages of Temporary Seeding
• Is generally inexpensive and easy to do
• Establishes plant cover fast when conditions are good
• Stabilizes soils well, is aesthetic, and can provide sedimentation controls for other site areas
• May help reduce costs of maintenance on other erosion controls (e.g., sediment basins may need to be cleaned out less often)

Disadvantages of Temporary Seeding
• Depends heavily on the season and rainfall rate for success
• May require extensive fertilizing of plants grown on some soils, which can cause problems with local water quality
• Requires protection from heavy use, once seeded
• May produce vegetation that requires irrigation and maintenance

Permanent Seeding and Planting

What Is It

Permanent seeding of grass and planting trees and brush provides stabilization to the soil by holding soil particles in place. Vegetation reduces sediments and runoff to downstream areas by slowing the velocity of runoff and permitting greater infiltration of the runoff. Vegetation also filters sediments, helps the soil absorb water, improves wildlife habitats, and enhances the aesthetics of a site.

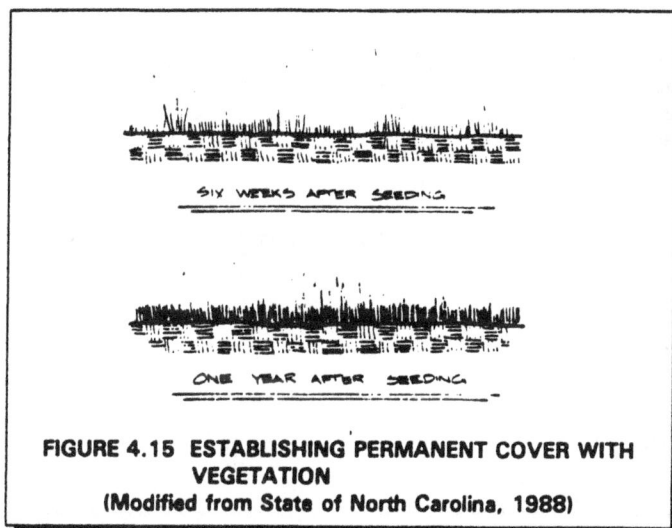

SIX WEEKS AFTER SEEDING

ONE YEAR AFTER SEEDING

FIGURE 4.15 ESTABLISHING PERMANENT COVER WITH
VEGETATION
(Modified from State of North Carolina, 1988)

When and Where to Use It

Permanent seeding and planting is appropriate for any graded or cleared area where long-lived plant cover is desired. Some areas where permanent seeding is especially important are filter strips, buffer areas, vegetated swales, steep slopes, and stream banks. This practice is effective on areas where soils are unstable because of their texture, structure, a high water table, high winds, or high slope. When seeding in northern areas during fall or winter, cover the area with mulch to provide a protective barrier against cold weather (see Mulching). Seeding should also be mulched if the seeded area slopes 4:1 or more, if soil is sandy or clayey, or if weather is excessively hot or dry. Plant when conditions are most favorable for growth. When possible, use low-maintenance local plant species. Install all other erosion control practices such as dikes, basins, and surface runoff control measures before planting.

What to Consider

For this practice to work, it is important to select appropriate vegetation, prepare a good seedbed, properly time planting, and water and fertilize. Planting local plants during their regular growing season will increase the chances for success and may lessen the need for watering. Check seeded areas frequently for proper watering and growth conditions.

Topsoil should be used on areas where topsoils have been removed, where the soils are dense or impermeable, or where mulching and fertilizers alone cannot improve soil quality. Topsoiling should be coordinated with the seeding and planting practices and should not be planned while the ground is frozen or too wet. Topsoil layers should be at least 2 inches deep (or similar to the existing topsoil depth).

To minimize erosion and sedimentation, remove as little existing topsoil as possible. All site controls should be in place before the topsoil is removed. If topsoils are brought in from another site, it is important that its texture is compatible with the subsoils onsite; for example, sandy topsoils are not compatible with clay subsoils.

Stockpiling of topsoils onsite requires good planning so soils will not obstruct other operations. If soil is to be stockpiled, consider using temporary seeding, mulching, or silt fencing to prevent or control erosion. Inspect the stockpiles frequently for erosion. After topsoil has been spread, inspect it regularly, and reseed or replace areas that have eroded.

Advantages of Permanent Seeding and Planting
• Improves the aesthetics of a site
• Provides excellent stabilization
• Provides filtering of sediments
• Provides wildlife habitat
• Is relatively inexpensive
Disadvantages of Permanent Seeding and Planting
• May require irrigation to establish vegetation
• Depends initially on climate and weather for success

Sodding

What Is It

Sodding stabilizes an area by establishing permanent vegetation, providing erosion and sedimentation controls, and providing areas where storm water can infiltrate the ground.

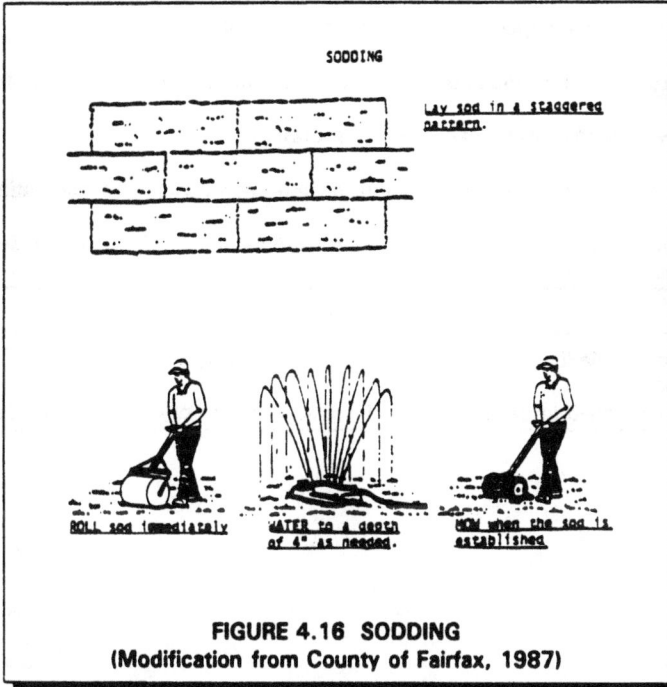

FIGURE 4.16 SODDING
(Modification from County of Fairfax, 1987)

When and Where to Use It

Sodding is appropriate for any graded or cleared area that might erode and where a permanent, long-lived plant cover is needed immediately. Examples of where sodding can be used are buffer zones, stream banks, dikes, swales, slopes, outlets, level spreaders, and filter strips.

What to Consider

The soil surface should be fine-graded before laying down the sod. Topsoil may be needed in areas where the soil textures are inadequate (see topsoil discussion in section on Permanent Seeding and Planting). Lime and fertilizers should be added to the soil to promote good growth conditions. Sodding can be applied in alternating strips or other patterns, or alternate areas can be seeded to reduce expense. Sod should not be planted during very hot or wet weather. Sod should not be placed on slopes that are greater than 3:1 if they are to be mowed. If placed on steep slopes, sod should be laid with staggered joints and/or be pegged. In areas such as steep slopes or next to running waterways, chicken wire, jute, or other netting can be placed over the sod for extra protection against lifting (see Mulching, Matting, and Netting). Rolled or compact immediately after installation to ensure firm contact with the underlying topsoil. Inspect the sod frequently after it is first installed, especially after large storm events, until it is established as permanent cover. Remove and replace dead sod. Watering may be necessary after planting and during periods of intense heat and/or lack of rain.

Advantages of Sodding
• Can provide immediate vegetative cover and erosion control
• Provides more stabilizing protection than initial seeding through dense cover formed by sod
• Produces lower weed growth than seeded vegetation
• Can be used for site activities within a shorter time than can seeded vegetation
• Can be placed at any time of the year as long as moisture conditions in the soil are favorable, except when the ground is frozen
Disadvantages of Sodding
• Purchase and installation costs are higher than for seeding
• May require continued irrigation if the sod is placed during dry seasons or on sandy soils

Chemical Stabilization

What Is It

Chemical stabilization practices, often referred to as a chemical mulch, soil binder, or soil palliative, are temporary erosion control practices. Materials made of vinyl, asphalt, or rubber are sprayed onto the surface of the soil to hold the soil in place and protect against erosion from storm water runoff and wind. Many of the products used for chemical stabilization are human-made, and many different products are on the market.

When and Where to Use It

Chemical stabilization can be used as an alternative in areas where temporary seeding practices cannot be used because of the season or climate. It can provide immediate, effective, and inexpensive erosion control anywhere erosion is occurring on a site.

What to Consider

The application rates and procedures recommended by the manufacturer of a chemical stabilization product should be followed as closely as possible to prevent the products from forming ponds and from creating large areas where moisture cannot get through.

Advantages of Chemical Stabilization
• Is easily applied to the surface of the soil
• Is effective in stabilizing areas where plants will not grow
• Provides immediate protection to soils that are in danger of erosion
Disadvantages of Chemical Stabilization
• Can create impervious surfaces (where water cannot get through), which may in turn increase the amount and speed of storm water runoff
• May cause harmful effects on water quality if not used correctly
• Is usually more expensive than vegetative cover

4.5.2 Structural Erosion Prevention and Sediment Control Practices

Structural practices used in sediment and erosion control divert storm water flows away from exposed areas, convey runoff, prevent sediments from moving offsite, and can also reduce the erosive forces of runoff waters. The controls can either be used as permanent or temporary measures. Practices discussed include the following:

- Interceptor Dikes and Swales

- Pipe Slope Drains

- Subsurface Drains

- Filter Fence

- Straw Bale Barrier

- Brush Barrier

- Gravel or Stone Filter Berm

- Storm Drain Inlet Protection

- Sediment Trap

- Temporary Sediment Basin

- Outlet Protection

- Check Dams

- Surface Roughening

- Gradient Terraces.

Interceptor Dikes and Swales

What Are They

Interceptor dikes (ridges of compacted soil) and swales (excavated depressions) are used to keep upslope runoff from crossing areas where there is a high risk of erosion. They reduce the amount and speed of flow and then guide it to a stabilized outfall (point of discharge) (see section on Outlet Protection) or sediment trapping area (see sections on Level Spreaders, Vegetated Filter Strips, Sediment Traps, and Temporary Sediment Basins). Interceptor dikes and swales divert runoff using a combination of earth dike and vegetated swale. Runoff is channeled away from locations where there is a high risk of erosion by placing a diversion dike or swale at the top of a sloping disturbed area. Dikes and swales also collect overland flow, changing it into concentrated flows (i.e., flows that are combined). Interceptor dikes and swales can be either temporary or permanent storm water control structures.

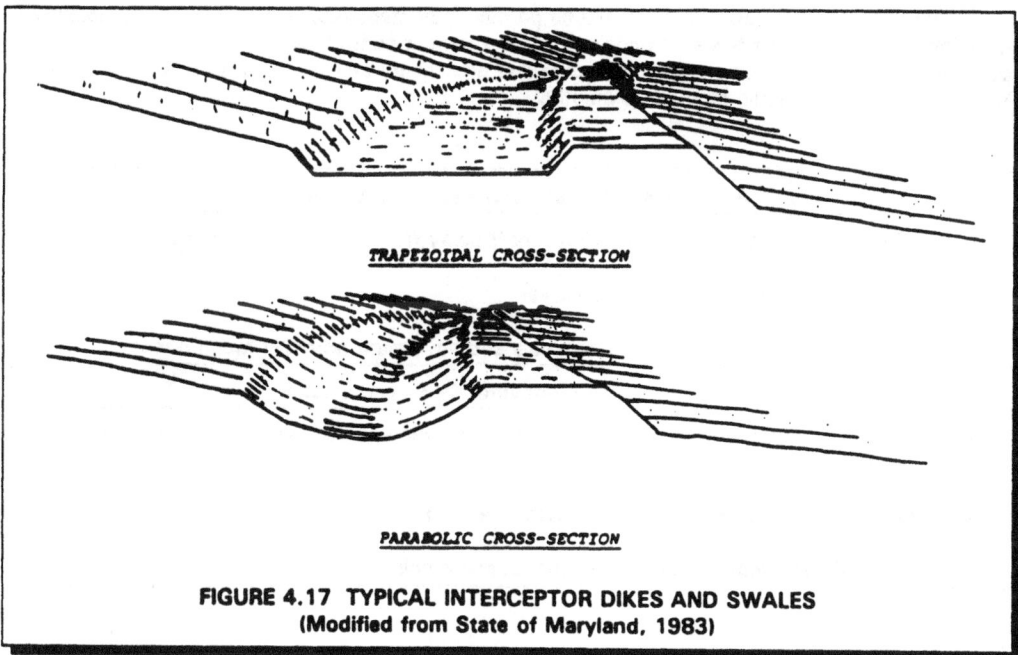

TRAPEZOIDAL CROSS-SECTION

PARABOLIC CROSS-SECTION

FIGURE 4.17 TYPICAL INTERCEPTOR DIKES AND SWALES
(Modified from State of Maryland, 1983)

When and Where to Use Them

Interceptor dikes and swales are generally built around the perimeter of a construction site before any major soil disturbing activity takes place. Temporary dikes or swales may also be used to protect existing buildings; areas, such as stockpiles; or other small areas that have not yet been fully stabilized. When constructed along the upslope perimeter of a disturbed or high-risk area (though not necessarily all the way around it), dikes or swales prevent runoff from uphill areas from crossing the unprotected slope. Temporary dikes or swales constructed on the down slope side of the disturbed or high-risk area will prevent runoff that contains sediment from leaving the site before sediment is removed. For short slopes, a dike or swale at the top of the slope reduces the amount of runoff reaching the disturbed area. For longer slopes, several dikes or swales are placed across the slope at intervals. This practice reduces the amount of runoff that accumulates on the face of the slope and carries the runoff safely down the slope. In all cases, runoff is guided to a sediment trapping area or a stabilized outfall before release.

What to Consider

Temporary dikes and swales are used in areas of overland flow; if they remain in place longer than 15 days, they should be stabilized. Runoff channeled by a dike or swale should be directed to an adequate sediment trapping area or stabilized outfall. Care should be taken to provide enough slope for drainage but not too much slope to cause erosion due to high runoff flow speed. Temporary interceptor dikes and swales may remain in place as long as 12 to 18 months (with proper stabilization) or be rebuilt at the end of each day's activities. Dikes or swales should remain in place until the area they were built to protect is permanently stabilized. Interceptor dikes and swales can be permanent controls. However, permanent controls: should be designed to handle runoff after construction is complete; should be permanently stabilized; and should be inspected and maintained on a regular basis. Temporary and permanent control measures should be inspected once each week on a regular schedule and after every storm. Repairs necessary to the dike and flow channel should be made promptly.

Advantages of Interceptor Dikes and Swales
• Are simple and effective for channeling runoff away from areas subject to erosion
• Can handle flows from large drainage areas
• Are inexpensive because they use materials and equipment normally found onsite
Disadvantages of Interceptor Dikes and Swales
• If constructed improperly, can cause erosion and sediment transport since flows are concentrated
• May cause problems to vegetation growth if water flow is too fast
• Require additional maintenance, inspections, and repairs

Pipe Slope Drains

What Are They

Pipe slope drains reduce the risk of erosion by discharging runoff to stabilized areas. Made of flexible or rigid pipe, they carry concentrated runoff from the top to the bottom of a slope that has already been damaged by erosion or is at high risk for erosion. They are also used to drain saturated slopes that have the potential for soil slides. Pipe slope drains can be either temporary or permanent depending on the method of installation and material used.

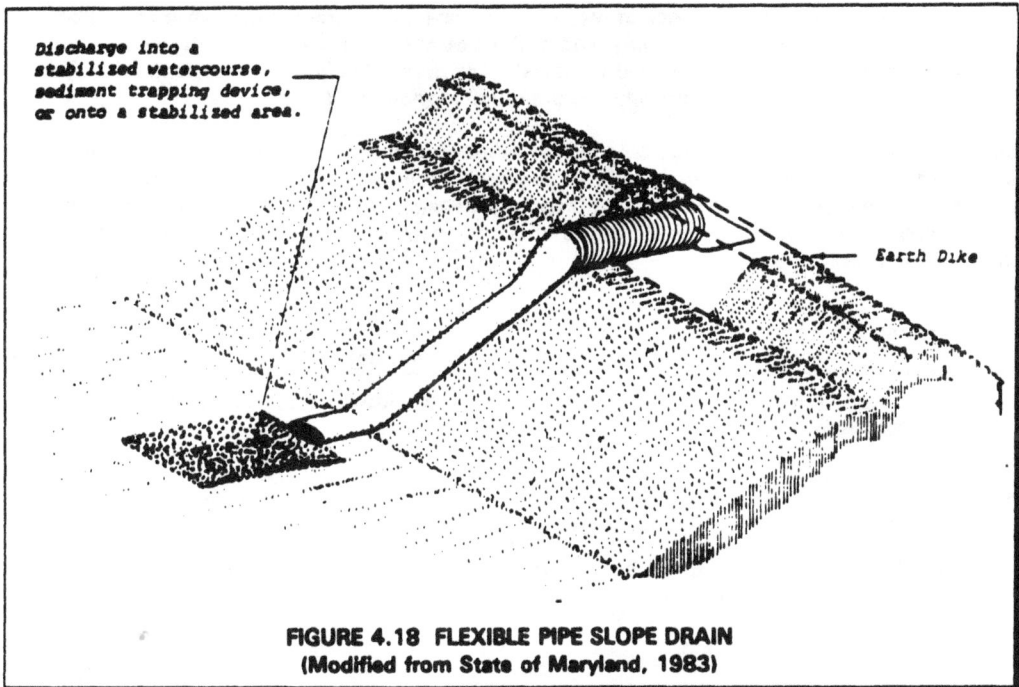

Discharge into a stabilized watercourse, sediment trapping device, or onto a stabilized area.

Earth Dike

FIGURE 4.18 FLEXIBLE PIPE SLOPE DRAIN
(Modified from State of Maryland, 1983)

When and Where to Use Them

Pipe slope drains are used whenever it is necessary to convey water down a slope without causing erosion. They are especially effective before a slope has been stabilized or before permanent drainage structures are ready for use. Pipe slope drains may be used with other devices, including diversion dikes or swales, sediment traps, and level spreaders (used to spread out storm water runoff uniformly over the surface of the ground). Temporary pipe slope drains, usually flexible tubing or conduit, may be installed prior to the construction of permanent drainage structures. Permanent slope drains may be placed on or beneath the ground surface; pipes, sectional downdrains, paved chutes, or clay tiles may be used.

Paved chutes may be covered with a surface of concrete or other impenetrable material. Subsurface drains can be constructed of concrete, PVC, clay tile, corrugated metal, or other permanent material.

What to Consider

The drain design should be able to handle the volume of flow. The effective life span of a temporary pipe slope drain is up to 30 days after permanent stabilization has been achieved. The maximum recommended drainage area for pipe slope drains is 10 acres (Washington State, 1992).

The inlets and outlets of a pipe slope drain should be stabilized. This means that a flared end section should be used at the entrance of the pipe. The soil around the pipe entrance should be fully compacted. The soil at the discharge end of the pipe should be stabilized with riprap (a combination of large stones, cobbles, and boulders). The riprap should be placed along the bottom of a swale which leads to a sediment trapping structure or another stabilized area.

Pipe slope drains should be inspected on a regular schedule and after any major storm. Be sure that the inlet from the pipe is properly installed to prevent bypassing the inlet and undercutting the structure. If necessary, install a headwall, riprap, or sandbags around the inlet. Check the outlet point for erosion and check the pipe for breaks or clogs. Install outlet protection if needed and promptly clear breaks and clogs.

Advantages of Pipe Slope Drains
• Can reduce or eliminate erosion by transporting runoff down steep slopes or by draining saturated soils
• Are easy to install and require little maintenance
Disadvantages of Pipe Slope Drains
• Require that the area disturbed by the installation of the drain should be stabilized or it, too, will be subject to erosion
• May clog during a large storm

Subsurface Drains

What Are They

A subsurface drain is a perforated pipe or conduit placed beneath the surface of the ground at a designed depth and grade. It is used to drain an area by lowering the water table. A high water table can saturate soils and prevent the growth of certain types of vegetation. Saturated soils on slopes will sometimes "slip" down the hill. Installing subsurface drains can help prevent these problems.

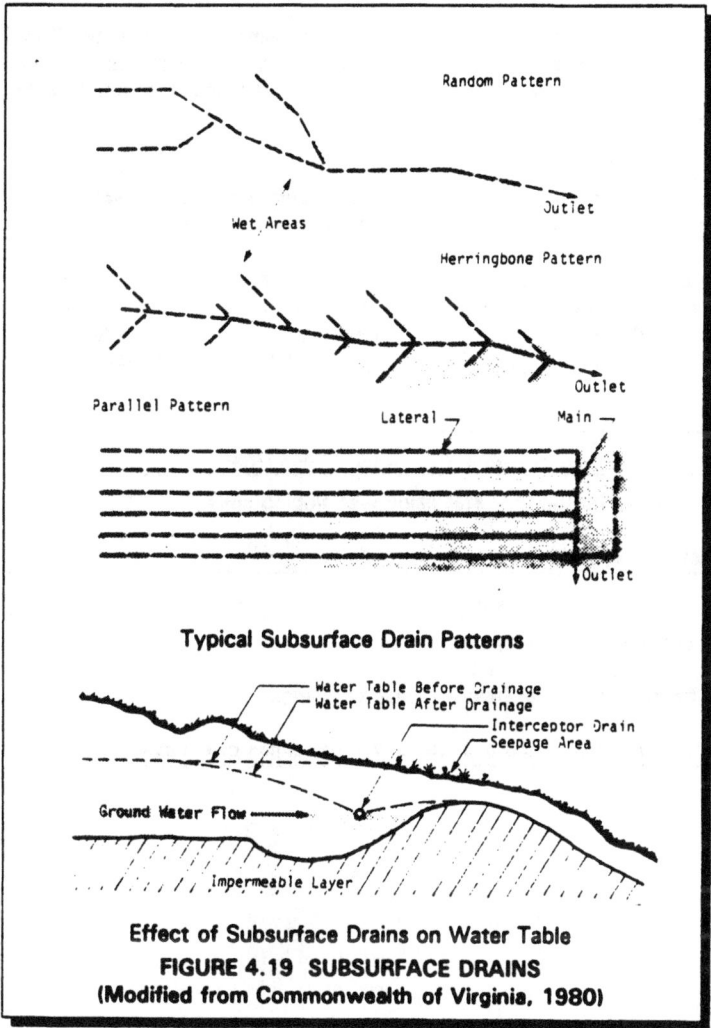

Typical Subsurface Drain Patterns

Effect of Subsurface Drains on Water Table
FIGURE 4.19 SUBSURFACE DRAINS
(Modified from Commonwealth of Virginia, 1980)

When and Where to Use Them

There are two types of subsurface drains: relief drains and interceptor drains. Relief drains are used to dewater an area where the water table is high. They may be placed in a gridiron, herringbone, or random pattern. Interceptor drains are used to remove water where sloping soils are excessively wet or subject to slippage. They are usually placed as single pipes instead of in patterns. Generally, subsurface drains are suitable only in areas where the soil is deep enough for proper installation. They are not recommended where they pass under heavy vehicle crossings.

What to Consider

Drains should be placed so that tree roots will not interfere with drainage pipes. The drain design should be adequate to handle the volume of flow. Areas disturbed by the installation of a drain should be stabilized or they, too, will be subject to erosion. The soil layer must be deep enough to allow proper installation.

Backfill immediately after the pipe is placed. Material used for backfill should be open granular soil that is highly permeable. The outlet should be stabilized and should direct sediment-laden storm water runoff to a sediment trapping structure or another stabilized area.

Inspect subsurface drains on a regular schedule and check for evidence of pipe breaks or clogging by sediment, debris, or tree roots. Remove blockage immediately, replace any broken sections, and restabilize the surface. If the blockage is from tree roots, it may be necessary to relocate the drain. Check inlets and outlets for sediment or debris. Remove and dispose of these materials properly.

Advantages of Subsurface Drains
• Provide an effective method for stabilizing wet sloping soils
• Are an effective way to lower the water table
Disadvantages of Subsurface Drains
• May be pierced and clogged by tree roots
• Should not be installed under heavy vehicle crossings
• Cost more than surface drains because of the expenses of excavation for installation

Filter Fence

What Is It

A silt fence, also called a "filter fence," is a temporary measure for sedimentation control. It usually consists of posts with filter fabric stretched across the posts and sometimes with a wire support fence. The lower edge of the fence is vertically trenched and covered by backfill. A silt fence is used in small drainage areas to detain sediment. These fences are most effective where there is overland flow (runoff that flows over the surface of the ground as a thin, even layer) or in minor swales or drainageways. They prevent sediment from entering receiving waters. Silt fences are also used to catch wind blown sand and to create an anchor for sand dune creation. Aside from the traditional wooden post and filter fabric method, there are several variations of silt fence installation including silt fence which can be purchased with pockets presewn to accept use of steel fence posts.

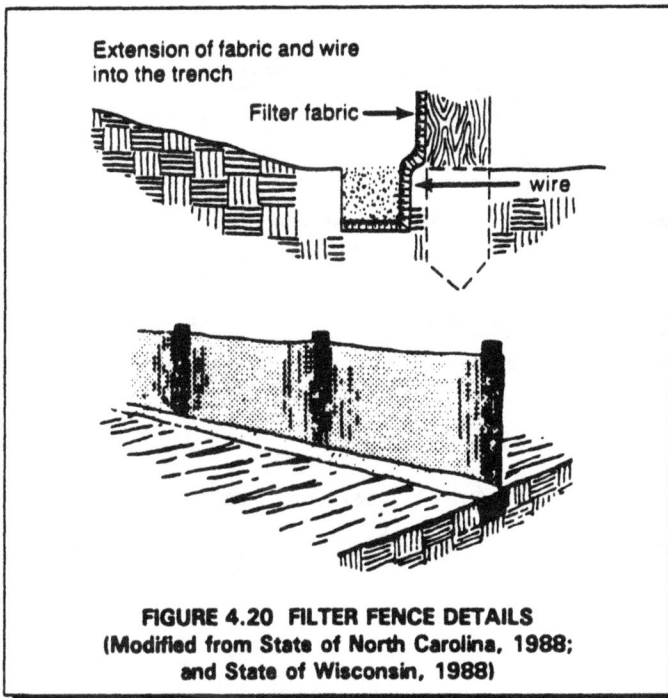

FIGURE 4.20 FILTER FENCE DETAILS
(Modified from State of North Carolina, 1988; and State of Wisconsin, 1988)

When and Where to Use It

A silt fence should be installed prior to major soil disturbance in the drainage area. Such a structure is only appropriate for drainage areas of 1 acre or less with velocities of 0.5 cfs or less (Washington State, 1992). The fence should be placed across the bottom of a slope or minor drainageway along a line of uniform elevation (perpendicular to the direction of flow). It can be used at the outer boundary of the work area. However, the fence does not have to surround the work area completely. In addition, a silt fence is effective where sheet and rill erosion may be a problem. Silt fences should not be constructed in streams or swales.

What to Consider

A silt fence is not appropriate for a large area or where the flow rate is greater than 0.5 cfs. This type of fence can be more effective than a straw bale barrier if properly installed and maintained. It may be used in combination with other erosion and sediment practices.

The effective life span for a silt fence is approximately 6 months. During this period, the fence requires frequent inspection and prompt maintenance to maintain its effectiveness. Inspect the fence after each rainfall. Check for areas where runoff eroded a channel beneath the fence, or where the fence was caused to sag or collapse by runoff flowing over the top. Remove and properly dispose of sediment when it is one-third to one-half the height of the fence or after each storm.

Advantages of a Filter Fence
• Removes sediments and prevents downstream damage from sediment deposits
• Reduces the speed of runoff flow
• Minimal clearing and grubbing required for installation
• Inexpensive

Disadvantages of a Filter Fence
• May result in failure from improper choice of pore size in the filter fabric or improper installation
• Should not be used in streams
• Is only appropriate for small drainage areas with overland flow
• Frequent inspection and maintenance is necessary to ensure effectiveness

Straw Bale Barrier

What Is It

Straw bales can be used as a temporary sediment barrier. They are placed end to end in a shallow excavated trench (with no gaps in between) and staked into place. If properly installed, they can detain sediment and reduce flow velocity from small drainage areas. A straw bale barrier prevents sediment from leaving the site by trapping the sediment in the barrier while allowing the runoff to pass through. It can also be used to decrease the velocity of sheetflow or channel flows of low-to-moderate levels.

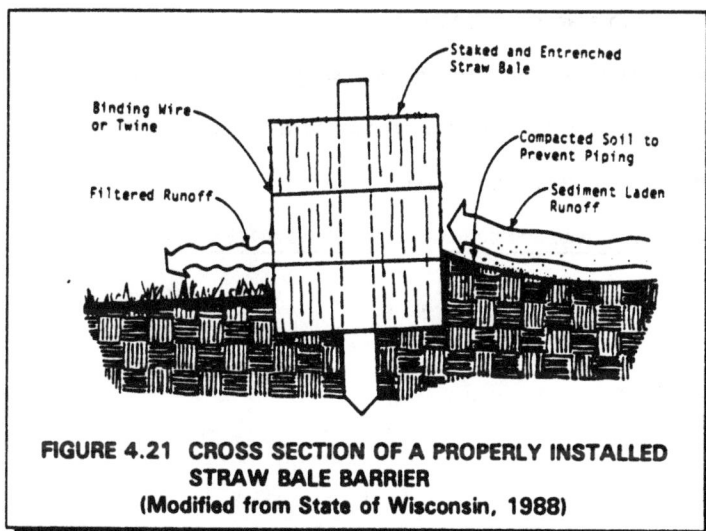

FIGURE 4.21   CROSS SECTION OF A PROPERLY INSTALLED
STRAW BALE BARRIER
(Modified from State of Wisconsin, 1988)

When and Where to Use It

A straw bale barrier should be installed prior to major soil disturbance in the drainage area. This type of barrier is placed perpendicular to the flow, across the bottom of a slope or minor drainageway where there is sheetflow. It can be used at the perimeter of the work area, although is does not have to surround it completely. It can also be very effective when used in combination with other erosion and sediment control practices. A straw bale barrier may be used where the length of slope behind the barrier is less than 100 feet and where the slope is less than 2:1.

What to Consider

The success of a straw bale barrier depends on proper installation. The bales must be firmly staked into the entrenchment and the entrenchment must be properly backfilled. To function effectively, the bales must be placed end to end and there can be no gaps between the bales.

Straw bale barriers are useful for approximately 3 months. They must be inspected and repaired immediately after each rainfall or daily if there is prolonged rainfall. Damaged straw bales require immediate replacement. After each storm, or on a regular basis, trapped sediments must be removed and disposed of properly.

Advantages of a Straw Bale Barrier
• Can prevent downstream damage from sediment deposits if properly installed, used, and maintained
• Can be an inexpensive way to reduce or prevent erosion
Disadvantages of a Straw Bale Barrier
• May not be used in streams or large swales
• Poses a risk of washouts if the barrier is installed improperly or a storm is severe
• Has a short life span and a high inspection and maintenance requirement
• Is appropriate for only small drainage areas
• Is easily subject to misuse and can contribute to sediment problems

Brush Barrier

What Is It

A brush barrier is a temporary sediment barrier constructed from materials resulting from onsite clearing and grubbing. It is usually constructed at the bottom perimeter of the disturbed area. Filter fabric is sometimes used as an anchor over the barrier to increase its filtering efficiency. Brush barriers are used to trap and retain small amounts of sediment by intercepting the flow from small areas of soil disturbance.

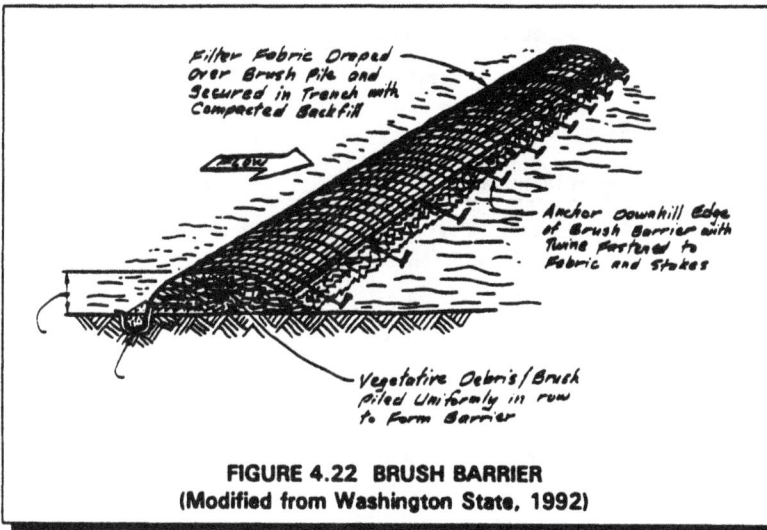

FIGURE 4.22 BRUSH BARRIER
(Modified from Washington State, 1992)

When and Where to Use It

A brush barrier should only be used to trap sediment from runoff which is from a small drainage area. The slope which the brush barrier is placed across should be very gentle. Do not place a brush barrier in a swale or any other channel. Brush barriers should be constructed below areas subject to erosion.

What to Consider

The construction of a brush barrier should be started as soon as clearing and grubbing has produced enough material to make the structure. Wood chips should not be included in the material used for the barrier because of the possibility of leaching. When the site has been stabilized and any excess sediment has been disposed of properly, the filter fabric can be removed. Over time, natural vegetation will establish itself within the barrier, and the barrier itself will decompose.

You will not have to maintain the brush barrier unless there is a very large amount of sediment being deposited. If used, the filter fabric anchor should be checked for tears and the damaged sections replaced promptly. The barrier should be inspected after each rainfall and checked for areas breached by concentrated flow. If necessary, repairs should be made promptly and excess sediment removed and disposed of properly.

Advantages of a Brush Barrier
• Can help prevent downstream damage from sediment deposits
• Is constructed of cleared onsite materials and, thus, is inexpensive
• Usually requires little maintenance, unless there are very heavy sediment deposits
Disadvantages of a Brush Barrier
• Does not replace a sediment trap or basin
• Is appropriate for only small drainage areas
• Has very limited sediment retention

Gravel or Stone Filter Berm

What Is It

A gravel or stone filter berm is a temporary ridge constructed of loose gravel, stone, or crushed rock. It slows and filters flow, diverting it from an exposed traffic area. Diversions constructed of compacted soil may be used where there will be little or no construction traffic within the right-of way. They are also used for directing runoff from the right-of-way to a stabilized outlet.

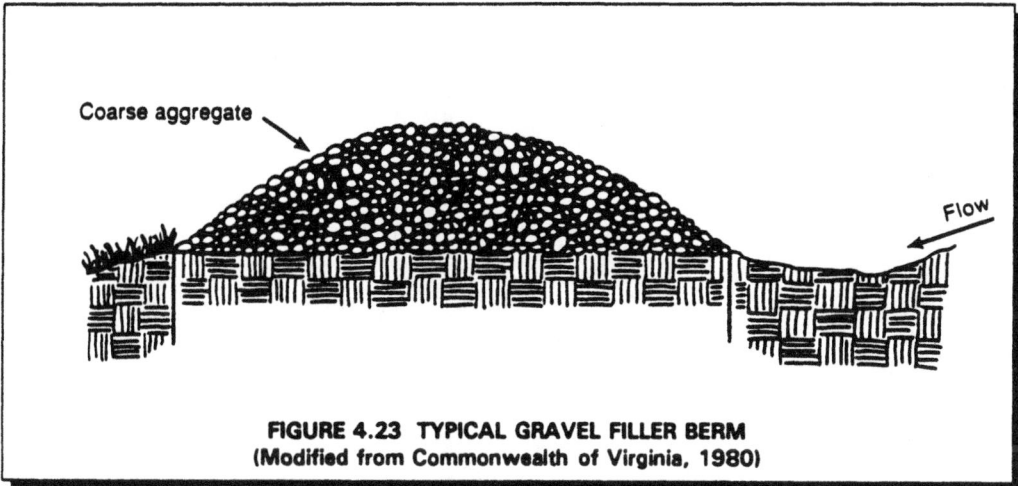

FIGURE 4.23 TYPICAL GRAVEL FILLER BERM
(Modified from Commonwealth of Virginia, 1980)

When and Where to Use It

This method is appropriate where roads and other rights-of-way under construction should accommodate vehicular traffic. Berms are meant for use in areas with shallow slopes. They may also be used at traffic areas within the construction site.

What to Consider

Berm material should be well graded gravel or crushed rock. The spacing of the berms will depend on the steepness of the slope: berms should be placed closer together as the slope increases. The diversion should be inspected daily, after each rainfall, or if breached by construction or other vehicles. All needed repairs should be performed immediately. Accumulated sediment should be removed and properly disposed of and the filter material replaced, as necessary.

Advantages of a Gravel or Stone Filter Berm
• Is a very efficient method of sediment control
Disadvantages of a Gravel or Stone Filter Berm
• Is more expensive than methods that use onsite materials
• Has a very limited life span
• Can be difficult to maintain because of clogging from mud and soil on vehicle tires

Storm Drain Inlet Protection

What Is It

Storm drain inlet protection is a filtering measure placed around any inlet or drain to trap sediment. This mechanism prevents the sediment from entering inlet structures. Additionally, it serves to prevent the silting-in of inlets, storm drainage systems, or receiving channels. Inlet protection may be composed of gravel and stone with a wire mesh filter, block and gravel, filter fabric, or sod.

Sod Inlet Protection

Excavated Gravel Inlet Protection

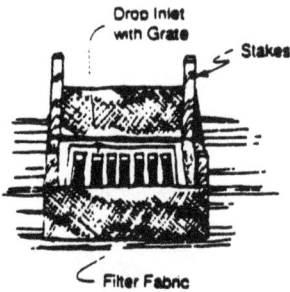

Filter Fabric Inlet Protection

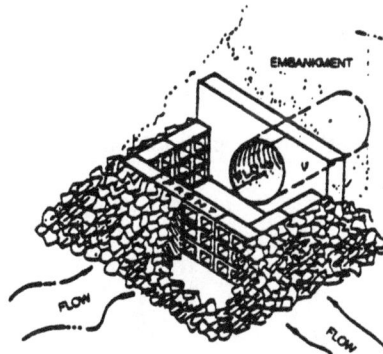

Block and Gravel Inlet Protection

FIGURE 4.24 EXAMPLES OF STORM DRAIN INLET PROTECTION
(Modified from State of North Carolina, 1988; Washington State, 1992; and County of Fairfax, 1987)

When and Where to Use It

This type of protection is appropriate for small drainage areas where storm drain inlets will be ready for use before final stabilization. Storm drain inlet protection is also used where a permanent storm drain structure is being constructed onsite. Straw bales are not recommended for this purpose. Filter fabric is used for inlet protection when storm water flows are relatively small with low velocities. This practice cannot be used where inlets are paved because the filter fabric should be staked. Block and gravel filters can be used where velocities are higher. Gravel and mesh filters can be used where flows are higher and subject to disturbance by site traffic. Sod inlet filters are generally used where sediments in the storm water runoff are low.

What to Consider

Storm drain inlet protection is not meant for use in drainage areas exceeding 1 acre or for large concentrated storm water flows. Installation of this measure should take place before any soil disturbance in the drainage area. The type of material used will depend on site conditions and the size of the drainage area. Inlet protection should be used in combination with other measures, such as small impoundments or sediment traps, to provide more effective sediment removal. Inlet protection structures should be inspected regularly, especially after a rainstorm. Repairs and silt removal should be performed as necessary. Storm drain inlet protection structures should be removed only after the disturbed areas are completely stabilized.

Advantages of Storm Drain Inlet Protection
• Prevents clogging of existing storm drainage systems and the siltation of receiving waters
• Reduces the amount of sediment leaving the site
Disadvantages of Storm Drain Inlet Protection
• May be difficult to remove collected sediment
• May cause erosion elsewhere if clogging occurs
• Is practical only for low sediment, low volume flows

Sediment Trap

What Is It

A sediment trap is formed by excavating a pond or by placing an earthen embankment across a low area or drainage swale. An outlet or spillway is constructed using large stones or aggregate to slow the release of runoff. The trap retains the runoff long enough to allow most of the silt to settle out.

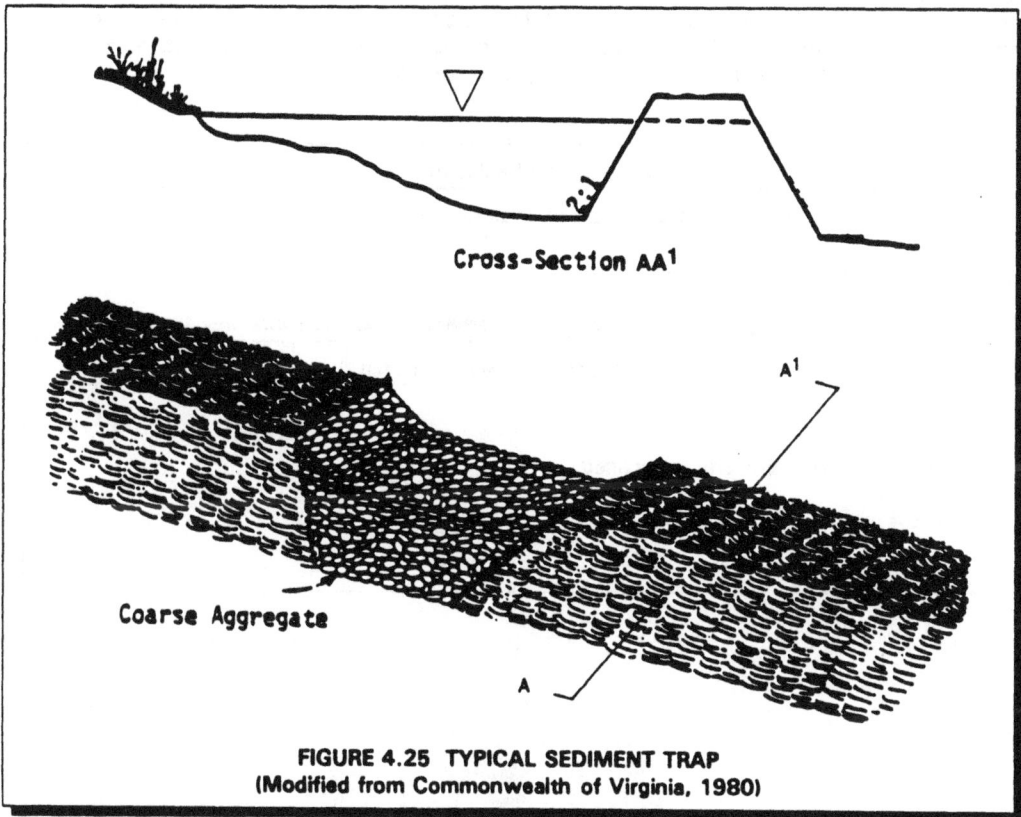

Cross-Section AA[1]

Coarse Aggregate

FIGURE 4.25 TYPICAL SEDIMENT TRAP
(Modified from Commonwealth of Virginia, 1980)

When and Where to Use It

A temporary sediment trap may be used in conjunction with other temporary measures, such as gravel construction entrances, vehicle wash areas, slope drains, diversion dikes and swales, or diversion channels. This device is appropriate for sites with short time schedules.

What to Consider

Sediment traps are suitable for small drainage areas, usually no more than 10 acres, that have no unusual drainage features. The trap should be large enough to allow the sediments to settle and should have a capacity to store the collected sediment until it is removed. The volume of storage required depends upon the amount and intensity of expected rainfall and on estimated quantities of sediment in the storm water runoff. Check your Permit to see if it specifies a minimum storage volume for sediment traps.

A sediment trap is effective for approximately 18 months. During this period, the trap should be readily accessible for periodic maintenance and sediment removal. Traps should be inspected after each rainfall and cleaned when no more than half the design volume has been filled with collected sediment. The trap should remain in operation and be properly maintained until the site area is permanently stabilized by vegetation and/or when permanent structures are in place.

Advantages of a Sediment Trap
• Protects downstream areas from clogging or damage due to sediment deposits
• Is inexpensive and simple to install
• Can simplify the design process by trapping sediment at specific spots onsite
Disadvantages of a Sediment Trap
• Is suitable only for a limited area
• Is effective only if properly maintained
• Will not remove very fine silts and clays
• Has a short life span

Temporary Sediment Basin

What Is It

A temporary sediment basin is a settling pond with a controlled storm water release structure used to collect and store sediment produced by construction activities. A sediment basin can be constructed by excavation or by placing an earthen embankment across a low area or drainage swale. Sediment basins can be designed to maintain a permanent pool or to drain completely dry. The basin detains sediment-laden runoff from larger drainage areas long enough to allow most of the sediment to settle out.

The pond has a gravel outlet or spillway to slow the release of runoff and provide some sediment filtration. By removing sediment, the basin helps prevent clogging of offsite conveyance systems and sediment-loading of receiving waterways. In this way, the basin helps prevent destruction of waterway habitats.

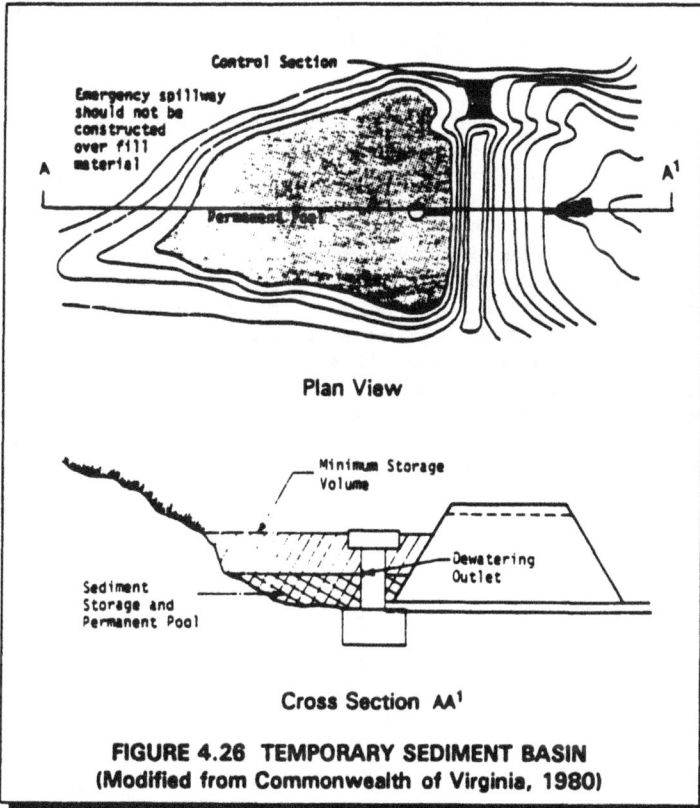

Plan View

Cross Section AA¹

FIGURE 4.26 TEMPORARY SEDIMENT BASIN
(Modified from Commonwealth of Virginia, 1980)

When and Where to Use It

A temporary sediment basin should be installed before clearing and grading is undertaken. It should not be built on an embankment in an active stream. The creation of a dam in such a site may result in the destruction of aquatic habitats. Dam failure can also result in flooding. A temporary sediment basin should be located only where there is sufficient space and appropriate topography. The basin should be made large enough to handle the maximum expected amount of site drainage. Fencing around the basin may be necessary for safety or vandalism reasons.

A temporary sediment basin used in combination with other control measures, such as seeding or mulching, is especially effective for removing sediments.

What to Consider

Temporary sediment basins are usually designed for disturbed areas larger than 5 acres. The pond should be large enough to hold runoff long enough for sediment to settle. Sufficient space should be allowed for collected sediments. Check the requirements of your permit to see if there is a minimum storage requirement for sediment basins. The useful life of a temporary sediment basin is about 12 to 18 months.

Sediment trapping efficiency is improved by providing the maximum surface area possible. Because finer silts may not settle out completely, additional erosion control measures should be used to minimize release of fine silt. Runoff should enter the basin as far from the outlet as possible to provide maximum retention time.

Sediment basins should be readily accessible for maintenance and sediment removal. They should be inspected after each rainfall and be cleaned out when about half the volume has been filled with sediment. The sediment basin should remain in operation and be properly maintained until the site area is permanently stabilized by vegetation and/or when permanent structures are in place. The embankment forming the sedimentation pool should be well compacted and stabilized with vegetation. If the pond is located near a residential area, it is recommended for safety reasons that a sign be posted and that the area be secured by a fence. A well built temporary sediment basin that is large enough to handle the post construction runoff volume may later be converted to use as a permanent storm water management structure.

Advantages of a Temporary Sediment Basin
• Protects downstream areas from clogging or damage due to sediment deposits generated during construction activities
• Can trap smaller sediment particles than sediment traps can because of the longer detention time

Disadvantages of a Temporary Sediment Basin
• Is generally suitable for small areas
• Requires regular maintenance and cleaning
• Will not remove very fine silts and clays unless used in conjunction with other measures
• Is a more expensive way to remove sediment than several other methods
• Requires careful adherence to safety practices since ponds are attractive to children

Outlet Protection

What Is It

Outlet protection reduces the speed of concentrated storm water flows and therefore it reduces erosion or scouring at storm water outlets and paved channel sections. In addition, outlet protection lowers the potential for downstream erosion. This type of protection can be achieved through a variety of techniques, including stone or riprap, concrete aprons, paved sections and settling basins installed below the storm drain outlet.

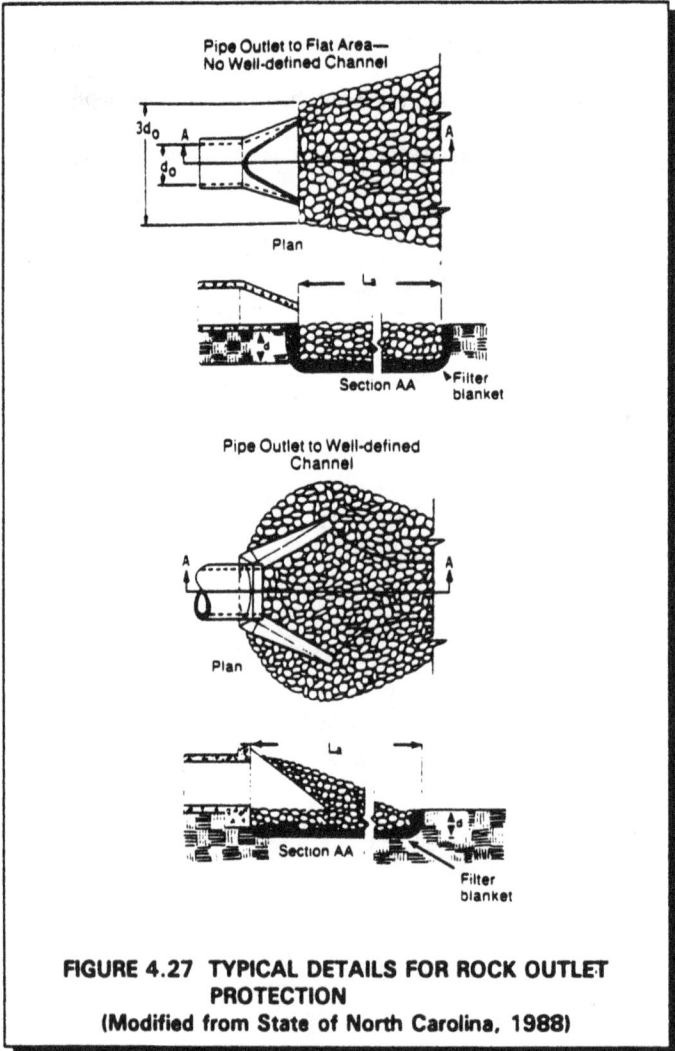

FIGURE 4.27 TYPICAL DETAILS FOR ROCK OUTLET PROTECTION
(Modified from State of North Carolina, 1988)

When and Where to Use It

Outlet protection should be installed at all pipe, interceptor dike, swale, or channel section outlets where the velocity of flow may cause erosion at the pipe outlet and in the receiving channel. Outlet protection should also be used at outlets where the velocity of flow at the design capacity may result in plunge pools (small permanent pools located at the inlet to or the outfall from BMPs). Outlet protection should be installed early during construction activities, but may be added at any time, as necessary.

What to Consider

The exit velocity of the runoff as it leaves the outlet protection structure should be reduced to levels that minimize erosion. Outlet protection should be inspected on a regular schedule to look for erosion and scouring. Repairs should be made promptly.

Advantages of Outlet Protection
• Provides, with riprap-line apron (the most common outlet protection), a relatively low cost method that can be installed easily on most sites
• Removes sediment in addition to reducing flow speed
• Can be used at most outlets where the flow speed is high
• Is an inexpensive but effective measure
• Requires less maintenance than many other measures
Disadvantages of Outlet Protection
• May be unsightly
• May cause problems in removing sediment (without removing and replacing the outlet protection structure itself)
• May require frequent maintenance for rock outlets with high velocity flows

Check Dams

What Are They

A check dam is a small, temporary or permanent dam constructed across a drainage ditch, swale, or channel to lower the speed of concentrated flows. Reduced runoff speed reduces erosion and gullying in the channel and allows sediments and other pollutants to settle out.

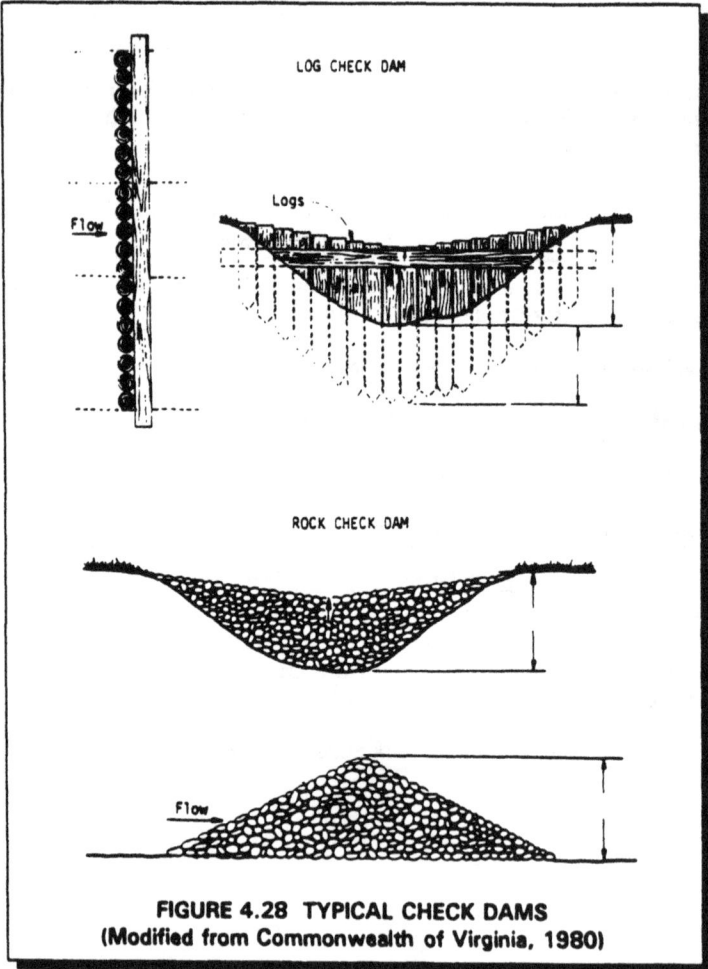

FIGURE 4.28 TYPICAL CHECK DAMS
(Modified from Commonwealth of Virginia, 1980)

When and Where to Use Them

A check dam should be installed in steeply sloped swales, or in swales where adequate vegetation cannot be established. A check dam may be built from logs, stone, or pea gravel-filled sandbags.

What to Consider

Check dams should be used only in small open channels that drain 10 acres <u>or less</u>. The dams should not be placed in streams (unless approved by appropriate State authorities). The center section of the check dam should be lower than the edges. Dams should be spaced so that the toe of the upstream dam is at the same elevation as the top of the downstream dam.

After each significant rainfall, check dams should be inspected for sediment and debris accumulation. Sediment should be removed when it reaches one half the original dam height. Check for erosion at edges and repair promptly as required. After construction is complete, all stone and riprap should be removed if vegetative erosion controls will be used as a permanent erosion control measure. It will be important to know the expected erosion rates and runoff flow rate for the swale in which this measure is to be installed. Contact the State/local storm water program agency or a licensed engineer for assistance in designing this measure.

Advantages of Check Dams
• Are inexpensive and easy to install
• May be used permanently if designed properly
• Allow a high proportion of sediment in the runoff to settle out
• Reduce velocity and provide aeration of the water
• May be used where it is not possible to divert the flow or otherwise stabilize the channel
Disadvantages of Check Dams
• May kill grass linings in channels if the water level remains high after it rains or if there is significant sedimentation
• Are useful only for drainage areas of 10 acres or less

Surface Roughening

What Is It

Surface roughening is a temporary erosion control practice. The soil surface is roughened by the creation of horizontal grooves, depressions, or steps that run parallel to the contour of the land. Slopes that are not fine-graded and that are left in a roughened condition can also control erosion. Surface roughening reduces the speed of runoff, increases infiltration, and traps sediment. Surface roughening also helps establish vegetative cover by reducing runoff velocity and giving seed an opportunity to take hold and grow.

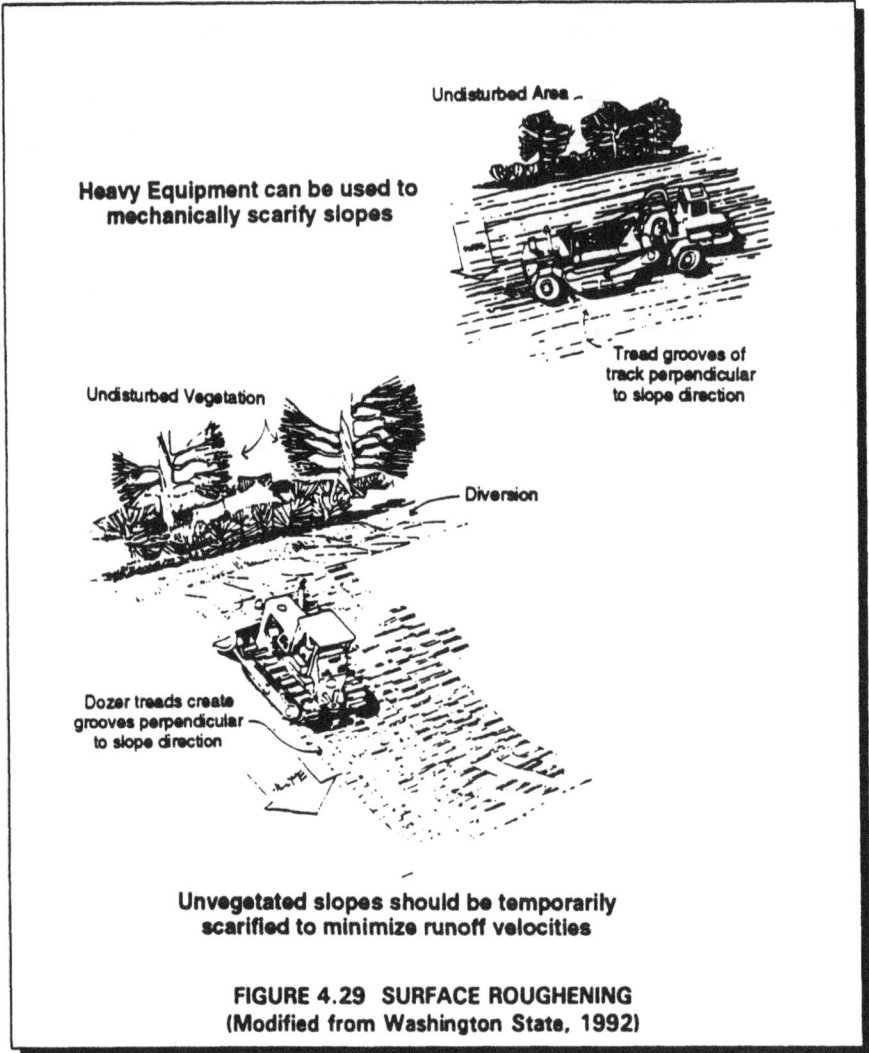

Undisturbed Area

Heavy Equipment can be used to mechanically scarify slopes

Tread grooves of track perpendicular to slope direction

Undisturbed Vegetation

Diversion

Dozer treads create grooves perpendicular to slope direction

Unvegetated slopes should be temporarily scarified to minimize runoff velocities

FIGURE 4.29 SURFACE ROUGHENING
(Modified from Washington State, 1992)

When and Where to Use It

Surface roughening is appropriate for all slopes. To slow erosion, roughening should be done as soon as possible after the vegetation has been removed from the slope. Roughening can be used with both seeding and planting and temporary mulching to stabilize an area. For steeper slopes and slopes that will be left roughened for longer periods of time, a combination of surface roughening and vegetation is appropriate.

What to Consider

Different methods can be used to roughen the soil surface on slopes. They include stair-step grading, grooving (using disks, spring harrows, or teeth on a front-end loader), and tracking (driving a crawler tractor up and down a slope, leaving the cleat imprints parallel to the slope contour). The selection of an appropriate method depends on the grade of the slope, mowing requirements after vegetative cover is established, whether the slope was formed by cutting or filling, and type of equipment available.

Cut slopes with a gradient steeper than 3:1 but less than 2:1 should be stair-step graded or groove cut. Stair-step grading works well with soils containing large amounts of small rock. Each step catches material discarded from above and provides a level site where vegetation can grow. Stairs should be wide enough to work with standard earth moving equipment. Grooving can be done by any implement that can be safely operated on the slope, including those described above. Grooves should not be less than 3 inches deep nor more than 15 inches apart. Fill slopes with a gradient steeper than 3:1 but less than 2:1 should be compacted every 9 inches of depth. The face of the slope should consist of loose, uncompacted fill 4 to 6 inches deep that can be left rough or can be grooved as described above, if necessary.

Any cut or filled slope that will be mowed should have a gradient less than 3:1. Such a slope can be roughened with shallow grooves parallel to the slope contour by using normal tilling. Grooves should be close together (less than 10 inches) and not less than 1 inch deep. Any gradient with a slope greater than 2:1 should be stair-stepped.

It is important to avoid excessive compacting of the soil surface, especially when tracking, because soil compaction inhibits vegetation growth and causes higher runoff speed. Therefore, it is best to limit roughening with tracked machinery to sandy soils that do not compact easily and to avoid tracking on clay soils. Surface roughened areas should be seeded as quickly as possible. Also, regular inspections should be made of all surface roughened areas, especially after storms. If rills (small watercourses that have steep sides and are usually only a few inches deep) appear, they should be filled, graded again, and reseeded immediately. Proper dust control procedures should be followed when surface roughening.

Advantages of Surface Roughening
• Provides a degree of instant erosion protection for bare soil while vegetative cover is being established
• Is inexpensive and simple for short-term erosion control
Disadvantages of Surface Roughening
• Is of limited effectiveness in anything more than a gentle rain
• Is only temporary; if roughening or vegetative cover is washed away in a heavy storm or the vegetation does not take hold, the surface will have to be re-roughened and new seed laid

Gradient Terraces

What Are They

Gradient terraces are earth embankments or ridge-and-channels constructed with suitable spacing and with an appropriate grade. They reduce erosion damage by capturing surface runoff and directing it to a stable outlet at a speed that minimizes erosion.

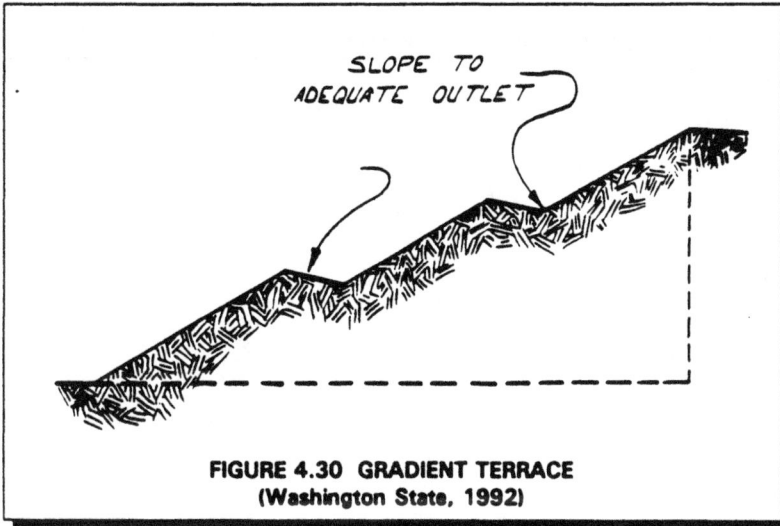

FIGURE 4.30 GRADIENT TERRACE
(Washington State, 1992)

When and Where to Use Them

Gradient terraces are usually limited to use on land that has no vegetation and that has a water erosion problem, or where it is anticipated that water erosion will be a problem. Gradient terraces should not be constructed on slopes with sandy or rocky soils. They will be effective only where suitable runoff outlets are or will be made available.

What to Consider

Gradient terraces should be designed and installed according to a plan determined by an engineering survey and layout. It is important that gradient terraces are designed with adequate outlets, such as a grassed waterway, vegetated area, or tile outlet. In all cases, the outlet should direct the runoff from the terrace system to a point where the outflow will not cause erosion or other damage. Vegetative cover should be used in the outlet where possible. The design elevation of the water surface of the terrace should not be lower than the design elevation of the water surface in the outlet at their junction, when both are operating at design flow. Terraces should be inspected regularly at least once a year and after major storms. Proper dust control procedures should be followed while constructing these features.

Advantages of Gradient Terraces
• Reduce runoff speed and increase the distance of overland runoff flow
• Hold moisture better than do smooth slopes and minimize sediment loading of surface runoff
Disadvantages of Gradient Terraces
• May significantly increase cut and fill costs and cause sloughing if excessive water infiltrates the soil
• Are not practical for sandy, steep, or shallow soils

4.6 INFILTRATION PRACTICES

Infiltration practices are surface or subsurface measures that allow for quick infiltration of storm water runoff. Rapid infiltration is possible because the structures or soils used in these practices are very porous. Infiltration practices offer an advantage over other practices in that they provide some treatment of runoff, preserve the natural flow in streams, and recharge ground water. Many of the infiltration practices also can reduce the velocity of the runoff so that it will not cause damaging erosion. Another benefit of infiltration practices is that they reduce the need for expensive storm water conveyance systems. Construction and maintenance of these practices may, however, require some level of expertise to prevent clogging and to retain high effectiveness. The infiltration practices in this section have been divided into two categories: vegetative infiltration practices and infiltration structures.

Infiltration BMPs are not practical in all cases. These practices should not be used in areas where runoff is contaminated with pollutants other than sediment or oil and grease. Excessively drained (i.e., very sandy) soils may provide inadequate treatment of runoff, which could result in ground water contamination. Other site-specific conditions, such as depth to bedrock or depth to the water table, could limit their use or make it impossible to use infiltration BMPs. Also, infiltration practices should not be installed near wells, foundations, septic tank drainfields, or on unstable slopes.

Vegetative infiltration practices rely on vegetated soils that are well drained to provide storage for the infiltration of storm water. Soils used for this practice generally have not previously been disturbed or compacted so that they more easily allow infiltration. Once vegetation has been planted, use of the area must be limited or the practice may not operate efficiently. The practices that are discussed include vegetated filter strips, grassed swales, and level spreaders.

Infiltration structures are built over soils to aid in collection of storm water runoff and are designed to allow storm water to infiltrate into the ground. These structures generally require a level of expertise for both their design and construction so that they function properly. Maintenance activities are very important because infiltration structures are easily damaged by high sediment loads. Often, infiltration structures are used with other structures that pretreat the storm water runoff for sediments, oil, and grease. These pretreatment structures may be as simple as a buffer zone (see Buffer Zones) or may be something more complex, such as an oil and grease separator. The types of infiltration structures discussed include infiltration trenches, porous pavements, concrete grids, and modular pavements.

Vegetated Filter Strips

What Are They

Vegetated filter strips are gently sloping areas of natural vegetation or are graded and artificially planted areas used to provide infiltration, remove sediments and other pollutants, and reduce the flow and velocity of the storm water moving across the terrain. Vegetated filter strips function similarly to vegetated or grassed swales. The filter strips, however, are fairly level and treat sheetflow, whereas grassed swales are indentations (see section on Grassed Swales) and treat concentrated flows. Vegetated filter strips provide permanent storm water control measures on a site.

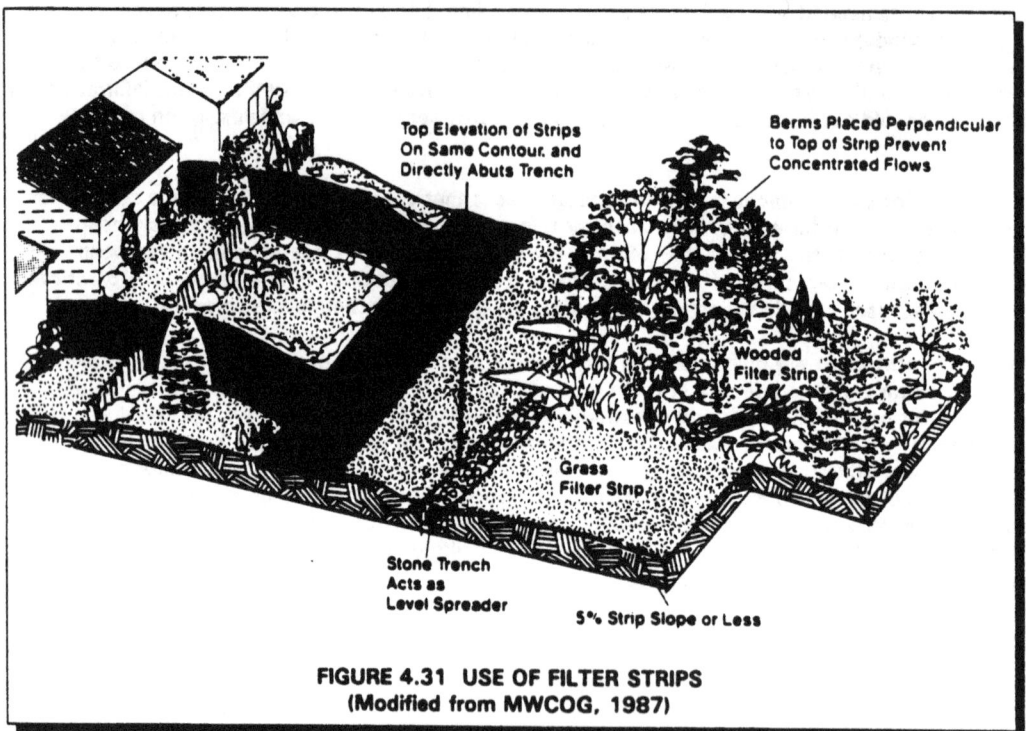

FIGURE 4.31 USE OF FILTER STRIPS
(Modified from MWCOG, 1987)

When and Where to Use Them

Vegetated filter strips are suited for areas where the soils are well drained or moderately well drained and where the bedrock and the water table are well below the surface. Vegetated filter strips will not function well on steep slopes, in hilly areas, or in highly paved areas because of the high velocity of runoff. Sites with slopes of 15 percent or more may not be suitable for filtering storm water flows. However, they should still be vegetated (MWCOG, 1987). This practice can be put into place at any time, provided that climatic conditions allow for planting.

What to Consider

At a minimum, a filter strip must be approximately 20 feet wide to function well. The length of the strip should be approximately 50 to 75 feet. Where slopes become steeper, the length of the strip must be increased. Forested strips are always preferred to vegetated strips, and existing vegetation is preferred to planted vegetation. In planning for vegetated strips, consider climatic conditions, since vegetation may not take hold in especially dry and/or cold regions.

Regular inspections are necessary to ensure the proper functioning of the filter strips. Removing sediments and replanting may be necessary on a regular basis. The entire area should be examined for damage due to equipment and vehicles. Vegetation should be dense. Also, the portions of the strip where erosion may have created ponding of runoff should be inspected. This situation can be eliminated by grading.

Advantages of Vegetated Filter Strips
• Provide low to moderate treatment of pollutants in storm water while providing a natural look to a site
• Can provide habitat for wildlife
• Can screen noise and views if trees or high shrubs are planted on the filter strips
• Are easily constructed and implemented
• Are inexpensive
Disadvantages of Vegetated Filter Strips
• Are not effective for high velocity flows (large paved areas or steep slopes)
• Require significant land space
• May have a short useful life due to clogging by sediments and oil and grease

Grassed Swales

What Are They

Grassed swales are vegetated depressions used to transport, filter, and remove sediments. Grassed swales control high runoff rates by reducing the speed of the runoff and by reducing the volume of the runoff through infiltration of the storm water. Pollutants are removed because runoff travels slowly and infiltrates into the soil and because the vegetation in the grassed swale works as a filter or strainer.

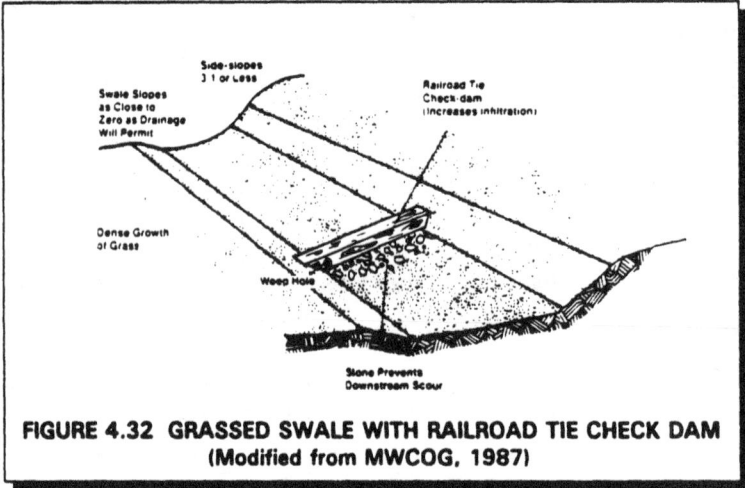

FIGURE 4.32 GRASSED SWALE WITH RAILROAD TIE CHECK DAM
(Modified from MWCOG, 1987)

When and Where to Use Them

Grassed swales are suitable for most areas where storm water runoff is low. Certain factors will affect the operation of grassed swales, including soil type, land features, and the depth of the soil from the surface to the water table (i.e., the top of the drenched portion of the soil or bedrock layer). The soil must be permeable for runoff to be able to infiltrate well. Sandy soils will not hold vegetation well nor form a stable channel structure. Steep slopes will increase runoff rates and create greater potential for erosion. Storm water flows will not be easily absorbed where the water table is near the surface. Swales are most useful for sites smaller than 10 acres (MWCOG, 1987). Even without highly permeable soils, swales reduce velocity and thus are useful.

Grassed swales usually do not work well for construction runoff because the runoff has high sediment loads.

What to Consider

The channel of the swale should be as level as possible to maximize infiltration. Side slopes in the swale should be designed to no steeper than 3:1 to minimize channel erosion (MWCOG, 1987). Plans should consider (1) the use of existing topography and existing drainage patterns and (2) the highest flow rate that is expected from a typical storm to determine the most practical size for the swale (in keeping with State or local requirements).

The swale should be tilled before grass is planted, and a dense cover of grasses should be planted in the swale. The location of the swale will determine the best type of vegetation (e.g., if the swale runs next to a road, then the grass chosen should be resistant to the use of de-icing salts in northern states).

Check dams (i.e., earthen or log structures) may be installed in the swales to reduce runoff speed and increase infiltration. Planners should also consider the design of the outlet at the end of the swale so that the runoff is released from the swale at a low rate (see section on Outlet Protection).

Maintenance activities for the swales include those practices needed to maintain healthy, dense vegetation and to retain efficient infiltration and movement of the storm water into and through the swale. Periodic mowing, reseeding, and weed control are required to maintain pollutant removal efficiency. The swale and channel outlet should be kept free from sediment buildup, litter, brush, or fallen tree limbs.

Periodic inspections will identify erosion problems or damaged areas. Damaged or eroded areas of the channel should be repaired. Areas with damaged vegetation should be reseeded immediately.

Advantages of Grassed Swales
• Are easily designed and constructed
• Provide moderate removal of sediments if properly constructed and maintained
• May provide a wildlife habitat
• Are inexpensive
• Can replace curb and gutter systems
• Can last for long periods of time if well maintained
Disadvantages of Grassed Swales
• Cannot control runoff from very large storms
• If they do not drain properly between storms, can encourage nuisance problems such as mosquitos, ragweed, dumping, and erosion
• Are not capable of removing significant amounts of soluble nutrients
• Cannot treat runoff with high sediment loadings

Level Spreaders

What Are They

Level spreaders are devices used at storm water outlets to spread out collected storm water flows into sheetflow (runoff that flows over ground surface in a thin, even layer). Typically, a level spreader consists of a depression in the soil surface that spreads the flow onto a flat area across a gentle slope. Level spreaders then release the storm water flow onto level areas stabilized by vegetation to reduce speed and increase infiltration.

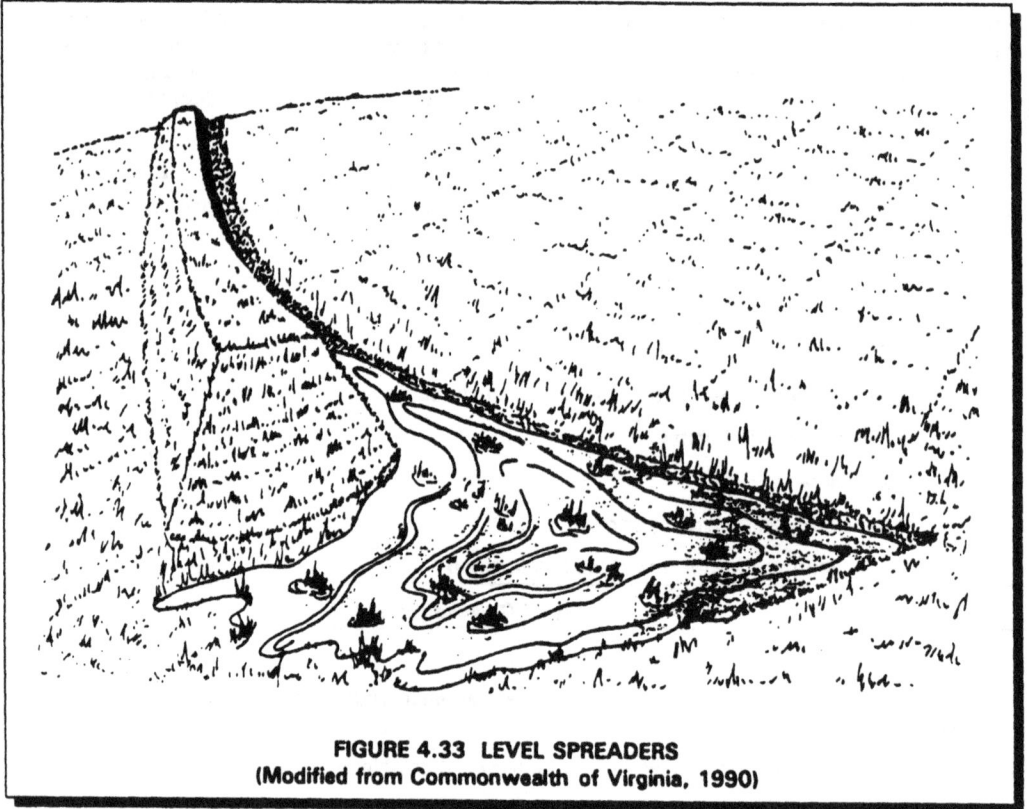

FIGURE 4.33 LEVEL SPREADERS
(Modified from Commonwealth of Virginia, 1990)

When and Where to Use Them

Level spreaders are most often used as an outlet for temporary or permanent storm water conveyances or dikes. Runoff that contains high sediment loads should be treated in a sediment trapping device prior to release into a level spreader.

What to Consider

The length of the spreader depends upon the amount of water that flows through the conveyance. Larger volumes of water need more space to even out. Level spreaders are generally used with filter strips (see Vegetated Filter Strips). The depressions are seeded with vegetation (see Permanent Seeding).

Level spreaders should not be used on soil that might erode easily. They should be constructed on natural soils and not on fill material. The entrance to the spreader should be level so that the flow can spread out evenly.

The spreader should be inspected after every large storm event to check for damage. Heavy equipment and other traffic should be kept off the level spreader because these vehicles may compact the soil or disturb the grade of the slope. If ponding or erosion channels develop, the spreader should be regraded. Dense vegetation should be maintained and damaged areas reseeded as needed.

Advantages of Level Spreaders
• Reduce storm water flow velocity, encourage sedimentation and infiltration
• Are relatively inexpensive to install
Disadvantages of Level Spreaders
• Can easily develop "short circuiting" (concentration of flows into small streams instead of sheetflow over the spreader) because of erosion or other disturbance
• Cannot handle large quantities of sediment-laden storm water

Infiltration Trenches

What Are They

An infiltration trench usually consists of a long, narrow excavation ranging from 3 to 12 feet deep. The trench is filled with stone, which allows for temporary storage of storm water runoff in the open spaces between the stones. The stored storm water infiltrates into the surrounding soil or drains into underground pipes through holes and is then routed to an outflow point. Infiltration trenches are designed to remove both fine sediments and soluble pollutants rather than larger, coarse pollutants.

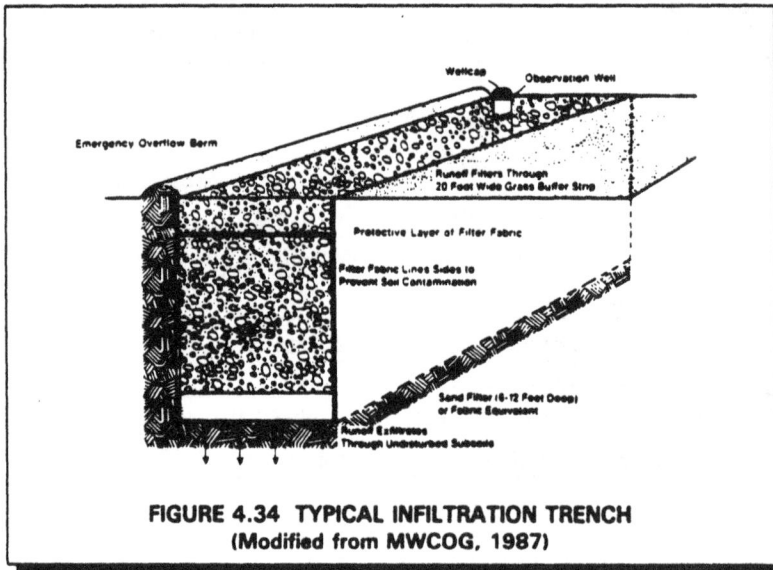

Wellcap
Observation Well
Emergency Overflow Berm
Runoff Filters Through
20 Foot Wide Grass Buffer Strip
Protective Layer of Filter Fabric
Filter Fabric Lines Sides to
Prevent Soil Contamination
Sand Filter (6-12 Feet Deep)
or Fabric Equivalent
Runoff Exfiltrates
Through Undisturbed Subsoils

FIGURE 4.34 TYPICAL INFILTRATION TRENCH
(Modified from MWCOG, 1987)

When and Where to Use Them

Infiltration trenches should be restricted to areas with certain soil, ground water, slope, area, and pollutant conditions. For example, infiltration trenches will not operate well in soils that have high clay contents, silt/clay soils, sandy/clay loams, or soils that have been compacted. Trenches should not be sited over fill soils because such soils are unstable. Hardened soils are often not suitable for infiltration trenches because these types of soils do not easily absorb water. Infiltration practices in general should not be used to manage contaminated storm water.

The drainage area contributing runoff to a single trench should not exceed 5 acres (State of Maryland, 1983). Construction of trenches should not start until after all land-disturbing activities have ceased so that runoff with high levels of sediment does not fill in the structure.

If slopes draining into the trench are steeper than 5 percent, the runoff will enter the trench too fast and will overwhelm the infiltration capacity of the soil, causing overflow. The depth from the bottom of the trench to the bedrock layer and the seasonal high water table must be at least three feet. Infiltration trenches may not be suitable in areas where there are cold winters and deep frost levels.

What to Consider

Pretreatment of runoff before it is channeled to the trench is important to efficient operation because pretreatment removes sediment, grit, and oil. Reducing the pollutant load in the runoff entering the trench lengthens trench life. One method of pretreatment is to install a buffer zone just above the trench to act as a filter (see Buffer Zones). In addition, a layer of filter fabric 1 foot below the bottom of the trench can be used to trap the sediments that get through the buffer strip. If excavation around the trenches is necessary, the use of light duty equipment will avoid compacting, which could cause a loss of infiltration capability.

Infiltration trenches should be inspected at least once per year and after major rainfall events. Debris should be removed from all areas of the trench, especially the inlets and overflow channels. Dense vegetative growth should be maintained in buffer areas surrounding the trench.

Test wells can be installed in every trench to monitor draining times and provide information on how well the system is operating. Daily test well monitoring is necessary, especially after large storm events. If the trench does not drain after 3 days, it usually means that the trench is clogged.

Advantages of Infiltration Trenches
• Preserve the natural water balance of the site
• Are effective for small sites
• Remove pollutants effectively
Disadvantages of Infiltration Trenches
• Require high maintenance when sediment loads are heavy
• Have short life span, especially if not maintained properly
• May be expensive (cost of excavation and fill material)

Porous Pavements/Concrete Grids and Modular Pavements

What Are They

Porous pavement, concrete grids, and modular pavements allow storm water to infiltrate so that the speed and amount of runoff from a site can be reduced.

Porous Pavement—Can be either asphalt or concrete. With porous asphalt pavement, runoff infiltrates through a porous asphalt layer into a stone "reservoir" layer. Storm water runoff filters through the stone reservoir into the underlying subsoil or drains into underground pipes through holes and is routed away. The bottom and sides of the stone reservoir are lined with filter fabric to prevent the movement of soils into the reservoir area.

Porous Concrete Pavement—Is made out of a special concrete mix that has a high number of open spaces between the particles and a coarse surface texture. These open spaces allow runoff to pass through the surface to lower levels. This type of pavement can be placed directly on graded soils. When a subbase is used for stability, 6 inches of sand is placed under the concrete mixture. Up to 6 inches of storm water can be held on the surface of the pavement and within the concrete.

Concrete Grids and Modular Pavement—Are made out of precast concrete, poured-in-place concrete, brick, or granite. These types of pavements can also reduce the loading and concentration of pollutants in the runoff. Concrete grids and modular pavements are designed and/or constructed so that they have open spaces within the pavement through which storm water can infiltrate into the ground. These open spaces can be filled with gravel or sand or have vegetation growing out of them.

When and Where to Use Them

These structures are usually only suitable for low-volume parking areas (1/4 acre to 10 acres) (State of Maryland, 1983) and lightly used access roads. However, areas that are expected to get moderate or high volumes of traffic or heavy equipment can use conventional pavements (for the heavy traffic areas) that are sloped to drain to areas with the porous pavements. These pavements are not effective in drainage areas that receive runoff containing high levels of sediment.

The soil types over which concrete grids and modular pavement are to be placed should allow for rapid drainage through the pores in the pavement. These pavements are not recommended for sites with slopes steeper than 5 percent (MWCOG, 1987) or sites with high water tables, shallow bedrock, fill soils, or localized clay lenses, which are conditions that would limit the ability of the runoff to infiltrate into surface soils. For example, the water table and bedrock should be at least 3 feet below the bottom of the stone reservoir. Porous pavement will not operate well in windy areas where sediment will be deposited on the porous pavement.

Construction of these pavements should be timed so that installation occurs on the site after other construction activities are finished and the site has been stabilized. Therefore, sediments are less likely to be tracked or carried on to the surface.

CONCRETE CURB
CALCULATED WATER STORAGE

6"

PERVIOUS CONCRETE
PAVING

PERVIOUS SUBGRADE

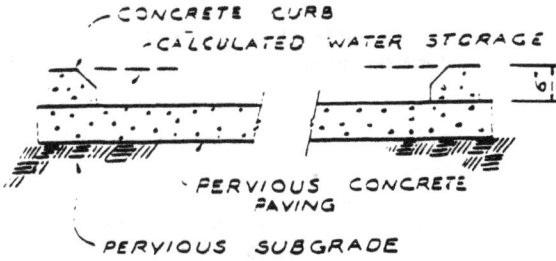

Detail of Pervious Concrete Pavement

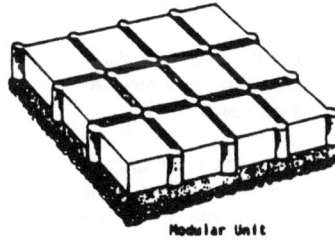

Poured-In-Place Slab

Castellated Unit

Lattice Unit

Modular Unit

Types of Grid and Modular Pavements

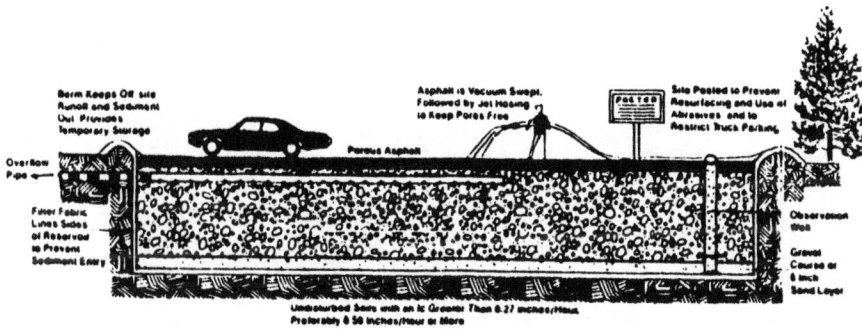

Berm Keeps Off site
Runoff and Sediment
Out Provides
Temporary Storage

Asphalt is Vacuum Swept,
Followed by Jet Hosing
to Keep Pores Free

Site Posted to Prevent
Resurfacing and Use of
Abrasives and to
Restrict Truck Parking

Overflow
Pipe

Porous Asphalt

Filter Fabric
Lines Sides
of Reservoir
to Prevent
Sediment Entry

Observation
Well

Gravel
Course or
Sand Layer

Undisturbed Soils with an fc Greater Than 0.27 inches/Hour
Preferably 0.50 inches/Hour or More

FIGURE 4.35 POROUS PAVEMENTS, CONCRETE GRIDS, AND MODULAR PAVEMENTS
(Modified from Commonwealth of Virginia, 1980; MWCOG, 1987; and Washington State, 1992)

What to Consider

Proper installation of these pavements requires a high level of construction expertise and workmanship. Only contractors who are familiar with the installation of these pavements should be used.

Designers of porous pavement areas should consider sediment and erosion control. Sediments must kept away from the pavement area because they can clog the pores. Controls to consider for sediments include a diversion berm (i.e., earthen mound) around the edge of the pavement area to block the flow of runoff from certain drainages onto the pavement, or other filtering controls such as silt fences. De-icing salt mixtures, sands, or ash also may clog pores and should not be used for snow removal. Signs should be posted to prohibit these activities.

Since the infiltration of storm water runoff may contaminate ground water sources, these pavements are not suitable for areas close to drinking water wells (at least 100 feet away is recommended) (State of Maryland, 1983).

Maintenance of the surface is very important. For porous pavements, this includes vacuum sweeping at least four times per year followed by high-pressure hosing to reduce the chance of sediments clogging the pores of the top layer. Potholes and cracks can be filled with typical patching mixes unless more than 10 percent of the surface area needs repair. Spot clogging may be fixed by drilling half-inch holes through the porous pavement layer every few feet.

The pavement should be inspected several times the first few months after installation and then annually. Inspections after large storms are necessary to check for pools of water. These pools may indicate clogging. The condition of adjacent vegetated filter strips, silt fences, or diversion dikes should also be inspected.

Concrete grids and modular pavements should be designed in accordance with manufacturers' recommendations. Designers also need information on soils, depth to the water table, and storm water runoff quantity and quality.

Maintenance of concrete grids and modular pavements is similar to that of the porous pavements; however, turf maintenance such as mowing, fertilizing, and irrigation may be needed where vegetation is planted in the open spaces.

Advantages of Porous Pavements/Concrete Grids and Modular Pavements
• Provide erosion control by reducing the speed and quantity of the storm water runoff from the site
• Provide some treatment to the water by removing pollutants
• Reduce the need for curbing and storm sewer installation and expansion
• Improve road safety by providing a rougher surface
• Provide some recharge to local aquifers
• Are cost effective because they take the place of more expensive and complex treatment systems
Disadvantages of Porous Pavements/Concrete Grids and Modular Pavements
• Can be more expensive than typical pavements
• Are easily clogged with sediment and/or oil; however, pretreatment and proper maintenance will prevent this problem
• May cause ground water contamination
• Are not structurally suited for high-density traffic or heavy equipment
• Asphalt pavements may break down if gasoline is spilled on the surface
• Are less effective when the subsurface is frozen

APPENDIX A

REFERENCES

REFERENCES

API, "Suggested Procedure for Development of Spill Prevention Control and Countermeasure Plans," American Petroleum Institute Bulletin D16, Second Edition. August 1, 1989.

APWA, "Urban Stormwater Management, Special Report No. 49," American Public Works Association Research Foundation. 1981.

Arapahoe County, "Erosion Control Standards," prepared by Kiowa Engineering Corporation. April 8, 1988.

Available EPA Pollution Control Manuals (see Appendix D)

Commonwealth of Pennsylvania, "Erosion and Sediment Pollution Control Program Manual," Pennsylvania Department of Environmental Resources, Bureau of Soil and Water Conservation. April 1990.

Commonwealth of Virginia, "Virginia Erosion and Sediment Control Handbook," Virginia Department of Conservation and Historical Preservation, Division of Soil & Water Conservation, Second Edition. 1980.

County of Fairfax, "Check List For Erosion and Sediment Control Fairfax County, Virginia." 1990 and 1987 Editions.

MWCOG, "Controlling Urban Runoff: A Practical Manual for Planning and Designing Urban BMPs," Department of Environmental Programs, Metropolitan Washington Council of Governments. July 1987.

Northern Virginia Planning District Commission, "BMP Handbook for the Occoquan Watershed," prepared for Occoquan Basin Nonpoint Pollution Management Program. August 1987.

Salt Institute, "The Salt Storage Handbook, A Practical Guide for Storing and Handling Deicing Salt," Alexandria, Virginia. 1987.

Santa Clara Valley Nonpoint Source Pollution Control Program, "Automotive-Related Industries, BMPs for Industrial Sanitary Sewer Discharges and Storm Water Pollution Control." No date.

State of Maryland, "1983 Maryland Standards and Specifications for Soil and Erosion and Sediment Control," Maryland Water Resources Administration, Soil Conservation Service, and State Soil Conservation Committee. April 1983.

State of North Carolina, "Erosion and Sediment control Planning and Design Manual," North Carolina Sedimentation Control Commission, Department of Natural Resources and Community Development, and Agricultural Extension Service. September 1, 1988.

State of Wisconsin, "Wisconsin Construction Site Best Management Practice Handbook," Wisconsin Department of Natural Resources, Bureau of Water Resources Management, Nonpoint Source and Land Management Section. June 1990.

Thron, H. and Rogashewski, O.J., "Useful Tools for Cleaning Up." Hazardous Material & Spills Conference, 1982.

U.S. Environmental Protection Agency, CZARA NPS Guidance.

U.S. Environmental Protection Agency, "Draft - A Current Assessment of Urban Best Management Practices. Techniques for Reducing Non-point Source Pollution in the Coastal Zone," EPA Office of Wetlands, Oceans and Watersheds, prepared by Metropolitan Washington Council of Governments. December, 1991.

U.S. Environmental Protection Agency, "Draft Construction Site Stormwater Discharge Control, An Inventory of Current Practices," EPA Office of Water Enforcement and Permits, prepared by Kamber Engineering. June 26, 1991.

U.S. Environmental Protection Agency, "Draft Report on Best Management Practices for the Control of Storm Water From Urbanized Areas," Science Applications International Corporation. June 1987.

U.S. Environmental Protection Agency, "Draft Sediment and Erosion Control, An Inventory of Current Practices," EPA Office of Water Enforcement and Permits, prepared by Kamber Engineering. April 20, 1990.

U.S. Environmental Protection Agency, "NPDES Best Management Practices Guidance Document," Industrial Environmental Research Laboratory, Cincinnati, Ohio, prepared by Hydroscience, Inc., EPA 600/9-79-0451. December 1979.

U.S. Environmental Protection Agency, "Pollution Prevention in Printing and Allied Industries: Saving Money Through Pollution Prevention," ORD, Pollution Prevention Office. October 1989.

U.S. Environmental Protection Agency, "Pollution Prevention Training Opportunities in 1992," EPA/560/8-92-002. January 1992. A comprehensive listing of pollution prevention resources, documents, courses, and programs, including names and phone numbers, is contained in a new annual EPA publication. Copies of this document may be obtained by calling the PPIC/PIES support number at (703) 821-4800.

U.S. Environmental Protection Agency, "Process, Procedure, and Methods to Control Pollution Resulting from All Construction Activity," EPA Office of Air and Water Programs, PB-257-318. October 1973.

U.S. Environmental Protection Agency, "Staff Analysis," Storm Water Section. July 1991.

U.S. Environmental Protection Agency, "Waste Minimization Opportunity Assessment Manual," Hazardous Waste Engineering Research Laboratory. July 1988.

Washington State, "Draft Stormwater Management Manual for the Puget Sound Basin," Washington State Department of Ecology. January 23, 1992.

Washington State, "Standards for Storm Water Management for the Puget Sound Basin," Chapter 173-275 WAC, Washington State Department of Ecology. July 29, 1991.

APPENDIX B

GLOSSARY

GLOSSARY

Aeration: A process which promotes biological degradation of organic matter. The process may be passive (as when waste is exposed to air) or active (as when a mixing or bubbling device introduces the air).

Backfill: Earth used to fill a trench or an excavation.

Baffles: Fin-like devices installed vertically on the inside walls of liquid waste transport vehicles that are used to reduce the movement of the waste inside the tank.

Berm: An earthen mound used to direct the flow of runoff around or through a structure.

Best Management Practice (BMP): Schedules of activities, prohibitions of practices, maintenance procedures, and other management practices to prevent or reduce the pollution of waters of the United States. BMPs also include treatment requirements, operating procedures, and practices to control facility site runoff, spillage or leaks, sludge or waste disposal, or drainage from raw material storage.

Biodegradable: The ability to break down or decompose under natural conditions and processes.

Boom: 1. A floating device used to contain oil on a body of water. 2. A piece of equipment used to apply pesticides from ground equipment such as a tractor or truck.

Buffer Strip or Zone: Strips of grass or other erosion-resistant vegetation between a waterway and an area of more intensive land use.

By-product: Material, other than the principal product, that is generated as a consequence of an industrial process.

Calibration: A check of the precision and accuracy of measuring equipment.

CERCLA: Comprehensive Environmental Response, Compensation, and Liability Act.

Chock: A block or wedge used to keep rolling vehicles in place.

Clay Lens: A naturally occurring, localized area of clay that acts as an impermeable layer to runoff infiltration.

Concrete aprons: A pad of nonerosive material designed to prevent scour holes developing at the outlet ends of culverts, outlet pipes, grade stabilization structures, and other water control devices.

Conduit: Any channel or pipe for transporting the flow of water.

Conveyance: Any natural or manmade channel or pipe in which concentrated water flows.

Corrosion: The dissolving and wearing away of metal caused by a chemical reaction such as between water and the pipes that the water contacts, chemicals touching a metal surface, or contact between two metals.

Culvert: A covered channel or a large-diameter pipe that directs water flow below the ground level.

CWA: Clean Water Act (formerly referred to as the Federal Water Pollution Control Act or Federal Water Pollution Control Act Amendments of 1972).

Denuded: Land stripped of vegetation such as grass, or land that has had vegetation worn down due to impacts from the elements or humans.

Dike: An embankment to confine or control water, often built along the banks of a river to prevent overflow of lowlands; a levee.

Director: The Regional Administrator or an authorized representative.

Discharge: A release or flow of storm water or other substance from a conveyance or storage container.

Drip Guard: A device used to prevent drips of fuel or corrosive or reactive chemicals from contacting other materials or areas.

Emission: Pollution discharged into the atmosphere from smokestacks, other vents, and surface areas of commercial or industrial facilities and from motor vehicle, locomotive, or aircraft exhausts.

Erosion: The wearing away of land surface by wind or water. Erosion occurs naturally from weather or runoff but can be intensified by land-clearing practices related to farming, residential or industrial development, road building, or timber-cutting.

Excavation: The process of removing earth, stone, or other materials.

Fertilizer: Materials such as nitrogen and phosphorus that provide nutrients for plants. Commercially sold fertilizers may contain other chemicals or may be in the form of processed sewage sludge.

Filter Fabric: Textile of relatively small mesh or pore size that is used to (a) allow water to pass through while keeping sediment out (permeable), or (b) prevent both runoff and sediment from passing through (impermeable).

Filter Strip: Usually long, relatively narrow area of undisturbed or planted vegetation used to retard or collect sediment for the protection of watercourses, reservoirs, or adjacent properties.

Flange: A rim extending from the end of a pipe; can be used as a connection to another pipe.

Flow Channel Liner: A covering or coating used on the inside surface of a flow channel to prevent the infiltration of water to the ground.

Flowmeter: A gauge that shows the speed of water moving through a conveyance.

General Permit: A permit issued under the NPDES program to cover a certain class or category of storm water discharges. These permits allow for a reduction in the administrative burden associated with permitting storm water discharges associated with industrial activities. For example, EPA is planning to issue two general permits: NPDES General Permits for Storm Water Discharges From Construction Activities that are classified as "Associated with Industrial Activity" and NPDES General Permits for Storm Water Discharges from Industrial Activities that are classified as "Associated with Industrial Activities." EPA is also encouraging delegated States which have an approved general permits program to issue general permits.

Grading: The cutting and/or filling of the land surface to a desired slope or elevation.

Hazardous Substance: 1. Any material that poses a threat to human health and/or the environment. Hazardous substances can be toxic, corrosive, ignitable, explosive, or chemically reactive. 2. Any substance named required by EPA to be reported if a designated quantity of the substance is spilled in the waters of the United States or if otherwise emitted into the environment.

Hazardous Waste: By-products of human activities that can pose a substantial or potential hazard to human health or the environment when improperly managed. Possesses at least one of four characteristics (ignitability, corrosivity, reactivity, or toxicity), or appears on special EPA lists.

Holding Pond: A pond or reservoir, usually made of earth, built to store polluted runoff for a limited time.

Illicit Connection: Any discharge to a municipal separate storm sewer that is not composed entirely of storm water except discharges authorized by an NPDES permit (other than the NPDES permit for discharges from the municipal separate storm sewer) and discharges resulting from fire fighting activities.

Infiltration: 1. The penetration of water through the ground surface into sub-surface soil or the penetration of water from the soil into sewer or other pipes through defective joints, connections, or manhole walls. 2. A land application technique where large volumes of wastewater are applied to land, allowed to penetrate the surface and percolate through the underlying soil.

Inlet: An entrance into a ditch, storm sewer, or other waterway.

Intermediates: A chemical compound formed during the making of a product.

Irrigation: Human application of water to agricultural or recreational land for watering purposes.

Jute: A plant fiber used to make rope, mulch, netting, or matting.

Lagoon: A shallow pond where sunlight, bacterial action, and oxygen work to purify wastewater.

Land Application Units: An area where wastes are applied onto or incorporated into the soil surface (excluding manure spreading operations) for treatment or disposal.

Land Treatment Units: An area of land where materials are temporarily located to receive treatment. Examples include: sludge lagoons, stabilization pond.

Landfills: An area of land or an excavation in which wastes are placed for permanent disposal, and which is not a land application unit, surface impoundment, injection well, or waste pile.

Large and Medium Municipal Separate Storm Sewer System: All municipal separate storm sewers that are either: (i) located in an incorporated place (city) with a population of 100,000 or more as determined by the latest Decennial Census by the Bureau of Census (these cities are listed in Appendices F and G of 40 CFR Part 122); or (ii) located in the counties with unincorporated urbanized populations of 100,000 or more, except municipal separate storm sewers that are located in the incorporated places, townships, or towns within such counties (these counties are listed in Appendices H and I of 40 CFR Part 122); or (iii) owned or operated by a municipality other than those described in paragraph (i) or (ii) and that are designated by the Director as part of the large or medium municipal separate storm sewer system.

Leaching: The process by which soluble constituents are dissolved in a solvent such as water and carried down through the soil.

Level Spreader: A device used to spread out storm water runoff uniformly over the ground surface as sheetflow (i.e., not through channels). The purpose of level spreaders are to prevent concentrated, erosive flows from occurring and to enhance infiltration.

Liming: Treating soil with lime to neutralize acidity levels.

Liner: 1. A relatively impermeable barrier designed to prevent leachate from leaking from a landfill. Liner materials include plastic and dense clay. 2. An insert or sleeve for sewer pipes to prevent leakage or infiltration.

Liquid Level Detector: A device that provides continuous measures of liquid levels in liquid storage areas or containers to prevent overflows.

Material Storage Areas: Onsite locations where raw materials, products, final products, by-products, or waste materials are stored.

Mulch: A natural or artificial layer of plant residue or other materials covering the land surface which conserves moisture, holds soil in place, aids in establishing plant cover, and minimizes temperature fluctuations.

Noncontact Cooling Water: Water used to cool machinery or other materials without directly contacting process chemicals or materials.

Notice of Intent (NOI): An application to notify the permitting authority of a facility's intention to be covered by a general permit; exempts a facility from having to submit an individual or group application.

NPDES: EPA's program to control the discharge of pollutants to waters of the United States. See the definition of "National Pollutant Discharge Elimination System" in 40 CFR 122.2 for further guidance.

NPDES Permit: An authorization, license, or equivalent control document issued by EPA or an approved State agency to implement the requirements of the NPDES program.

Oil and Grease Traps: Devices which collect oil and grease, removing them from water flows.

Oil Sheen: A thin, glistening layer of oil on water.

Oil/Water Separator: A device installed, usually at the entrance to a drain, which removes oil and grease from water flows entering the drain.

Organic Pollutants: Substances containing carbon which may cause pollution problems in receiving streams.

Organic Solvents: Liquid organic compounds capable of dissolving solids, gases, or liquids.

Outfall: The point, location, or structure where wastewater or drainage discharges from a sewer pipe, ditch, or other conveyance to a receiving body of water.

Permeability: The quality of a soil that enables water or air to move through it. Usually expressed in inches/hour or inches/day.

Permit: An authorization, license, or equivalent control document issued by EPA or an approved State agency to implement the requirements of an environmental regulation; e.g., a permit to operate a wastewater treatment plant or to operate a facility that may generate harmful emissions.

Permit Issuing Authority (or Permitting Authority): The State agency or EPA Regional office which issues environmental permits to regulated facilities.

Plunge pool: A basin used to slow flowing water, usually constructed to a design depth and shape. The pool may be protected from erosion by various lining materials.

Pneumatic Transfer: A system of hoses which uses the force of air or other gas to push material through; used to transfer solid or liquid materials from tank to tank.

Point Source: Any discernible, confined, and discrete conveyance, including but not limited to any pipe, ditch, channel, tunnel, conduit, well, discrete fissure, container, rolling stock, concentrated animal feeding operation, or vessel or other floating craft, from which pollutants are or may be discharged. This term does not include return flows from irrigated agriculture or agricultural storm water runoff.

Pollutant: Any dredged spoil, solid waste, incinerator residue, filter backwash, sewage, garbage, sewage sludge, munitions, chemical wastes, biological materials, radioactive materials (except those regulated under the Atomic Energy Act of 1954, as amended (42 (U.S.C. 2011 et seq.)), heat, wrecked or discharged equipment, rock, sand, cellar dirt, and industrial, municipal, and agricultural waste discharged into water. It does not mean:
(i) Sewage from vessels; or
(ii) Water, gas, or other material which is injected into a well to facilitate production of oil or gas, or water derived in association with oil and gas production and disposed of in a well, if the well used either to facilitate production or for disposal purposes is approved by the authority of the State in which the well is located, and if the State determines that the injection or disposal will not result in the degradation of ground or surface water resources [Section 502(6) of the CWA].

Radioactive materials covered by the Atomic Energy Act are those encompassed in its definition of source, byproduct, or special nuclear materials. Examples of materials not covered include radium and accelerator-produced isotopes. See Train v. Colorado Public Interest Research Group, Inc., 426 U.S. 1 (1976).

Porous Pavement: A human-made surface that will allow water to penetrate through and percolate into soil (as in porous asphalt pavement or concrete). Porous asphalt pavement is comprised of irregular shaped crush rock precoated with asphalt binder. Water seeps through into lower layers of gravel for temporary storage, then filters naturally into the soil.

Precipitation: Any form of rain or snow.

Preventative Maintenance Program: A schedule of inspections and testing at regular intervals intended to prevent equipment failures and deterioration.

Process Wastewater: Water that comes into direct contact with or results from the production or use of any raw material, intermediate product, finished product, by-product, waste product, or wastewater.

PVC (Polyvinyl Chloride): A plastic used in pipes because of its strength; does not dissolve in most organic solvents.

Raw Material: Any product or material that is converted into another material by processing or manufacturing.

RCRA: Resource Conservation and Recovery Act.

Recycle: The process of minimizing the generation of waste by recovering usable products that might otherwise become waste. Examples are the recycling of aluminum cans, wastepaper, and bottles.

Reportable Quantity (RQ): The quantity of a hazardous substance or oil that triggers reporting requirements under CERCLA or the Clean Water Act. If a substance is released in amounts exceeding its RQ, the release must be reported to the National Response Center, the State Emergency Response Commission, and community emergency coordinators for areas likely to be affected (see Appendix I for a list of RQs).

Residual: Amount of pollutant remaining in the environment after a natural or technological process has taken place, e.g., the sludge remaining after initial wastewater treatment, or particulates remaining in air after the air passes through a scrubbing or other pollutant removal process.

Retention: The holding of runoff in a basin without release except by means of evaporation, infiltration, or emergency bypass.

Retrofit: The modification of storm water management systems in developed areas through the construction of wet ponds, infiltration systems, wetland plantings, stream bank stabilization, and other BMP techniques for improving water quality. A retrofit can consist of the construction of a new BMP in the developed area, the enhancement of an older storm water management structure, or a combination of improvement and new construction.

Rill Erosion: The formation of numerous, closely spread streamlets due to uneven removal of surface soils by storm water or other water.

Riparian Habitat: Areas adjacent to rivers and streams that have a high density, diversity, and productivity of plant and animal species relative to nearby uplands.

Runon: Storm water surface flow or other surface flow which enters property other than that where it originated.

Runoff: That part of precipitation, snow melt, or irrigation water that runs off the land into streams or other surface water. It can carry pollutants from the air and land into the receiving waters.

Sanitary Sewer: A system of underground pipes that carries sanitary waste or process wastewater to a treatment plant.

Sanitary Waste: Domestic sewage.

SARA: Superfund Amendments and Reauthorization Act.

Scour: The clearing and digging action of flowing water, especially the downward erosion caused by stream water in sweeping away mud and silt from the stream bed and outside bank of a curved channel.

Sealed Gate: A device used to control the flow of liquid materials through a valve.

Secondary Containment: Structures, usually dikes or berms, surrounding tanks or other storage containers and designed to catch spilled material from the storage containers.

Section 313 Water Priority Chemical: A chemical or chemical categories which are: (1) are listed at 40 CFR 372.65 pursuant to Section 313 of the Emergency Planning and Community Right-to-Know Act (EPCRA) [also known as Title III of the Superfund Amendments and Reauthorization Act (SARA) of 1986]; (2) are present at or above threshold levels at a facility subject to EPCRA Section 313 reporting requirements; and (3) that meet at least one of the following criteria: (i) are listed in Appendix D of 40 CFR Part 122 on either Table II (organic priority pollutants), Table III (certain metals, cyanides, and phenols), or Table V (certain toxic pollutants and hazardous substances); (ii) are listed as a hazardous substance pursuant to Section 311(b)(2)(A) of the CWA at 40 CFR 116.4; or (iii) are pollutants for which EPA has published acute or chronic water quality criteria. See Addendum B of this permit. (List is included as Appendix I.)

Sediment Trap: A device for removing sediment from water flows; usually installed at outfall points.

Sedimentation: The process of depositing soil particles, clays, sands, or other sediments that were picked up by flowing water.

Sediments: Soil, sand, and minerals washed from land into water, usually after rain. They pile up in reservoirs, rivers, and harbors, destroying fish-nesting areas and holes of water animals and cloud the water so that needed sunlight might not reach aquatic plants. Careless farming, mining, and building activities will expose sediment materials, allowing them to be washed off the land after rainfalls.

Sheet Erosion: Erosion of thin layers of surface materials by continuous sheets of running water.

Sheetflow: Runoff which flows over the ground surface as a thin, even layer, not concentrated in a channel.

Shelf Life: The time for which chemicals and other materials can be stored before becoming unusable due to age or deterioration.

Significant Materials: Include, but are not limited to: raw materials; fuels; materials such as solvents, detergents and plastic pellets; finished materials such as metallic products; raw materials used in food processing or production; hazardous substances designated under section 101(14) of the Comprehensive Environmental Response, Compensation, and Liability Act (CERCLA); any chemical the facility is required to report pursuant to section 313 of Title III of the Superfund Amendments and Reauthorization Act (SARA); fertilizers; pesticides; and waste products such as ashes, slag, and sludge that have a potential to be released with storm water discharges [122.26(b)(12)].

Significant Spills: Includes, but is not limited to: releases of oil or hazardous substances in excess of reportable quantities under Section 311 of the CWA (see 40 CFR 110.10 and CFR 117.21) or Section 102 of CERCLA (see 40 CFR 302.4).

Slag: Non-metal containing waste leftover from the smelting and refining of metals.

Slide Gate: A device used to control the flow of water through storm water conveyances.

Sloughing: The movement of unstabilized soil layers down a slope due to excess water in the soils.

Sludge: A semi-solid residue from any of a number of air or water treatment processes. Sludge can be a hazardous waste.

Soil: The unconsolidated mineral and organic material on the immediate surface of the earth that serves as a natural medium for the growth of plants.

Solids Dewatering: A process for removing excess water from solids to lessen the overall weight of the wastes.

Source Control: A practice or structural measure to prevent pollutants from entering storm water runoff or other environmental media.

Spent Solvent: A liquid solution that has been used and is no longer capable of dissolving solids, gases, or liquids.

Spill Guard: A device used to prevent spills of liquid materials from storage containers.

Spill Prevention Control and Countermeasures Plan (SPCC): Plan consisting of structures, such as curbing, and action plans to prevent and respond to spills of hazardous substances as defined in the Clean Water Act.

Stopcock Valve: A small valve for stopping or controlling the flow of water or other liquid through a pipe.

Storm Drain: A slotted opening leading to an underground pipe or an open ditch for carrying surface runoff.

Storm Water: Runoff from a storm event, snow melt runoff, and surface runoff and drainage.

Storm Water Discharge Associated with Industrial Activity: The discharge from any conveyance which is used for collecting and conveying storm water and which is directly related to manufacturing, processing or raw materials storage areas at an industrial plant. The term does not include discharges from facilities or activities excluded from the NPDES program under 40 CFR Part 122. For the categories of industries identified in subparagraphs (i) through (x) of this subsection, the term includes, but is not limited to, storm water discharges from industrial plant yards; immediate access roads and rail lines used or traveled by carriers of raw materials, manufactured products, waste material, or by-products used or created by the facility; material handling sites; refuse sites; sites used for the application or disposal of process waste waters (as defined at 40 CFR 401); sites used for the storage and maintenance of material handling equipment; sites used for residual treatment, storage, or disposal; shipping and receiving areas; manufacturing buildings; storage areas (including tank farms) for raw materials, and intermediate and finished products; and areas where industrial activity has taken place in the past and significant materials remain and are exposed to storm water. For the categories of industries identified in subparagraph (xi), the term includes only storm water discharges from all the areas (except access roads and rail lines) that are listed in the previous sentence where material handling equipment or activities, raw materials, intermediate products, final products, waste material, by-products, or industrial machinery are exposed to storm water. For the purposes of this paragraph, material handling activities include the: storage, loading and unloading, transportation, or conveyance of any raw material, intermediate product, finished product, by-product or waste product. The term excludes areas located on plant lands separate from the plant's industrial activities, such as office buildings and accompanying parking lots as long as the drainage from the excluded areas is not mixed with storm water drained from the above described areas. Industrial facilities (including industrial facilities that are Federally, State, or municipally owned or operated that meet the description of the facilities listed in this paragraph (i)-(xi) include those facilities designated under the provision of 122.26(a)(1)(v). The following categories of facilities are considered to be engaging in "industrial activity" for purposes of this subsection:

(i) Facilities subject to storm water effluent limitations guidelines, new source performance standards, or toxic pollutant effluent standards under 40 CFR Subchapter N (except facilities with toxic pollutant effluent standards which are excepted under category (xi) of this paragraph);

(ii) Facilities classified as Standard Industrial Classifications 24 (except 2434), 26 (except 265 and 267), 28 (except 283 and 285) 29, 311, 32 (except 323), 33, 3441, 372;

(iii) Facilities classified as Standard Industrial Classifications 10 though 14 (mineral industry) including active or inactive mining operations (except for areas of coal mining operations no longer meeting the definition of a reclamation area under 40 CFR 434.11(l) because the performance bond issued to the facility by the appropriate SMCRA authority has been released, or except for areas of non-coal mining operations which have been released from applicable State or Federal reclamation requirements after December 17, 1990 and oil and gas exploration, production, processing, or treatment operations, or transmission facilities that discharge storm water contaminated by contact with or that has come into contact with, any overburden, raw material, intermediate products, finished products, byproducts or waste products located on the site of such operations; (inactive mining operations are mining sites that are not being actively mined, but which have an identifiable owner/operator; inactive mining sites do not include sites where mining claims are being maintained prior to disturbances associated with the extraction, beneficiation, or processing of mined materials, nor sites where minimal activities are undertaken for the sole purpose of maintaining mining claim);

(iv) Hazardous waste treatment, storage, or disposal facilities, including those that are operating under interim status or a permit under Subtitle C of RCRA;

(v) Landfills, land application sites, and open dumps that receive or have received any industrial wastes (waste that is received from any of the facilities described under this subsection) including those that are subject to regulation under Subtitle D of RCRA;

(vi) Facilities involved in the recycling of materials, including metal scrapyards, battery reclaimers, salvage yards, and automobiles junkyards, including but limited to those classified as Standard Industrial Classification 5015 and 5093;

(vii) Steam electric power generating facilities, including coal handling sites;

(viii) Transportation facilities classified as Standard Industrial Classifications 40, 41, 42 (except 4221-25), 43, 44, 45, and 5171 which have vehicle maintenance shops, equipment cleaning operations, or airport deicing operations. Only those portions of the facility that are either involved in vehicle maintenance (including vehicle rehabilitation, mechanical repairs, painting, fueling, and lubrication), equipment cleaning operations, airport deicing operations, or which are otherwise identified under paragraphs (i)-(vii) or (ix)-(xi) of this subsection are associated with industrial activity;

(ix) Treatment works treating domestic sewage or any other sewage sludge or wastewater treatment device or system, used in the storage treatment, recycling, and reclamation of municipal or domestic sewage, including land dedicated to the disposal of sewage sludge that are located within the confines of the facility, with a design flow of 1.0 mgd or more, or required to have an approved pretreatment program under 40 CFR 403. Not included are farm lands, domestic gardens or lands used for sludge management where sludge is beneficially reused and which are not physically located in the confines of the facility, or areas that are in compliance with Section 405 of the CWA;

(x) Construction activity including clearing, grading and excavation activities except: operations that result in the disturbance of less than five acres of total land area which are not part of a larger common plan of development or sale;

(xi) Facilities under Standard Industrial Classification 20, 21, 22, 23, 2434, 25, 265, 267, 27, 283, 285, 30, 31 (except 311), 323, 34 (except 3441), 35, 36, 37 (except 373), 38, 39, 4221-25, (and which are not otherwise included within categories (ii)-(x));

Note: The Transportation Act of 1991 provides an exemption from storm water permitting requirements for certain facilities owned or operated by municipalities with a population of less than 100,000. Such municipalities must submit storm water discharge permit applications for only airports, power plants, and uncontrolled sanitary landfills that they own or operate, unless a permit is otherwise required by the permitting authority.

Subsoil: The bed or stratum of earth lying below the surface soil.

Sump: A pit or tank that catches liquid runoff for drainage or disposal.

Surface Impoundment: Treatment, storage, or disposal of liquid wastes in ponds.

Surface Water: All water naturally open to the atmosphere (rivers, lakes, reservoirs, streams, wetlands impoundments, seas, estuaries, etc.); also refers to springs, wells, or other collectors which are directly influenced by surface water.

Swale: An elongated depression in the land surface that is at least seasonally wet, is usually heavily vegetated, and is normally without flowing water. Swales direct storm water flows into primary drainage channels and allow some of the storm water to infiltrate into the ground surface.

Tarp: A sheet of waterproof canvas or other material used to cover and protect materials, equipment, or vehicles.

Topography: The physical features of a surface area including relative elevations and the position of natural and human-made features.

Toxic Pollutants: Any pollutant listed as toxic under Section 501(a)(1) or, in the case of "sludge use or disposal practices," any pollutant identified in regulations implementing Section 405(d) of the CWA. Please refer to 40 CFR Part 122 Appendix D.

Treatment: The act of applying a procedure or chemicals to a substance to remove undesirable pollutants.

Tributary: A river or stream that flows into a larger river or stream.

Underground Storage Tanks (USTs): Storage tanks with at least 10 percent or more of its storage capacity underground (the complete regulatory definition is at 40 CFR Part 280.12).

Waste: Unwanted materials left over from a manufacturing or other process.

Waste Pile: Any noncontainerized accumulation of solid, nonflowing waste that is used for treatment or storage.

Water Table: The depth or level below which the ground is saturated with water.

Waters of the United States:
"(a) All waters, which are currently used, were used in the past, or may be susceptible to use in interstate or foreign commerce, including all waters which are subject to the ebb and flow of the tide;
(b) All interstate waters, including interstate "wetlands;"
(c) All other waters such as intrastate lakes, rivers, streams (including intermittent streams), mudflats, sandflats, "wetlands," sloughs, prairie potholes, wet meadows, playa lakes, or natural ponds, the use, degradation, or destruction of which would affect or could affect interstate or foreign commerce including any such waters:
(1) Which are or could be used by interstate or foreign travelers for recreational or other purposes;

(2) From which fish or shellfish are or could be taken and sold in interstate or foreign commerce; or

(3) Which are used or could be used for industrial purposes by industries in interstate commerce;

(d) All impoundments of waters otherwise defined as waters of the United States under this definition;

(e) Tributaries of waters identified in paragraphs (a) through (d) of this definition;

(f) The territorial sea; and

(g) "Wetlands" adjacent to waters (other than waters that are themselves wetlands) identified in paragraphs (a) through (f) of this definition.

Waste treatment systems, including treatment ponds or lagoons designed to meet the requirements of CWA (other than cooling ponds as defined in 40 CFR 423.11(m) which also meet the criteria of this definition) are not waters of the United States. This exclusion applies only to manmade bodies of water which neither were originally created in waters of the United States (such as disposal area in wetlands) nor resulted from the impoundment of waters of the United States.

Waterway: A channel for the passage or flow of water.

Wet Well: A chamber used to collect water or other liquid and to which a pump is attached.

Wetlands: An area that is regularly saturated by surface or ground water and subsequently is characterized by a prevalence of vegetation that is adapted for life in saturated soil conditions. Examples include: swamps, bogs, fens, marshes, and estuaries.

Wind Break: Any device designed to block wind flow and intended for protection against any ill effects of wind.

APPENDIX C

MODEL STORM WATER POLLUTION PREVENTION PLAN

Double Scoop Ice Cream Company

40 Wonka Drive
Anytown, OK 12345

December 1992

Storm Water Pollution Prevention Plan	
Emergency Contact: **Cheryl Glenn**	Work Phone: **(101) 555-1234**
Title: **Plant Manager**	Emergency Phone: **(101) 555-6929**
Secondary Contact: **Rachel Meyers**	Work Phone: **(101) 555-3923**
Title: **Engineering Supervisor**	Emergency Phone: **(101) 555-6789**
Type of Manufacturer: **Ice Cream Manufacturer**	
Operating Schedule: **8:00 a.m. - 11:30 p.m.**	
Number of Employees: **The plant has 21 employees, including part time staff. Shifts overlap all day.**	
Average Wastewater Discharge: **5,000 gallons per week**	
NPDES Permit Number: **OK1234567**	

POLLUTION PREVENTION TEAM MEMBER ROSTER	Worksheet #1 Completed by: _Cheryl Glenn_ Title: _Plant Manager_ Date: _December 12, 1992_

Leader: _Cheryl Glenn_ **Title:** _Plant Manager_
Office Phone: _(101) 555-1234_
Responsibilities: _Signatory authority; coordinate all stages of plan development and implementation; coordinate employee training program; keep all records and ensure reports are submitted._

Members:

(1) _Stephen Michaels_ **Title:** _Production Supervisor_
Office Phone: _(101) 555-3923_
Responsibilities: _Note any process changes; help conduct inspections._

(2) _Rachel Meyers_ **Title:** _Engineering Dept. Supervisor_
Office Phone: _(101) 555-5890_
Responsibilities: _Responsible for implementing the Preventive Maintenance program; oversee inspections._

(3) _Isaac Feldman_ **Title:** _Maintenance Dept. Supervisor_
Office Phone: _(101) 555-0482_
Responsibilities: _Mr. Feldman is the spill response coordinator; Oversees "good housekeeping."_

(4) _Group Activities_ **Title:** _____
Office Phone: _____
Responsibilities: _Developing the plan elements, choosing storm water management options._

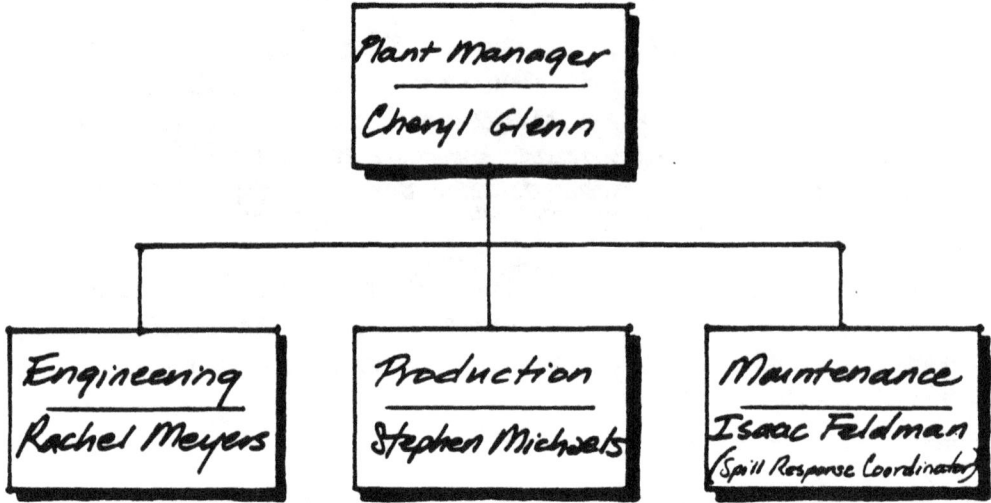

Pollution Prevention Team
Organization Chart

Plant Manager
Cheryl Glenn

Engineering
Rachel Meyers

Production
Stephen Michaels

Maintenance
Isaac Feldman
(Spill Response Coordinator)

Double Scoop Ice Cream Company

**Storm Water Pollution Prevention Plan
Comparison with SPCC Plan**

Double Scoop Ice Cream Plant has an SPCC plan in operation for its aboveground fuel storage tank. Overlaps are noted below:

- Isaac Feldman is the SPCC Coordinator and reports directly to Cheryl Glenn. He will be the Storm Water Spill Prevention and Response Coordinator.

- A complete description of potential for oil to contaminate storm water discharges including quantity of oil that could be discharged.

- Curbing around aboveground fuel storage tank identified on site map.

- Expanded SPCC schedules and procedures to include Storm Water Pollution Prevention Plan requirements.

- Incorporated SPCC plan training into storm water training programs on spill prevention and response.

- Relevant portions of the SPCC plan will be included in this plan.

DEVELOPING A SITE MAP	Worksheet #2 Completed by: _Cheryl Glenn_ Title: _Plant Manager_ Date: _December 12, 1992_

Instructions: Draw a map of your site including a footprint of all buildings, structures, paved areas, and parking lots. The information below describes additional elements required by EPA's General Permit (see example maps in Figures 2.3 and 2.4).

EPA's General Permit requires that you indicate the following features on your site map:

- All outfalls and storm water discharges

- Drainage areas of each storm water outfall

- Structural storm water pollution control measures, such as:

    - Flow diversion structures
    - Retention/detention ponds
    - Vegetative swales
    - Sediment traps

- Name of receiving waters (or if through a Municipal Separate Storm Sewer System)

- Locations of exposed significant materials (see Section 2.2.2)

- Locations of past spills and leaks (see Section 2.2.3)

- Locations of high-risk, waste-generating areas and activities common on industrial sites such as:

    - Fueling stations
    - Vehicle/equipment washing and maintenance areas
    - Area for unloading/loading materials
    - Above-ground tanks for liquid storage
    - Industrial waste management areas (landfills, waste piles, treatment plants, disposal areas)
    - Outside storage areas for raw materials, by-products, and finished products
    - Outside manufacturing areas
    - Other areas of concern (specify:_____)

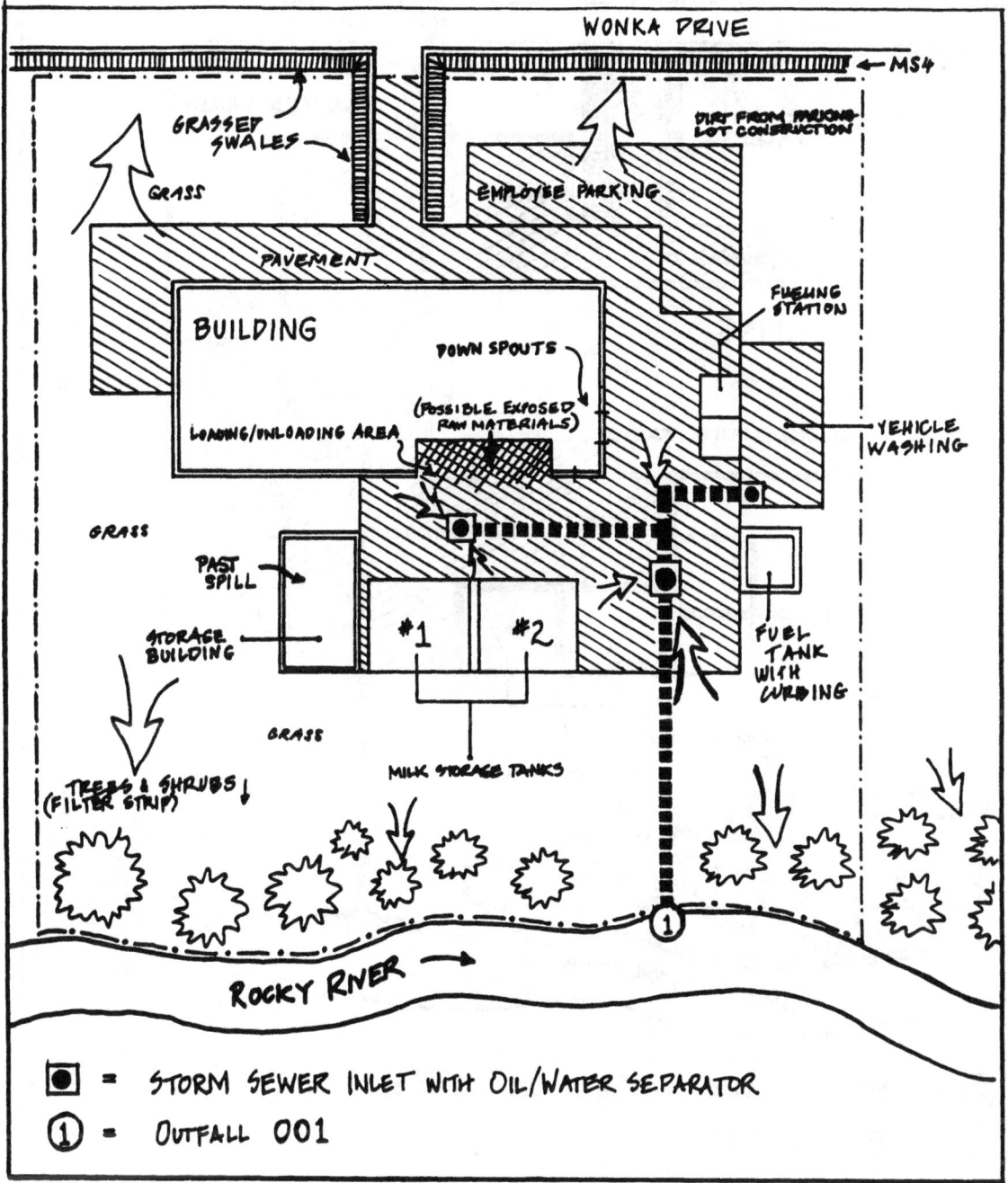

WONKA DRIVE

← MS4

GRASSED SWALES

GRASS

DIRT FROM PARKING LOT CONSTRUCTION

EMPLOYEE PARKING

PAVEMENT

FUELING STATION

BUILDING

DOWN SPOUTS

(POSSIBLE EXPOSED RAW MATERIALS)

LOADING/UNLOADING AREA

VEHICLE WASHING

GRASS

PAST SPILL

STORAGE BUILDING

#1   #2

FUEL TANK WITH CURBING

GRASS

TREES & SHRUBS (FILTER STRIP)

MILK STORAGE TANKS

ROCKY RIVER →

① = STORM SEWER INLET WITH OIL/WATER SEPARATOR
① = OUTFALL 001

DOUBLE SCOOP ICE CREAM COMPANY

PRE-BMP SITE MAP
MARCH 1, 1993

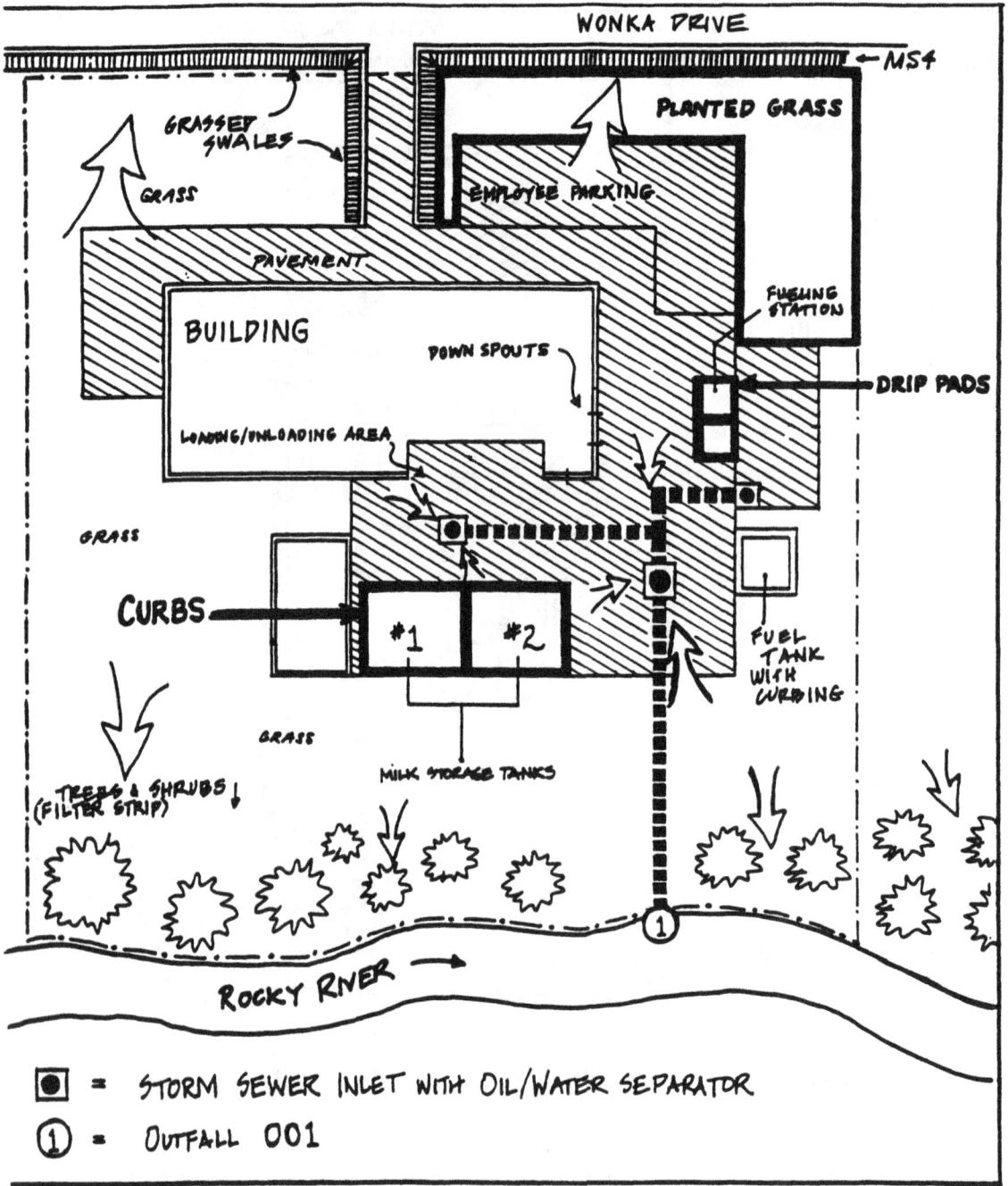

WONKA DRIVE

← MS4

PLANTED GRASS

GRASSED SWALES

GRASS

EMPLOYEE PARKING

PAVEMENT

FUELING STATION

BUILDING

DOWN SPOUTS

DRIP PADS

LOADING/UNLOADING AREA

GRASS

CURBS

#1   #2

FUEL TANK WITH CURBING

GRASS

TREES & SHRUBS (FILTER STRIP)

MILK STORAGE TANKS

①

ROCKY RIVER →

◉ = STORM SEWER INLET WITH OIL/WATER SEPARATOR

① = OUTFALL 001

DOUBLE SCOOP ICE CREAM COMPANY

POST-BMP SITE MAP
MARCH 1, 1993

Worksheet #3

MATERIAL INVENTORY

Completed by: _Cheryl Glenn_
Title: _Plant Manager_
Date: _December 12, 1992_

Instructions: List all materials used, stored, or produced onsite. Assess and evaluate these materials for their potential to contribute pollutants to storm water runoff. Also complete Worksheet 3A if the material has been exposed during the last three years.

Material	Purpose/Location	Quantity (units) Used	Quantity (units) Produced	Quantity (units) Stored	Quantity Exposed in Last 3 Years	Likelihood of contact with storm water. If yes, describe reason.	Past Significant Spill or Leak Yes	Past Significant Spill or Leak No
Butter fat	truck unloading area during transfer to liquid ingredient storage and milk vat storage.	72,600 gal/wk	—	2,000 gal/wk	NO	Truck loading area outside and possible exposure with ruptured tanks.		✓
Milk solids								
Whey solids								
Corn Syrup	Truck unloading area during transfer to sweetener storage.	7,100 gal/wk	—	—	yes	Truck loading area outside with possible exposure as a result of leaking tanks.	✓	
Liquid sugar								
Ice cream	Inside freezers for final product shipping.	35-40,000 lbs.			NO	NO		✓
Cleansers: Granular Chloorshire-0	Dry cleansers in dry storage area (indoors)	400 lb/wk	—	—	NO	Yes. Possible storage exposure during transfer to dry storage area.		✓
HDC-38								
Power Spray-R								

MATERIAL INVENTORY
Page 2

Worksheet #3
Completed by: _Cheryl Glenn_
Title: _Plant Manager_
Date: _December 13, 1992_

Instructions: List all materials used, stored, or produced onsite. Assess and evaluate these materials for their potential to contribute pollutants to storm water runoff. Also complete Worksheet 3A if the material has been exposed during the last three years.

Material	Purpose/Location	Quantity (units)			Quantity Exposed in Last 3 Years	Likelihood of contact with storm water. If yes, describe reason.	Past Significant Spill or Leak	
		Used	Produced	Stored			Yes	No
Cleansers:								
liquid	Cleansers are stored outside	100 ga/mo	—	—	NO	Yes-if material tanks stored outside		✓
M.R.S.-200-0	stored outside							
Acidize-O	water cover.							
Microsan								
Fuels:								
gasoline	above ground 250 gallon storage tank	250 gal/mo	—	—	No	Yes-possible exposure in the event of defective tanks or transfer of materials from tanks to containers		✓
motor oil	750 gallon storage tank	20 gal/mo	—	—	No			✓
soaps		40 gal/mo	—	—	No			✓
detergents								

DESCRIPTION OF EXPOSED SIGNIFICANT MATERIAL

Worksheet #3A
Completed by: _Cheryl Glenn_
Title: _Plant Manager_
Date: _December 12, 1992_

Instructions: Based on your material inventory, describe the significant materials that were exposed to storm water during the past three years and/or are currently exposed. For the definition of "significant materials" see Appendix B of the manual.

Description of Exposed Significant Material	Period of Exposure	Quantity Exposed (units)	Location (as indicated on the site map)	Method of Storage or Disposal (e.g., pile, drum, tank)	Description of Material Management Practice (e.g., pile covered, drum sealed)
liquid sugar	1/21/92	10 gal.	Storage building Tank #2	50 gal. tanks (2)	Leak was contained and mopped up. Remainder of liquid sugar transferred to tank that did not have a leaky valve.

LIST OF SIGNIFICANT SPILLS AND LEAKS

Worksheet #4	
Completed by:	Cheryl Glenn
Title:	Plant Manager
Date:	December 12, 1992

Directions: Record below all significant spills and significant leaks of toxic or hazardous pollutants that have occurred at the facility in the three years prior to the effective date of the permit.

Definitions: Significant spills include, but are not limited to, releases of <u>oil</u> or <u>hazardous substances in excess of reportable quantities.</u>

1st Year Prior

Date (month/day/year)	Spill	Leak	Location (as indicated on site map)	Description				Response Procedure			Preventive Measures Taken
				Type of Material	Quantity	Source, If Known	Reason	Amount of Material Recovered	Material No Longer Exposed to Storm Water (True/False)		
				N	O	N	E				

2nd Year Prior

Date (month/day/year)	Spill	Leak	Location (as indicated on site map)	Description				Response Procedure			Preventive Measures Taken
				Type of Material	Quantity	Source, If Known	Reason	Amount of Material Recovered	Material No Longer Exposed to Storm Water (True/False)		
1/21/91		✓	STORAGE BLDG.	LIQUID SUGAR	10 ga.	Tank #2	leaky valve	contained and mopped spill — transferred	true (circled)	installing curbing around tank-to be completed → 2/93

3rd Year Prior

Date (month/day/year)	Spill	Leak	Location (as indicated on site map)	Description				Response Procedure			Preventive Measures Taken
				Type of Material	Quantity	Source, If Known	Reason	Amount of Material Recovered	Material No Longer Exposed to Storm Water (True/False)		
				N	O	N	E				

NON-STORM WATER DISCHARGE ASSESSMENT AND CERTIFICATION

Worksheet #5
Completed by: _Rachel Meyers_
Title: _Engineering Department Supervisor_
Date: _3/1/93_

Date of Test or Evaluation	Outfall Directly Observed During the Test (identify as indicated on the site map)	Method Used to Test or Evaluate Discharge	Describe Results from Test for the Presence of Non-Storm Water Discharge	Identify Potential Significant Sources	Name of Person Who Conducted the Test or Evaluation
12/24/92	001	visual inspection	No discharge observed		R. Meyers and S. Goodhope
1/19/93	001	visual inspection	Significant Flow; oil	vehicle wash ongoing at time	R. Meyers and S. Goodhope
2/5/93	001	visual inspection	Small amount of discharge observed; clear	suspected to be delayed storm water discharge from storm that occurred 2/1/93	R. Meyers and S. Goodhope
			✱ See details in attached field notebook.		

CERTIFICATION

I, _Cheryl Glenn_ _____ (responsible corporate official), certify under penalty of law that this document and all attachments were prepared under my direction or supervision in accordance with a system designed to assure that qualified personnel properly gather and evaluate the information submitted. Based on my inquiry of the person or persons who manage the system or those persons directly responsible for gathering the information, the information submitted is, to the best of my knowledge and belief, true, accurate, and complete. I am aware that there are significant penalties for submitting false information, including the possibility of fine and imprisonment for knowing violations.

A. Name & Official Title (type or print)
Cheryl Glenn

B. Area Code and Telephone No.
_(101) 555-1234_

C. Signature
Cheryl Glenn

D. Date Signed
_3/2/93_

```
┌─────────────────────────────────────────┐
│  FIELD NOTEBOOK                          │
│  For non-storm water discharge inspections │
└─────────────────────────────────────────┘
```

INSPECTION TEAM:
R. Meyers
S. Goodhope

Completed by: Rachel Meyers
Date: 12/24/92
Time: 10:50 am

Time since last rain: 42 hours
Quantity of last rain: 0.12 inches
Flow observed: NO

SIGNATURE: Rachel Meyer

Completed by: Rachel Meyers
Date: 1/19/93
Time: 3:20 pm

Time since last rain: 5 days
Quantity of last rain: 0.5 inches
Flow observed: YES
DESCRIPTION: No odor; clear color
             (soap suds); oily sheen; some
             sediment.
Temperature: cold (37.5°F)
Volume: collected ten gallons/minute in buckets
Comments: Vehicle wash ongoing at time of inspection.
          This was the source of the flow.

SIGNATURE: Russel Meyers

Completed by: Rachel Meyers
Date: 2/5/93
Time: 12:15 pm.

Time since last rain: 96 hours
Quantity of last rain: 2.5 inches
Flow observed: YES
DESCRIPTION: No odor; clear; some sediments;
             few small pieces of paper (trash)
Temperature: cold (42.3°F)
Volume: Collected one gallon in 5 minutes.
Comments: We suspect that the flow was left over from
          storm that occurred on 2/1/93 (4 days ago)

SIGNATURE: Rachel Meyer

Double Scoop Ice Cream Company

Site Assessment Inspection

February 10, 1993

Evaluate the site for pollutants.

There are five areas where material handling and storage activities take place.

- The storage building contains tanks of corn syrup, liquid sugar, and the granular cleansers. The tanks were examined for possible leaks. We found that the valve on the liquid sugar tank #2 was faulty and had leaked approximately 10 gallons of liquid sugar. Although this leak occurred on 1/21/92, the faulty valve was not discovered until now. All other tanks are secure. Areas around the tanks were swept clean to determine if leaks or spills were prevalent.

- The milk storage tanks were then examined for leaks or exposure. Upon closer examination, it was found that the number 1 tank was leaking a small amount of milk to the drainage system. This leak may be the reason for the high concentration of biochemical oxygen demand found in the sample taken from the storm water discharge. The tank was temporarily fixed to ensure that no further contamination would result. A replacement tank was ordered on February 6, 1993, and was expected to arrive within 5 business days. The milk storage tanks shall be examined on a daily basis to further prevent possible exposure to the storm water collection system and receiving stream.

- We inspected the fueling station to see if there were any leaks. The general area surrounding the fueling station was clean but we observed that gasoline and motor oil falls during fueling. In accordance with standard operating conditions, facility personnel hose down the area during vehicle washing and the drain is connected to the storm sewer. We detected this connection on 1/19/93 during one of the non-storm water discharge assessment visual inspections. Since this discharge is not allowed under our general permit, we are in the process of submitting a separate permit application specifically for the discharge of vehicle wash water.

- We examined the fueling station which is adjacent to the vehicle washing area. Vehicle washing cleaners are used here and any empty or open containers were removed from the area.

- We next looked at the loading and unloading docks where raw materials and various cleansers are delivered. The transfer of goods from incoming trucks to storage areas is a source of pollution. Although no problems were noticed, the pollution prevention team has developed a spill prevention and response plan to clean up spills quickly and report them if necessary.

- The last area we inspected was the runoff field below the employee parking lot. Here we noticed a significant amount of erosion resulting from recent construction to expand the parking lot.

Describe existing management practices.

Grass was lightly planted around the parking lot after recent construction. The fuel storage tank has curbing around it in accordance with our SPCC plan. Also, the maintenance crew regularly picks up trash and empty containers from around the storage tanks, loading and unloading areas, and the vehicle washing areas. Used oils are collected in containers and taken to a recycling facility. In addition, we installed two oil/water separators at the drains into our underground storm sewer leading to the Rocky River. These separators are indicated on the site map.

Double Scoop Ice Cream Company

Existing Monitoring Data

Although our NPDES permit for process wastewater does not require storm water sampling, we sampled our storm water on one occasion in response to a questionnaire we received from the National Association of Ice Cream Makers. They were collecting information to submit as part of their comments on EPA's proposed general permit.

Date of Sampling	8/30/91
Outfall Sampled	001
Type of Storm	1 inch light rainfall (lasted 2 days)
Type of Samples	Grab samples taken during first hour of flow

Data		
Parameter	Quantity	Sample Type
BOD	250 mg/l	Grab
TSS	100 mg/l	Grab
pH	7.2 s.u.	Grab
Oil and grease	5.0 mg/l	Grab

Based upon the high concentration of BOD in the storm water samples collected, pollution prevention team is considering possible potential sources of BOD. We will look at storage areas housing butter fat, milk, and whey solids tanks.

Double Scoop Ice Cream Company

Summary of Pollutant Sources

March 5, 1993

Based on the site assessment inspection conducted on 12/1/92, the pollution prevention team identified four potential sources of pollutants:

- Oil and grease stains on the pavement in the fueling area indicate oil and grease may be picked up by storm water draining to the storm sewer. This area drains into the storm sewer leading to the Rocky River.

- Sediment and erosion potential in the field below the employee parking lot because of thinly planted grass.

- Potential for spills or leaks from liquid storage tanks, including the fuel storage tank, based on a spill that occurred on 1/21/92 and the leak that was detected in the milk storage tank. These pollutants would drain into the piped outfall into the Rocky River.

- Use of a toxic cleaning agent may result in a pollution problem if handled improperly.

Double Scoop Ice Cream Company

**Description of Storm Water Management Measures Taken
Based on Site Assessment Phase**

March 5, 1993

These measures correspond to the pollutant sources identified on the preceding page.

Oil and grease from fueling area.

We installed drip pads around the fuel pumps to pick up spilled gas and oil during truck refueling. These will be inspected regularly to make sure they are working well.

Sediment and erosion in the field below the employee parking lot.

We planted grass in this area to reduce potential for erosion.

Leaks/spills from liquid storage tanks.

We are in the process of installing curbing around the outdoor liquid storage tanks that will contain the volume of he largest tank in case a spill should occur. The spill response team has developed procedures to clean up this area should a spill occur. We are incorporating spill response procedures from our SPCC plan.

Toxic cleaning agent.

We have discontinued the use of this agent and are replacing it with a non-toxic cleaning agent.

POLLUTANT SOURCE IDENTIFICATION
(Section 2.2.6)

Worksheet #7
Completed by: _Cheryl Glenn_
Title: _Plant Manager_
Date: _3/5/93_

Instructions: List all identified storm water pollutant sources and describe existing management practices that address those sources. In the third column, list BMP options that can be incorporated into the plan to address remaining sources of pollutants.

Storm Water Pollutant Sources	Existing Management Practices	Description of New BMP Options
1. Oil and grease on pavement in fueling area	Oil and water separators installed in storm water drain	Install drip pads
2. Erosion in field below employee parking lot	Planted some grass after construction; grassed swales along Wonka Drive	Plant more grass
3. Potential for spills from liquid storage tanks (leak detected in milk tank #1 - past spill on 12/1/92)	Curbing around fuel storage tank (see SPCC plan)	Replace milk tank #1, replace valve on liquid sugar tank #2; install curbing around other outside tanks; spill prevention response plan (see tech?)
4. Use of toxic cleaning agent		Use non-toxic cleaning agent
5. Trash in loading/unloading fueling areas	Regular trash pickup (daily) by maintenance crew; collect and recycle used oil	Train staff in good housekeeping practices
6.		
7.		
8.		
9.		
10.		

BMP IDENTIFICATION
(Section 2.3.1)

Worksheet #7a
Completed by: _Cheryl Glen_
Title: _Plant Manager_
Date: _3/5/93_

Instructions: Describe the Best Management Practices that you have selected to include in your plan. For each of the baseline BMPs, describe actions that will be incorporated into facility operations. Also describe any additional BMPs (activity-specific (Chapter 3) and site-specific BMPs (Chapter 4)) that you have selected. Attach additional sheets if necessary.

BMPs	Brief Description of Activities
Good Housekeeping	Collect and recycle used oil; regular trash pick up; train staff in basic clean up procedures (sweeping loading & unloading areas, etc.)
Preventive Maintenance	Daily inspection of outside milk tanks; replace faulty valve on sugar tank #2; replace leaking milk tank #1
Inspections	Daily inspection of outside milk tanks; bi-monthly inspections of drip pads, curbing, loading/unloading areas, grassed areas, drainage system.
Spill Prevention Response	Install curbing around outside liquid storage tanks; fuel tank has curbing; install drip pads at fueling station
Sediment and Erosion Control	Plant grass around new parking area.
Management of Runoff	Grassed swales along Wonka Drive; (2) oil/water separators in storm drain system
Additional BMPs (Activity-specific and Site-specific)	Order non-toxic cleaning agent.

Double Scoop Ice Cream Company

Employee Training Program

Who:

Line Workers
Maintenance Crew
Shipping and Receiving Crew

When:

Employee meetings held the first Monday of each month to discuss:

- Any environmental/health and safety incidents

- Upcoming training sessions

- Brief reminders on good housekeeping, spill prevention and response procedures, and material handling practices

- Announce any changes to the plan

- Announce any new management practices

In-depth pollution prevention training for new employees

Refresher courses held every 6 months (October and March) addressing:

- Good housekeeping

- Spill prevention and response procedures

- Materials handling and storage

Employee Training Program Topics:

Good Housekeeping

- Review and demonstrate basic cleanup (sweeping and vacuuming) procedures.

- Clearly indicate proper disposal locations.

- Post signs in materials handling areas reminding staff of good housekeeping procedures.

- Be sure employees know where routine clean-up equipment is located.

Spill Prevention and Response

- Clearly identify potential spill areas and drainage routes

- Familiarize employees with past spill events -- why they happened and the environmental impact (use slides)

- Post warning signs in spill areas with emergency contacts and telephone numbers

- Introduce Isaac Feldman as the Spill Response Coordinator and introduce his "team"

- Drill on spill clean-up procedures

- Post the locations of spill clean-up equipment and the persons responsible for operating the equipment

Materials Handling and Storage

- Be sure employees are aware which materials are hazardous and where those materials are stored

- Point out container labels

- Tell employees to use the oldest materials first

- Explain recycling practices

- Demonstrate how valves are tightly closed and how drums should be sealed

- Show how to fuel vehicles and avoid "topping off"

Worksheet #8
Completed by: _Cheryl Glenn_
Title: _Plant Manager_
Date: _3/30/93_

IMPLEMENTATION
(Section 2.4.1)

Instructions: Develop a schedule for implementing each BMP. Provide a brief description of each BMP, the steps necessary to implement the BMP (i.e., any construction or design), the schedule for completing those steps (list dates) and the person(s) responsible for implementation.

BMPs	Description of Action(s) Required for Implementation	Scheduled Completion Date(s) for Req'd. Action	Person Responsible for Action	Notes
Good Housekeeping	1. Develop training program	3/10/93	Glenn	
	2. Conduct training	6/1/93	Glenn	
	3.			
Preventive Maintenance	1. Replace valve on sugar tank #2	3/1/93	Feldman	
	2. Install new milk tank #2	2/15/93	Feldman	
	3.			
Inspections	1. Develop inspections schedule	4/1/93	Glenn	
	2.			
	3.			
Spill Prevention and Response	1. Install curbing around milk storage tanks	4/30/95	Meyers	
	2. Install drip pads	4/1/93	Feldman	
	3. Develop/Implement Spill Prevention/Response Training	4/1/93 - DEVELOP 6/1/93 - TRAIN	Feldman	
Sediment and Erosion Control	1. Plant grass around parking area	4/15/93	Feldman	
	2.			
	3.			
Management of Runoff	1. BMPs already in place			
	2.			
	3.			
Additional BMPs (Actively-specific and site-specific)	1. Substitute non-toxic cleaning agent	2/28/93	Michaels	
	2.			
	3.			

EMPLOYEE TRAINING
(Section 2.4.2)

Worksheet #9
Completed by: _Cheryl Glenn_
Title: _Plant Manager_
Date: _3/2/93_

Instructions: Describe the employee training program for your facility below. The program should, at a minimum, address spill prevention and response, good housekeeping, and material management practices. Provide a schedule for the training program and list the employees who attend training sessions.

Training Topics	Brief Description of Training Program/Materials (e.g., film, newsletter course)	Schedule for Training (list dates)	Attendees						
Spill Prevention and Response	locate spill areas by signs; drill spill response procedures; show slides of past spills.	October/March	Maintenance/ shipping & receiving						
Good Housekeeping	Demonstration; post signs at disposal sites.	October/March	Maintenance/ shipping & receiving						
Material Management Practices	Introduce hazardous materials labels; discuss recycling.	October/March	Lineworkers/shipping and receiving/maintenance						
Other Topics	Environmental/health incidents; reminders of pollution prevention plan issues.	1st Monday of each month	All employees.						

APPENDIX D

**STORM WATER AND POLLUTION PREVENTION CONTACTS
AND ADDITIONAL POLLUTION PREVENTION INFORMATION**

STATE STORM WATER AND POLLUTION PREVENTION CONTACTS		
State	**Storm Water Contact**	**Pollution Prevention Contact**
*Alabama	John Poole 205-271-7852	Daniel E. Cooper 205-271-7939
Alaska	Michael Menge 907-465-5260	David Wigglesworth 907-465-5275
Arizona	See Region IX Contact	Stephanie Wilson 602-257-2318
*Arkansas	Marysia Jastrzebski 501-562-7444	Robert J. Finn 501-570-2861
*California	Don Parrin 916-657-1288	Kim Wilhelm 916-324-1807
*Colorado	Patricia Nelson 303-331-4590	Kate Kramer 303-331-4510
*Connecticut	Dick Mason 203-566-7167	Rita Lomasney (ConnTap) 203-241-0777
*Delaware	Sarah Cooksey 302-739-5731	Andrea Farrell 302-739-3822
District of Columbia	James Collier 202-404-1120	Hampton Cross 202-939-7116
Florida	Eric Livingston 904-488-0782	Janet A. Campbell 904-488-0300
*Georgia	Mike Creason 404-656-4887	Susan Hendricks 404-656-2833
*Hawaii	Steve Chang 808-586-4309	Jane Dewell 808-586-4226
Idaho	Jerry Yoder 208-334-5898	Joy Palmer 208-334-5879
*Illinois	Tim Kluge 217-782-0610	Mike Hayes 217-782-8700
*Indiana	Lonnie Brumfield 317-232-8705	Joanna Joyce 317-232-8172
*Iowa	Monica Wnuk 515-281-7017	John Konefes 319-273-2079
*Kansas	Don Carlson 913-296-5555	Tom Gross 913-296-1603
*Kentucky	Douglas Allgeier 502-564-3410	Joyce St. Clair 502-588-7260

*Approved NPDES Program

STATE STORM WATER AND POLLUTION PREVENTION CONTACTS

State	Storm Water Contact	Pollution Prevention Contact
Louisiana	Jim Delahoussaye 504-765-0525	Gary Johnson 504-765-0720
Maine	Norm Marcotte 207-289-3901	Scott Whittier 207-289-2651
*Maryland	Vince Berg 410-631-3553	Harry Benson 301-631-3315
Massachusetts	Cynthia Hall 617-292-5656	Barbara Kelly 617-727-3260
*Michigan	Gary Boersen 517-373-1982	Larry E. Hartwig 517-335-1178
*Minnesota	Scott Thompson 612-296-7203	Cindy McComas (MNTAP) 612-296-4646
*Mississippi	Jerry Cain 601-961-5171	Caroline Hill 601-325-8454
*Missouri	Bob Hentges 314-751-6825	Becky Shannon 314-751-3176
*Montana	Fred Shewman 406-444-2406	Bill Potts 406-444-2821
*Nebraska	Clark Smith 402-471-4239	Teri Swarts 402-471-4217
*Nevada	Rob Saunders 702-687-4670	Kevin Dick 702-784-1717
New Hampshire	Jeff Andrews 603-271-2457	Vincent R. Perelli 603-271-2902
*New Jersey	Sandra Cohen 609-633-7021	Jean Herb 609-777-0518
New Mexico	Glen Saums 505-827-2827	Alex Puglisi 505-827-2804
*New York	Ken Stevens 518-457-1157	John Ianotti 518-457-7267
*North Carolina	Coleen Sullins 919-733-5083	Gary Hunt 919-571-4100
*North Dakota	Sheila McClenatahan 701-221-5210	Neil Knatterud 703-221-5166
*Ohio	Robert Phelps 614-644-2034	Mike Kelly 614-644-3492

*Approved NPDES Program

STATE STORM WATER AND POLLUTION PREVENTION CONTACTS		
State	Storm Water Contact	Pollution Prevention Contact
Oklahoma	Brooks Kirlin 504-231-2500	Chris Varga 405-271-7047
*Oregon	Ranei Nomura 503-229-5256	Roy W. Brower 503-229-6585
*Pennsylvania	R.B. Patel 717-787-8184	Greg Harder 717-772-2724
*Rhode Island	Ed Symanski 401-244-3931	Janet Keller 401-277-3434
*South Carolina	Brigit McDade 803-734-5300	Jeffrey DeBossonet 803-734-4715
South Dakota	Glenn Pieritz 605-773-3351	Vonnie Kallmeyn 605-773-3153
*Tennessee	Robert Haley 615-741-2275	James Ault 615-742-6547
Texas	Randy Wilburn 512-463-8446	Priscilla Seymour 512-463-7761
*Utah	Harry Campbell 801-538-6146	Sonja Wallace 801-538-6170
*Vermont	Brian Kooiker 802-244-5674	Gary Gulka 802-244-8702
*Virgin Islands	Marc Pacifico 809-773-0565	See Region II Contact
*Virginia	Martin Ferguson, Jr. 804-527-5030	Sharon Kenneally-Baxter 804-371-8716
*Washington	Peter Birch 206-438-7076	Stan Springer 206-438-7541
*West Virginia	Jerry Ray 304-348-0375	Dale Moncer 304-348-4000
*Wisconsin	Ann Mauel 608-267-7634	Lynn Persson 608-267-3763
*Wyoming	John Wagner 307-777-7082	David Finley 307-777-7752

*Approved NPDES Program

EPA REGIONAL STORM WATER AND POLLUTION PREVENTION CONTACTS		
State	**Storm Water Contact**	**Pollution Prevention Contact**
REGION I	Veronica Harrington 617-565-3525	Mark Mahoney 617-565-1155
REGION II	Jose Rivera 212-264-2911	Janet Sapadin 212-264-1925
REGION III	Kevin Magerr 215-597-1651	Roy Denmark 215-597-8327
REGION IV	Roosevelt Childress 404-347-3379	Carol Monell 404-347-7109
REGION V	Peter Swenson 312-886-0236	Louis Blume 312-353-4135
REGION VI	Brent Larsen 214-655-7175	Laura Townsend 214-655-6525
REGION VII	Ralph Summers 913-551-7418	Alan Wehmeyer 913-551-7336
REGION VIII	Vern Berry 303-293-1630	Sharon Childs 303-293-1456
REGION IX	Eugene Bromley 415-744-1906	Jesse Baskir 415-744-2189
REGION X	Steve Bubnick 206-553-8399	Carolyn Gangmark 206-553-4072

ADDITIONAL POLLUTION PREVENTION INFORMATION

State pollution prevention programs have people who are knowledgeable about pollution prevention and are willing to provide information and sometimes technical assistance on pollution prevention. The EPA has pollution prevention experts located in a number of different program offices, laboratories, and EPA Regional offices. These experts can provide information on starting a pollution prevention program or on specific waste reduction BMPs. This Appendix lists State and Federal pollution prevention contacts above. Trade associations are another good source of pollution prevention information. Trade associations can often provide you with pollution prevention assistance directly or refer you to someone who can.

A comprehensive listing of pollution prevention resources, documents, courses, and programs, including names and phone numbers, is contained in a new annual EPA publication. Copies of this document -- *Pollution Prevention Training Opportunities in 1992* -- may be obtained by calling the PPIC/PIES support number at (703) 821-4800.

One good source of information on pollution prevention is EPA's Pollution Prevention Information Clearinghouse (PPIC). PPIC contains technical, policy, programmatic, legislative, and financial information on pollution prevention efforts in the United States and abroad. The PPIC may be reached by personal computer modem, telephone hotline, or mail. The PIES, or Pollution Prevention Information Exchange System, is a free 24-hour electronic bulletin board consisting of message centers, technical data bases, issue-specific "mini-exchanges," and a calendar of pollution prevention events. The PIES allows a user to access the full range of infirmation in the PPIC. For information on how to use the PPIC/PIES, call (703) 821-4800. To log on to the PIES system using a modem and a PC, call (703) 506-1025 (set your communication software at 8 bits and no parity).

EPA and State programs have developed manuals and fact sheets containing specific pollution prevention information. These manuals and fact sheets listed below can be ordered free of charge by calling the EPA Pollution Prevention Information Clearinghouse at (703) 821-4800.

> A *Pollution Prevention Technology Handbook*, edited by Robert Noyes, is available from Noyes Data Corporation, Park Ridge, NJ (1993, 683 pp, $98). The handbook describes specific pollution prevention technology applications in 36 industries.

INDUSTRY-SPECIFIC POLLUTION PREVENTION GUIDANCE MANUALS
AVAILABLE FROM THE PPIC

Guides to Pollution Prevention:	*Automotive Refinishing Industry*	EPA/625/7-91/016
Guides to Pollution Prevention:	*Auto Repair Industry*	EPA/625/7-91/013
Guides to Pollution Prevention:	*The Commercial Printing Industry*	EPA/625/7-90/008
Guides to Pollution Prevention:	*The Fabricated Metal Industry*	EPA/625/7-90/006
Guides to Pollution Prevention:	*Fiberglass Reinforced and Composite Plastics*	EPA/625/7-91/014
Guides to Pollution Prevention:	*Marine Maintenance and Repair*	EPA/625/7-91/015
Guides to Pollution Prevention:	*The Paint Manufacturing Industry*	EPA/625/7-90/005
Guides to Pollution Prevention:	*The Pesticide Formulating Industry*	EPA/625/7-90/004
Guides to Pollution Prevention:	*Pharmaceutical Preparation*	EPA/625/7-91/017
Guides to Pollution Prevention:	*Photoprocessing Industry*	EPA/625/7-91/012
Guides to Pollution Prevention:	*The Printed Circuit Board Manufacturing Industry*	EPA/625/7-90/00
Guides to Pollution Prevention:	*Research and Educational Institutions*	EPA/625/7-90/010
Guides to Pollution Prevention:	*Selected Hospital Waste Streams*	EPA/625/7-90-009

FACT SHEETS AVAILABLE FROM PPIC

General/Introductory Information

- Conservation Tips for Business
- General Guidelines
- Getting More Use Out of What We Have
- Glossary of Waste Reduction Terms
- Guides to Pollution Prevention
- Hazardous Waste Fact Sheet for Minnesota Generators
- Hazardous Waste Minimization
- How Business Organizations Can Help
- Increase Your Corporate and Product Image
- Industrial Hazardous Wastes in Minnesota
- Local Governments and Pollution Prevention
- Pollution Prevention (General)
- Pollution Prevention Fees
- Pollution Prevention Training and Education
- Pollution Prevention Through Waste Reduction
- Recent Publications
- Reduce Hazardous Waste
- Reuse Strategies for Local Government
- Source Reduction Techniques for Local Government
- U.S. EPA's Pollution Prevention Program
- Video Tapes Available from the Virginia Waste Minimization Program
- Waste Exchange: Everybody Wins!
- Waste Exchange Services
- Waste Minimization Fact Sheet

- Waste Minimization in the Workplace
- Waste Reduction Can Work For You
- Waste Reduction Overview
- Waste Reduction/Pollution Prevention: Getting Started
- Waste Reduction Tips for All Businesses
- Waste Source Reduction Checklist
- What is Pollution Prevention?
- Why Reduce Waste?

Legislative Information/ EPA and State Initiatives

- About Minnesota's "But Recycled Campaign"
- Alaska State Agency Waste Reduction and Recycling
- EPA's 2% Set Aside Pollution Prevention Projects
- EPA's "List of Lists" Projects
- EPA's Pollution Prevention Enforcement Settlement Policy
- EPA's Pollution Prevention Incentives for States
- EPA's Pollution Prevention Strategy
- Introducing the Colorado Pollution Prevention Program
- Michigan's Solid Waste Reduction Strategy
- Minnesota's Toxic Pollution Prevention Act
- New Form R Reporting Requirements
- Oregon's Toxic Use Reduction Act
- Pollution Prevention Act of 1990
- Promoting Pollution Prevention in Minnesota State Government

Setting Up A Program

- 1991 Small Business Pollution Prevention Grants

- An Organization Strategy for Pollution Prevention

- Considerations in Selecting a Still for Onsite Recycling

- Colorado Technical Information Center

- Onsite Assistance (Colorado only)

- Pollution Prevention Grant Program Summaries and Reports

- Procuring Recycled Products

- Recycling Market Development Program

- Selecting a Supplier, Hauler, and Materials Broker

- Solid Waste Management Financial Assistance Program

- Source Reduction at Your Facility

- Starting Your Own Waste Reduction Program

- The Alexander Motor's Success Story

- The Eastside Plating Success Story

- The Tektronics Payoff

- The Wacker Payoff

- Waste Reduction Checklists:

  - General

  - Cleaning

  - Coating/Painting

  - Formulating

  - Machining

  - Operating Procedures

  - Plating/Metal Finishing

- Waste Source Reduction: Implementing a Program

Process/Material Specific

- Aerosol Containers

- Aircraft Rinsewater Disposal

- Acids/Bases

- Chemigation Practices to Prevent Ground Water Contamination

- Corrugated Cardboard Waste Reduction

- Demolition

- Empty Containers

- Gunwasher Maintenance

- Lead Acid Batteries

- Machine Coolants:

  - Prolonging Coolant Life

  - Waste Reduction

- Metal Recovery:

  - Dragout Reduction

  - Ion Exchange/Electrolytic Recovery

  - Etchant Substitution

- Metals Recycling

- Office Paper Waste Reduction

- Old Paints, Inks, Residuals, and Related Materials

- Pesticides:

  - Disposal of Unused Pesticides, Tank Mixes, and Rinsewater

  - In-Filled Sprayer Rinse System to Reduce Pesticide Wastes

  - Pesticide Container Disposal

  - Preventing Pesticide Pollution of Surface and Ground Water

- Preventing Well Contamination by Pesticides

- Protecting Mountain Springs from Pesticide Contamination

- Reducing and Saving Money Using Integrated Pest Management

• Plastics:

- The Facts About Production, Use, and Disposal

- The Facts on Degradable Plastics

- The Facts on Recycling Plastics

- The Facts on Source Reduction

• Printing Equipment

• Refrigerant Reclamation Equipment/ Services

• Reverse Osmosis

• Safety Kleen, Inc., Users

• Shop Rags from Printers

• Small Silver Recovery Units

• Solvents:

- Alternatives to CFC-113 Used in the Cleaning of Electronic Circuit Boards

- Onsite Solvent Reclamation

- Reducing Shingle Waste at a Manufacturing Facility

- Reducing Solvent Emissions from Vapor Degreasers

- Small Solvent Recovery Systems

- Solvent Loss Control

- Solvent Management: Printing Press

- Solvent Recovery: Fiber Production Plant

- Solvent Reduction in Metal Parts Cleaning

- Solvent Reuse: Technical Institute

- Trichloroethylene and Stoddard Solvent Reduction Alternatives

• Ultrafiltration

• Used Containers: Management

• Used Oil Recycling

• Waste Management Guidance for Oil Clean-Up

• Water and Chemical Reduction for Cooling Towers

• Waste Water Treatment Opportunities

Industry-Specific Information

• Aerospace Industry

• Auto Body Shops

• Automotive Painting

• Automotive/Vehicle Repair Shops

• Auto Salvage Yards

• Asbestos Handling, Transport, and Disposal

• Chemical Production

• Coal Mining

• Concrete Panel Manufacturers

• Dairy Industry:

- Cut Waste and Reduce Surcharges for Your Dairy Plant

- Dairy CEOs: Do You Have a $500 Million Opportunity?

- Liquid Assets for Your Dairy Plant

- Water and Wastewater Management in a Dairy Processing Plant

• Dry Cleaners

• Electrical Power Generators

- Electroplating Industry:

  - Dragout Management for Electroplaters

  - Plating with Trivalent Chrome Instead of Cr + 6

  - Water Conservation Using Counter Current Rinsing

  - Water Conservation: Tank Design

  - Water Conservation: Rinsewater Reuse

  - What Should I Do With My Electroplating Sludge?

- Fabricated Metal Manufacturers

- Fiberglass Fabricators: Volatile Emissions Reduction

- Machine Toolers

- Metal Finishers:

  - General

  - Effluent Minimization

  - Rinsewater Reduction

- Oil Refiners

- Paint Formulators

- Paper Manufacturers

- Pesticide Formulating Industry

- Photofinishers/Photographic Processors

- Poultry Industry:

  - Poultry CEOs: You May Have a $60 Million Opportunity

  - Poultry Processors: You Can Reduce Waste Load and Cut Sewer Surcharges

  - Survey Shows That Poultry Processors Can Save Money By Conserving Water

  - Systems for Recycling Water in Poultry Processing

- Printed Circuit Board Manufacturers

- Printing Industry

- Radiator Service Firms

- Shrimp Processors

- Steel Manufacturers

- Textile Industry:

  - Dye Bath and Bleach Bath Reconstitution

  - Water Conservation

- Wire Milling Operations: Process Water Reduction

APPENDIX E

BMP FACT SHEETS

SILT FENCE

September 1992

Design Criteria

▲ Silt fences are appropriate at the following general locations:

   ▲ Immediately upstream of the point(s) of runoff discharge from a site before flow becomes concentrated (maximum design flow rate should not exceed 0.5 cubic feet per second).
   ▲ Below disturbed areas where runoff may occur in the form of overland flow.

▲ Ponding should not be allowed behind silt fences since they will collapse under high pressure; the design should provide sufficient outlets to prevent overtopping.
▲ The drainage area should not exceed 0.25 acre per 100 feet of fence length.
▲ For slopes between 50:1 and 5:1, the maximum allowable upstream flow path length to the fence is 100 feet; for slopes of 2:1 and steeper, the maximum is 20 feet.
▲ The maximum upslope grade perpendicular to the fence line should not exceed 1:1.
▲ Synthetic silt fences should be designed for 6 months of service; burlap is only acceptable for periods of up to 60 days.

Materials

▲ Synthetic filter fabric should be a pervious sheet of polypropylene, nylon, polyester, or polyethylene yarn conforming to the requirements in Table 1 below.

TABLE 1. SYNTHETIC FILTER FABRIC REQUIREMENTS

Physical Property	Requirements
Filtering Efficiency	75% - 85% (minimum)
Tensile Strength at 20% (maximum) Elongation	Standard Strength - 30 lb/linear inch (minimum)
	Extra Strength - 50 lb/linear inch (minimum)
Slurry Flow Rate	0.3 gal/ft^2/min (minimum)

▲ Synthetic filter fabric should contain ultraviolet ray inhibitors and stabilizers to provide a minimum of 6 months of expected usable construction life at a temperature range of 0 to 120°F.
▲ Burlap of 10 ounces per square yard of fabric can also be used.
▲ The filter fabric should be purchased in a continuous roll to avoid joints.
▲ While not required, wire fencing may be used as a backing to reinforce standard strength filter fabric. The wire fence (14 gauge minimum) should be at 22-48 inches wide and should have a maximum mesh spacing of 6 inches.
▲ Posts should be 2-4 feet long and should be composed of either 2" x 2-4" pine (or equivalent) or 1.00 to 1.33 lb/linear ft steel. Steel posts should have projections for fastening wire and fabric to them.

Construction Specifications

▲ The maximum height of the filter fence should range between 18 and 36 inches above the ground surface (depending on the amount of upslope ponding expected).

SILT FENCE

- ▲ Posts should be spaced 8 to 10 feet apart when a wire mesh support fence is used and no more than 6 feet apart when extra strength filter fabric (without a wire fence) is used. The posts should extend 12 to 30 inches into the ground.
- ▲ A trench should be excavated 4 to 8 inches wide and 4 to 12 inches deep along the upslope side of the line of posts.
- ▲ If standard strength filter fabric is to be used, the optional wire mesh support fence may be fastened to the upslope side of the posts using 1 inch heavy duty wire staples, tie wires, or hog rings. Extend the wire mesh support to the bottom of the trench. The filter fabric should then be stapled or wired to the fence, and 8 to 20 inches of the fabric should extend into the trench (Figure 1).
- ▲ Extra strength filter fabric does not require a wire mesh support fence. Staple or wire the filter fabric directly to the posts and extend 8 to 20 inches of the fabric into the trench (Figure 1).
- ▲ Where joints in the fabric are required, the filter cloth should be spliced together only at a support post, with a minimum 6-inch overlap, and securely sealed.
- ▲ Do not attach filter fabric to trees.
- ▲ Backfill the trench with compacted soil or 0.75 inch minimum diameter gravel placed over the filter fabric.

Maintenance

- ▲ Inspect filter fences daily during periods of prolonged rainfall, immediately after each rainfall event, and weekly during periods of no rainfall. Make any required repairs immediately.
- ▲ Sediment must be removed when it reaches one-third to one-half the height of the filter fence. Take care to avoid damaging the fence during cleanout.
- ▲ Filter fences should not be removed until the upslope area has been permanently stabilized. Any sediment deposits remaining in place after the filter fence has been removed should be dressed to conform with the existing grade, prepared, and seeded.

Cost

- ▲ Silt fence installation costs approximately $6.00 per linear foot.

Sources

- ▲ Commonwealth of Virginia - County of Fairfax, 1987. 1987 Check List For Erosion And Sediment Control - Fairfax County, Virginia.
- ▲ State of North Carolina, 1988. Erosion and Sediment Control Planning and Design Manual. North Carolina Sedimentation Control Commission, Department of Natural Resources and Community Development.
- ▲ Maryland Department of the Environment, 1991. 1991 Maryland Standards And Specifications For Soil Erosion And Sediment Control - Draft.

PIPE SLOPE DRAIN

September 1992

Design Criteria

▲ Pipe Slope Drains (PSD) are appropriate in the following general locations:

  ▲ On cut or fill slopes before permanent storm water drainage structures have been installed.
  ▲ Where earth dikes or other diversion measures have been used to concentrate flows.
  ▲ On any slope where concentrated runoff crossing the face of the slope may cause gullies, channel erosion, or saturation of slide-prone soils.
  ▲ As an outlet for a natural drainageway.

▲ The drainage area may be up to 10 acres; however, many jurisdictions consider 5 acres the recommended maximum.
▲ The PSD design should handle the peak runoff for the 10-year storm. Typical relationships between area and pipe diameter are shown in Table 2 below.

TABLE 2. RELATIONSHIP BETWEEN AREA AND PIPE DIAMETER

Maximum Drainage Area (Acres)	Pipe Diameter (D) (Inches)
0.5	12
0.75	15
1.0	18

Materials

▲ Pipe may be heavy duty flexible tubing designed for this purpose, e.g., nonperforated, corrugated plastic pipe, corrugated metal pipe, bituminous fiber pipe, or specially designed flexible tubing.
▲ A standard flared end section secured with a watertight fitting should be use for the inlet. A standard T-section fitting may also be used.
▲ Extension collars should be 12-inch long sections of corrugated pipe. All fittings must be watertight.

Construction Specifications

▲ Place the pipe slope drain on undisturbed or well-compacted soil.
▲ Soil around and under the entrance section must be hand-tamped in 4-inch to 8-inch lifts to the top of the dike to prevent piping failure around the inlet.
▲ Place filter cloth under the inlet and extend 5 feet in front of the inlet and be keyed in 6-inches on all sides to prevent erosion. A 6-inch metal toe plate may also be used for this purpose.
▲ Ensure firm contact between the pipe and the soil at all points by backfilling around and under the pipe with stable soil material hand compacted in lifts of 4-inches to 8-inches.
▲ Securely stake the PSD to the slope using grommets provided for this purpose at intervals of 10 feet or less.
▲ Ensure that all slope drain sections are securely fastened together and have watertight fittings.

PIPE SLOPE DRAIN

- ▲ Extend the pipe beyond the toe of the slope and discharge at a nonerosive velocity into a stabilized area (e.g., rock outlet protection may be used) or to a sedimentation trap or pond.
- ▲ The PSD should have a minimum slope of 3 percent or steeper.
- ▲ The height at the centerline of the earth dike should range from a minimum of 1.0 foot over the pipe to twice the diameter of the pipe measured from the invert of the pipe. It should also be at least 6 inches higher than the adjoining ridge on either side.
- ▲ At no point along the dike will the elevation of the top of the dike be less than 6 inches higher than the top of the pipe.
- ▲ Immediately stabilize all areas disturbed by installation or removal of the PSD.

Maintenance

- ▲ Inspect regularly and after every storm. Make any necessary repairs.
- ▲ Check to see that water is not bypassing the inlet and undercutting the inlet or pipe. If necessary, install headwall or sandbags.
- ▲ Check for erosion at the outlet point and check the pipe for breaks or clogs. Install additional outlet protection if needed and immediately repair the breaks and clean any clogs.
- ▲ Do not allow construction traffic to cross the PSD and do not place any material on it.
- ▲ If a sediment trap has been provided, clean it out when the sediment level reaches 1/3 to 1/2 the design volume.
- ▲ The PSD should remain in place until the slope has been completely stabilized or up to 30 days after permanent slope stabilization.

Cost

- ▲ Pipe slope drain costs are generally based upon the pipe type and size (generally, flexible PVC at $5.00 per linear foot). Also adding to this cost are any expenses associated with inlet and outlet structures.

Sources

- ▲ Commonwealth of Virginia - County of Fairfax, 1987. 1987 Check List For Erosion And Sediment Control - Fairfax County, Virginia.
- ▲ State of North Carolina, 1988. Erosion and Sediment Control Planning and Design Manual. North Carolina Sedimentation Control Commission, Department of Natural Resources and Community Development.
- ▲ Maryland Department of the Environment, 1991. 1991 Maryland Standards And Specifications For Soil Erosion And Sediment Control - Draft.
- ▲ Storm Water Management Manual for the Puget Sound Basin. State of Washington, Department of Ecology, 1991.
- ▲ Cost Data:

  - ▲ Draft Sediment and Erosion Control, An Inventory of Current Practices, April 20, 1990. Prepared by Kamber Engineering for the U.S. Environmental Protection Agency, Office of Water Enforcement and Permits, Washington, D.C. 20460.

FILTER FABRIC INLET PROTECTION

September 1992

Design Criteria

▲ Inlet protection is appropriate in the following locations:

  ▲ In small drainage areas (less than 1 acre) where the storm drain inlet is functional before the drainage area has been permanently stabilized.
  ▲ Where there is danger of sediment silting in an inlet which is in place prior to permanent stabilization.

▲ Filter fabric inlet protection is appropriate for most types of inlets where the drainage area is one acre or less.
▲ The drainage area should be fairly flat with slopes of 5 % or less and the area immediately surrounding the inlet should not exceed a slope of 1%.
▲ Overland flow to the inlet should be no greater than 0.5 cfs.
▲ This type of inlet protection is not appropriate for use in paved areas because the filter fabric requires staking.
▲ To avoid failure caused by pressure against the fabric when overtopping occurs, it is recommended that the height of the filter fabric be limited to 1.5 feet above the crest of the drop inlet.
▲ It is recommended that a sediment trapping sump of 1 to 2 feet in depth with side slopes of 2:1 be provided.

Materials

▲ Filter fabric (see the fabric specifications for silt fence).
▲ Wooden stakes 2" x 2" or 2"x 4" with a minimum length of 3 feet.
▲ Heavy-duty wire staples at least ½ inch in length.
▲ Washed gravel ¾ inches in diameter.

Construction Specifications

▲ Place a stake at each corner of the inlet and around the edges at no more than 3 feet apart. Stakes should be driven into the ground 18 inches or at a minimum 8 inches.
▲ For stability a framework of wood strips should be installed around the stakes at the crest of the overflow area 1.5 feet above the crest of the drop inlet.
▲ Excavate a trench of 8 inches to 12 inches in depth around the outside perimeter of the stakes. If a sediment trapping sump is being provided then the excavation may be as deep as 2 feet.
▲ Staple the filter fabric to the wooden stakes with heavy-duty staples, overlapping the joints to the next stake. Ensure that between 12 inches to 32 inches of filter fabric extends at the bottom so it can be formed into the trench.
▲ Place the bottom of the fabric in the trench and backfill the trench all the way around using washed gravel to a minimum depth of 4 inches.

FILTER FABRIC INLET PROTECTION

Maintenance

- ▲ Inspect regularly and after every storm. Make any repairs necessary to ensure the measure is in good working order.
- ▲ Sediment should be removed and the trap restored to its original dimensions when sediment has accumulated to ½ the design depth of the trap.
- ▲ If the filter fabric becomes clogged it should be replaced immediately.
- ▲ Make sure that the stakes are firmly in the ground and that the filter fabric continues to be securely anchored.
- ▲ All sediments removed should be properly disposed.
- ▲ Inlet protection should remain in place and operational until the drainage area is completely stabilized or up to 30 days after the permanent site stabilization is achieved.

Cost

- ▲ The cost of storm drain inlet protection varies dependent upon the size and type of inlet to be protected but generally is about $300.00 per inlet.

Sources

- ▲ Commonwealth of Virginia - County of Fairfax, 1987. 1987 Check List For Erosion And Sediment Control - Fairfax County, Virginia.
- ▲ State of North Carolina, 1988. Erosion and Sediment Control Planning and Design Manual. North Carolina Sedimentation Control Commission, Department of Natural Resources and Community Development.
- ▲ Maryland Department of the Environment, 1991. 1991 Maryland Standards And Specifications For Soil Erosion And Sediment Control - Draft.
- ▲ Storm Water Management Manual for the Puget Sound Basin. State of Washington, Department of Ecology, 1991.
- ▲ Cost Data:

  - ▲ Draft Sediment and Erosion Control, An Inventory of Current Practices, April 20, 1990. Prepared by Kamber Engineering for the U.S. Environmental Protection Agency, Office of Water Enforcement and Permits, Washington, D.C. 20460.

EXCAVATED GRAVEL INLET PROTECTION

September 1992

Design Criteria

▲ Inlet protection is appropriate in the following locations:

  ▲ In small drainage areas (less than 1 acre) where the storm drain inlet is functional before the drainage area has been permanently stabilized.
  ▲ Where there is danger of sediment silting in an inlet which is in place prior to permanent stabilization.
  ▲ Where ponding around the inlet structure could be a problem to traffic on site.

▲ Excavated gravel and mesh inlet protection may be used with most inlets where overflow capability is needed and in areas of heavy flows, 0.5 cfs or greater.
▲ The drainage area should not exceed 1 acre.
▲ The drainage area should be fairly flat with slopes of 5% or less.
▲ The trap should have a sediment trapping sump of 1 to 2 feet measured from the crest of the inlet. Side slopes should be 2:1. The recommended volume of excavation is 35 yd³/acre disturbed.
▲ To achieve maximum trapping efficiency the longest dimension of the basin should be oriented toward the longest inflow area.

Materials

▲ Hardware cloth or wire mesh with ½ inch openings.
▲ Filter fabric (see the fabric specifications for silt fence).
▲ Washed gravel ¾ inches to 4 inches in diameter.

Construction Specifications

▲ Remove any obstructions to excavating and grading. Excavate sump area, grade slopes and properly dispose of soil.
▲ The inlet grate should be secured to prevent seepage of sediment laden water.
▲ Place wire mesh over the drop inlet so that the wire extends a minimum of 1 foot beyond each side of the inlet structure. Overlap the strips of mesh if more than one is necessary.
▲ Place filter fabric over the mesh extending it at least 18 inches beyond the inlet opening on all sides. Ensure that weep holes in the inlet structure are protected by filter fabric and gravel.
▲ Place stone/gravel over the fabric/wire mesh to a depth of at least 1 foot.

EXCAVATED GRAVEL INLET PROTECTION

Maintenance

- ▲ Inspect regularly and after every storm. Make any repairs necessary to ensure the measure is in good working order.
- ▲ Sediment should be removed and the trap restored to its original dimensions when sediment has accumulated to ½ the design depth of the trap.
- ▲ Clean or remove and replace the stone filter or filter fabric if they become clogged.
- ▲ Inlet protection should remain in place and operational until the drainage area is completely stabilized or up to 30 days after the permanent site stabilization is achieved.

Cost

- ▲ The cost of storm drain inlet protection varies dependent upon the size and type of inlet to be protected but generally is about $300.00 per inlet.

Sources

- ▲ Commonwealth of Virginia - County of Fairfax, 1987. 1987 Check List For Erosion And Sediment Control - Fairfax County, Virginia.
- ▲ State of North Carolina, 1988. Erosion and Sediment Control Planning and Design Manual. North Carolina Sedimentation Control Commission, Department of Natural Resources and Community Development.
- ▲ Maryland Department of the Environment, 1991. 1991 Maryland Standards And Specifications For Soil Erosion And Sediment Control - Draft.
- ▲ Storm Water Management Manual for the Puget Sound Basin. State of Washington, Department of Ecology, 1991.
- ▲ Cost Data:

  - ▲ Draft Sediment and Erosion Control, An Inventory of Current Practices, April 20, 1990. Prepared by Kamber Engineering for the U.S. Environmental Protection Agency, Office of Water Enforcement and Permits, Washington, D.C. 20460.

BLOCK AND GRAVEL INLET PROTECTION

September 1992

Design Criteria

▲ Inlet protection is appropriate in the following locations:

    ▲ In drainage areas (less than 1 acre) where the storm drain inlet is functional before the drainage area has been permanently stabilized.

    ▲ Where there is danger of sediment silting in an inlet which is in place prior to permanent stabilization.

▲ Block and gravel inlet protection may be used with most types of inlets where overflow capability is needed and in areas of heavy flows 0.5 cfs or greater.

▲ The drainage area should not exceed 1 acre.

▲ The drainage area should be fairly flat with slopes of 5% or less.

▲ To achieve maximum trapping efficiency the longest dimension of the basin should be oriented toward the longest inflow area.

▲ Where possible the trap should have sediment trapping sump of 1 to 2 feet in depth with side slopes of 2:1.

▲ There are several other types of inlet protection also used to prevent siltation of storm drainage systems and structures during construction, they are:

    ▲ Filter Fabric Inlet Protection
    ▲ Excavated Gravel Inlet Protection

Materials

▲ Hardware cloth or wire mesh with ½ inch openings

▲ Filter fabric (see the fabric specifications for silt fence)

▲ Concrete block 4 inches to 12 inches wide.

▲ Washed gravel ¾ inches to 4 inches in diameter

Construction Specifications

▲ The inlet grate should be secured to prevent seepage of sediment laden water.

▲ Place wire mesh over the drop inlet so that the wire extends a minimum of 12 inches to 18 inches beyond each side of the inlet structure. Overlap the strips of mesh if more than one is necessary.

▲ Place filter fabric (optional) over the mesh and extend it at least 18 inches beyond the inlet structure.

▲ Place concrete blocks over the filter fabric in a single row lengthwise on their sides along the sides of the inlet. The foundation should be excavated a minimum of 2 inches below the crest of the inlet and the bottom row of blocks should be against the edge of the structure for lateral support.

▲ The open ends of the block should face outward not upward and the ends of adjacent blocks should abut. Lay one block on each side of the structure on its side to allow for dewatering of the pool.

▲ The block barrier should be at least 12 inches high and may be up to a maximum of 24 inches high and may be from 4 inches to 12 inches in depth depending on the size of block used.

▲ Prior to backfilling, place wire mesh over the outside vertical end of the blocks so that stone does not wash down the inlet.

▲ Place gravel against the wire mesh to the top of the blocks.

| BLOCK AND GRAVEL INLET PROTECTION |

Maintenance

- ▲ Inspect regularly and after every storm. Make any repairs necessary to ensure the measure is in good working order.
- ▲ Sediment should be removed and the trap restored to its original dimensions when sediment has accumulated to ½ the design depth of the trap.
- ▲ All sediments removed should be properly disposed of.
- ▲ Inlet protection should remain in place and operational until the drainage area is completely stabilized or up to 30 days after the permanent site stabilization is achieved.

Cost

- ▲ The cost of storm drain inlet protection varies dependent upon the size and type of inlet to be protected but generally is about $300.00 per inlet.

Sources

- ▲ Commonwealth of Virginia - County of Fairfax, 1987. 1987 Check List For Erosion And Sediment Control - Fairfax County, Virginia.
- ▲ State of North Carolina, 1988. Erosion and Sediment Control Planning and Design Manual. North Carolina Sedimentation Control Commission, Department of Natural Resources and Community Development.
- ▲ Maryland Department of the Environment, 1991. 1991 Maryland Standards And Specifications For Soil Erosion And Sediment Control - Draft.
- ▲ Storm Water Management Manual for the Puget Sound Basin. State of Washington, Department of Ecology, 1991.
- ▲ Cost Data:

  - ▲ Draft Sediment and Erosion Control, An Inventory of Current Practices, April 20, 1990. Prepared by Kamber Engineering for the U.S. Environmental Protection Agency, Office of Water Enforcement and Permits, Washington, D.C. 20460.

TEMPORARY SEDIMENT TRAP

September 1992

Design Criteria

▲ Temporary sediment traps are appropriate in the following locations:

  ▲ At the outlet of the perimeter controls installed during the first stage of construction.
  ▲ At the outlet of any structure which concentrates sediment-laden runoff, e.g. at the discharge point of diversions, channels, slope drains, or other runoff conveyances.
  ▲ Above a storm water inlet that is in line to receive sediment-laden runoff.

▲ Temporary sediment traps may be constructed by excavation alone or by excavation in combination with an embankment.
▲ Temporary sediment traps are often used in conjunction with a diversion dike or swale.
▲ The drainage area for the sediment trap should not exceed 5 disturbed acres.
▲ The trap must be accessible for ease of regular maintenance which is critical to its functioning properly.
▲ Sediment traps are temporary measures and should not be planned to remain in place longer than between 18 and 24 months.
▲ The capacity of the sedimentation pool should provide storage volume for 3,600 cubic feet/acre drainage area.
▲ The outlet should be designed to provide a 2 foot settling depth and an additional sediment storage area 1½ feet deep at the bottom of the trap.
▲ The embankment may not exceed 5 feet in height.
▲ The recommended minimum width at the top of the embankment is between 2 feet and 5 feet.
▲ The minimum recommended length of the weir is between 3 feet and 4 feet, and the maximum is 12 feet in length.
▲ Table 5 illustrates the typical relationship between the embankment height, the height of the outlet (H_o), and the width (W) at the top of the embankment.

TABLE 5. EMBANKMENT HEIGHT vs. OUTLET HEIGHT AND WIDTH

H	H_o	W
1.5	0.5	2.0
2.0	1.0	2.0
2.5	1.5	2.5
3.0	2.0	2.5
3.5	2.5	3.0
4.0	3.0	3.0
4.5	3.5	4.0
5.0	4.0	4.5

Materials

▲ Filter fabric (see fabric requirement for silt fence)
▲ Coarse aggregate or riprap 2 inches to 14 inches in diameter
▲ Washed gravel ¾ to 1½ inches in diameter
▲ Seed and mulch for stabilization

TEMPORARY SEDIMENT TRAP

Construction Specifications

▲ Clear the area of all trees, brush, stumps or other obstructions.
▲ Construct the embankment in 8 inch lifts compacting each lift with the appropriate earth moving equipment. Fill material must be free of woody vegetation, roots, or large stones.
▲ Keep cut and fill slopes between 3:1 and 2:1 or flatter.
▲ Line the outlet area with filter fabric prior to placing stone or gravel.
▲ Construct the gravel outlet using heavy stones between 6 inches and 14 inches in diameter and face the upstream side with a 12 inch layer of ¾ inch to 1½ inch washed gravel on the upstream side.
▲ Seed and mulch the embankment as soon as possible to ensure stabilization.

Maintenance

▲ Inspect regularly and after every storm. Make any repairs necessary to ensure the measure is in good working order.
▲ Frequent removal of sediment is critical to the functioning of this measure. At a minimum sediment should be removed and the trap restored to its original volume when sediment reaches ⅓ of the original volume.
▲ Sediment removed from the trap must be properly disposed.
▲ Check the embankment regularly to make sure it is structurally sound.

Cost

▲ Costs for a sediment trap vary widely based upon their size and the amount of excavation and stone required, they usually can be installed for $500 to $7,000.

Sources

▲ Commonwealth of Virginia - County of Fairfax, 1987. 1987 Check List For Erosion And Sediment Control - Fairfax County, Virginia.
▲ State of North Carolina, 1988. Erosion and Sediment Control Planning and Design Manual. North Carolina Sedimentation Control Commission, Department of Natural Resources and Community Development.
▲ Maryland Department of the Environment, 1991. 1991 Maryland Standards And Specifications For Soil Erosion And Sediment Control - Draft.
▲ Storm Water Management Manual for the Puget Sound Basin. State of Washington, Department of Ecology, 1991.
▲ Cost Data:

  ▲ Draft Sediment and Erosion Control, An Inventory of Current Practices, April 20, 1990. Prepared by Kamber Engineering for the U.S. Environmental Protection Agency, Office of Water Enforcement and Permits, Washington, D.C. 20460.

APPENDIX F

TESTS FOR NON-STORM WATER DISCHARGES

TESTS FOR NON-STORM WATER DISCHARGES

DYE TESTING

Dye testing can be used to establish positively if certain facilities or fixtures are connected to a storm water collection system. The dye is simply introduced into the suspected waste stream, and storm water outfalls are examined for detections of the dye. Specially manufactured dyes are available for this type of testing. Check with your local sewer authority before conducting this test—dyes can be toxic and thus harmful to the municipal sewage treatment plant

Equipment

Two types of safe and harmless but effective dyes are available for dye testing. Powder in cans or containers is measured by a spoon or small dipper. Tablets of the dye are slower to dissolve than the powder form, but are less messy and are sometimes more desirable than the powder for this reason. The dye is the only piece of equipment needed. Regardless of the type of dye, dissolve it in the flow. A tablet may sink into a sump or wet well and not circulate with the usual flow.

CAUTION: Some dyes may leave a stain if spilled. These stains can be very difficult to remove.

Contact the water pollution control agency to determine if there are any regulations regarding the use of dyes.

Operation

While one operator applies the dye to the suspected location, another operator maintains a watch at the next downstream manhole from the location.

- Where a plumbing fixture is used, such as a water closet bowl or basin, the water is turned on and the dye powder or tablet is dropped directly into the drain.

- Where there is no immediate supply of water, such as a roof gutter or storm drain in dry weather, pouring a bucket of water with the dye powder is suggested. The amount of water and dye needed depends on the distance to the next manhole and the existing flow.

- Based on the assumed velocity of flow, an estimate may be made of the expected flow time to the downstream manhole. Allow plenty of time because the dye often takes much longer than expected.

- Use of powdered dye can be difficult and messy on a windy day. When the wind blows, either pre-mix the dye in water or enclose a quantity of the powder dye in either tissue or toilet paper. Wind can scatter a powdered dye, the dye is impossible to collect. The dye may land on the property of nearby residents and businesses, and when wet, cause stains on buildings, autos, clothes, and landscaping.

- When a number of dye tests are to be conducted on the same line or section of a sewer system, the dye testing should start at the facility farthest downstream and progressively work upstream for the other dye tests. Otherwise, if you dye the facilities upstream first, the flow is then contaminated with dye, and you then must wait several hours or until the next day to conduct additional tests.

- When tests are completed, record whether or not the service is connected to the sewer.

APPENDIX G

COMPARISON OF OTHER ENVIRONMENTAL PLANS

POTENTIALLY RELEVANT ELEMENTS OF OTHER FACILITY ENVIRONMENTAL PLANS

Required Elements of Each Plan	Storm Water Pollution Prevention Plan	Preparedness Prevention and Contingency Plan (40 CFR 264 and 265)	Spill Control and Countermeasures (40 CFR 112)	NPDES Toxic Organic Management Plan (40 CFR 413, 433, 469)	OSHA Emergency Action Plan (29 CFR 1910)
Identification of Pollutants of Concern	• Description of potential pollutant sources • Risk identification • Material inventory • Test for illicit connections	• Requires identification of hazardous wastes handled at the facility and associated hazards	• Requires prediction of direction, rate of flow and total quantity of oil that could be discharged	• Requires identification of toxic organic compounds used	• Requires list of major workplace fire and emergency hazards
Coordinator	• Pollution prevention planner or team under supervision of plant manager	• Emergency coordinator at facility or on call at all times to coordinate emergency response.	• Designated person who is accountable for oil spill prevention and who reports to line management	Not specifically addressed	Not specifically addressed
Operational Controls	• Preventive maintenance program • Good housekeeping • Spill prevention and response procedures • Site-specific storm water BMPs • Activity-specific BMPs	• Requires that personnel involved in hazardous waste activities have access to emergency communication device	• Requires appropriate spill prevention and containment procedures	• Requires method of disposal used instead of dumping into drain be specified • Procedures for assuring that toxic organics do not routinely spill or leak into wastewater	• Requires employer to control accumulations of flammable and combustible waste • Maintain equipment and systems to prevent accidental ignition of combustible materials
Structural Controls	• Sediment and erosion control • Site-specific storm water BMPs • Activity-specific BMPs • BMPs for non-storm water discharges • Enclosure of salt storage piles • *Provide containment, drainage control, and/or diversionary structures to prevent contamination of storm water discharges associated with industrial activity from facilities subject to EPCRA Section 313* • *Security for EPCRA Section 313 facilities*	• Maintain aisle space for movement of emergency equipment and personnel • Specific requirements for storage tanks	• Appropriate containment and/or diversionary structures or equipment (detailed suggestions provided in reg.) • Security - including fences and gates, locks for flow and drain valves and pumps, and lighting	• Specify method of disposal used instead of dumping into drain • Procedures for assuring that toxic organics do not routinely spill or leak into wastewater	Not specifically addressed

POTENTIALLY RELEVANT ELEMENTS OF OTHER FACILITY ENVIRONMENTAL PLANS (Continued)

Required Elements of Each Plan	Storm Water Pollution Prevention Plan	Preparedness Prevention and Contingency Plan (40 CFR 264 and 265)	Spill Control and Countermeasures (40 CFR 112)	NPDES Toxic Organic Management Plan (40 CFR 413, 433, 469)	OSHA Emergency Action Plan (29 CFR 1910)
Inspections	• Routine visual inspection of designated equipment and plant areas, including materials handling, by qualified plant personnel who will also develop procedures to ensure follow up • Annual site inspection to verify the accuracy of pollutant source description, drainage map and controls	Not specifically addressed	• Testing and inspection of pollution prevention/control equipment by owner/operator on a scheduled, periodic basis • Inspections should be in accordance with written procedures developed for the facility by the owner/operator	Not specifically addressed	Not specifically addressed
Employee Training	• Training for employee at all levels in: - spill response - good housekeeping - materials management • Specify periodic training dates in plan	Not specifically addressed	• Owners/operators are responsible for properly training personnel on applicable regulations and in the operation and maintenance of equipment to prevent discharges • Owners/operators should schedule and conduct spill prevention briefings for operating personnel	Not specifically addressed	• Designate and train a sufficient number of persons to assist in safe evacuation
Coordinate with Local Authorities	• Facilities which discharge storm water to large or medium municipal separate storm sewer systems must comply with applicable conditions in municipal storm water management programs	• Familiarize local police and fire departments, hospitals and emergency response teams - layout of facility - properties of hazardous wastes - types of injuries • Coordinate arrangements for plan implementation authorities	• Follow contingency plan provisions of 40 CFR 109 including consultation with State and local governments	Not specifically addressed	Not specifically addressed

POTENTIALLY RELEVANT ELEMENTS OF OTHER FACILITY ENVIRONMENTAL PLANS (Continued)

Required Elements of Each Plan	Storm Water Pollution Prevention Plan	Preparedness Prevention and Contingency Plan (40 CFR 264 and 265)	Spill Control and Countermeasures (40 CFR 112)	NPDES Toxic Organic Management Plan (40 CFR 413, 433, 469)	OSHA Emergency Action Plan (29 CFR 1910)
Emergency/ Spill Response Equipment	• Necessary equipment to implement a spill clean up	• List describing emergency equipment and its location: - Internal communications (intercom or alarm) - Immediately accessible line of communication to summon emergency assistance (fire/police) - fire extinguishers - water supplies - decontamination equipment - spill control equipment • All equipment must be tested and maintained	• Appropriate containment and/or diversionary structures or equipment • If impractical, a written commitment of equipment and materials required to expeditiously control and remove any harmful quantities of oil discharged	Not specifically addressed	• Alarm system

POTENTIALLY RELEVANT ELEMENTS OF OTHER FACILITY ENVIRONMENTAL PLANS (Continued)

Required Elements of Each Plan	Storm Water Pollution Prevention Plan	Preparedness Prevention and Contingency Plan (40 CFR 264 and 265)	Spill Control and Countermeasures (40 CFR 112)	NPDES Toxic Organic Management Plan (40 CFR 413, 433, 469)	OSHA Emergency Action Plan (29 CFR 1910)

Notification/ Record Keeping Procedures	• Record spills and other discharges
	• Record storm water quality/ quantity information
	• Document inspection and maintenance activities
	• Certify that discharge has been tested for the presence of non-storm water discharges or certify where such testing is not feasible
		• In case of imminent or actual emergency situation:
		  - activate alarms/ communication systems to notify facility personnel
		  - notify State/local agencies
		  - identify the character, exact source, amount and areal extent of release
		  - assess hazards to human health and the environment and respond
		  - facilitate containment
		  - coordinate clean up
		  - submit incident report
			• Written procedures for and records of inspections should be made part of the SPCC and maintained for 3 years
			• Detailed notification requirements apply if a facility has a single spill event of more than 1000 gallons of oil or has discharged oil in harmful quantities in two spill events within the last 12 months
				Not specifically addressed	• Means of reporting fires and other emergencies

Evacuation Procedures	Not specifically addressed	• Evacuation plan describing:
		  - signals to begin evacuation
		  - primary and alternate routes
			Not specifically addressed	Not specifically addressed	• Emergency escape routes
					• Procedures to account for all employees
					• Procedures for employees who remain behind to perform critical functions

POTENTIALLY RELEVANT ELEMENTS OF OTHER FACILITY ENVIRONMENTAL PLANS (Continued)

Required Elements of Each Plan	Storm Water Pollution Prevention Plan	Preparedness Prevention and Contingency Plan (40 CFR 264 and 265)	Spill Control and Countermeasures (40 CFR 112)	NPDES Toxic Organic Management Plan (40 CFR 413, 433, 469)	OSHA Emergency Action Plan (29 CFR 1910)
Plan Location/ Distribution	• Maintained at facility unless requested by the director or the municipal operator	• Maintained at facility • Submitted to local police, fire, hospital, and State and local emergency response teams	• Maintain at facility if facility is normally attended at least 8 hours per day or at nearest field office if not so attended	• Submitted to permitting authority for approval	• Plan shall be written and kept at the workplace unless there are fewer than 10 employees, then oral communication is sufficient • Employer shall review the plan with each employee covered by the plan when: - Plan is initially developed - Plan changes - Employee's responsibility changes
Modification of Plan	• Plan fails to control pollutants in storm water • Change in design, construction, operation or maintenance • Requested by the director	• Facility permit revised • Plan fails during emergency • Facility changes • Emergency coordinator(s) change • Emergency equipment changes	• By the Regional Administrator where the plan does not meet requirements or is necessary to prevent and contain discharges of oil • By the owner/operator: - change in facility - if warranted by findings of 3 years evaluation	Not specifically addressed	Not specifically addressed

POTENTIALLY RELEVANT ELEMENTS OF OTHER FACILITY ENVIRONMENTAL PLANS (Continued)

Required Elements of Each Plan	Storm Water Pollution Prevention Plan	Preparedness Prevention and Contingency Plan (40 CFR 264 and 265)	Spill Control and Countermeasures (40 CFR 112)	NPDES Toxic Organic Management Plan (40 CFR 413, 433, 469)	OSHA Emergency Action Plan (29 CFR 1910)
Certification	• Certify that discharges have been tested for the presence of non-storm water discharges • Plans must be signed and certified in accordance with 40 CFR 122.22 • Spill prevention and response plan for facilities subject to EPCRA Section 313 must be reviewed and certified every three years by a registered professional engineer	Not specifically addressed	• Plan must be reviewed and certified by a registered professional engineer	No dumping of toxic organic compounds into the wastewater has occurred and the approved TOMP is being implemented	Not specifically addressed

APPENDIX H

LIST OF HAZARDOUS SUBSTANCES AND REPORTABLE QUANTITIES

LIST OF HAZARDOUS SUBSTANCES AND REPORTABLE QUANTITIES
40 CFR 302.4 and 117
Note: All comments are located at the end of this table.

Hazardous Substance	CASRN	Regulatory Synonyms	Statutory RQ	Code†	RCRA Waste #	Category	Final RQ Pounds (Kg)
Acenaphthene	83329		1*	2		B	100 (45.4)
Acenaphthylene	208968		1*	2		D	5000 (2270)
Acetaldehyde	75070	Ethanal	1000	1,4	U001	C	1000 (454)
Acetaldehyde, chloro-	107200	Chloroacetaldehyde	1*	4	P023	C	1000 (454)
Acetaldehyde, trichloro-	75876	Chloral	1*	4	U034	D	5000 (2270)
Acetamide, N-(aminothioxomethyl)-	591082	1-Acetyl-2-thiourea	1*	4	P002	C	1000 (454)
Acetamide, N-(4-ethoxyphenyl)-	62442	Phenacetin	1*	4	U187	B	100 (45.4)
Acetamide, 2-fluoro-	640197	Fluoroacetamide	1*	4	P057	B	100 (45.4)
Acetamide, N-9H-fluoren-2-yl-	53963	2-Acetylaminofluorene	1*	4	U005	X	1 (0.454)
Acetic acid	64197		1000	1		D	5000 (2270)
Acetic acid (2,4-dichlorophenoxy)-	94757	2,4-D Acid 2,4-D, salts and esters	100	1,4	U240	B	100 (45.4)
Acetic Acid, lead(2+) salt	301042	Lead acetate	5000	1,4	U144		#
Acetic acid, thallium(1+) salt	563688	Thallium(I) acetate	1*	4	U214	B	100 (45.4)
Acetic acid (2,4,5-trichlorophenoxy)-	93765	2,4,5-T 2,4,5-T acid	100	1,4	U232	C	1000 (454)
Acetic acid, ethyl ester	141786	Ethyl acetate	1*	4	U112	D	5000 (2270)
Acetic acid, fluoro-, sodium salt	62748	Fluoroacetic acid, sodium salt	1*	4	P058	A	10 (4.54)
Acetic anhydride	108247		1000	1		D	5000 (2270)
Acetone	67641	2-Propanone	1*	4	U002	D	5000 (2270)
Acetone cyanohydrin	75865	Propanenitrile, 2-hydroxy-2-methyl-2-Methyllactonitrile	10	1,4	P069	A	10 (4.54)
Acetonitrile	75058		1*	4	U003	D	5000 (2270)
Acetophenone	98862	Ethanone, 1-phenyl-	1*	4	U004	D	5000 (2270)
2-Acetylaminofluorene	53963	Acetamide, N-9H-fluoren-2-yl-	1*	4	U005	X	1 (0.454)
Acetyl bromide	506967		5000	1		D	5000 (2270)
Acetyl chloride	75365		5000	1,4	U006	D	5000 (2270)
1-Acetyl-2-thiourea	591082	Acetamide, N-(aminothioxomethyl)-	1*	4	P002	C	1000 (454)
Acrolein	107028	2-Propenal	1	1,2,4	P003	X	1 (0.454)
Acrylamide	79061	2-Propenamide	1*	4	U007	D	5000 (2270)

Hazardous Substance	CASRN	Regulatory Synonyms	Statutory			Final RQ	
			RQ	Code†	RCRA Waste #	Category	Pounds (Kg)
Acrylic acid	79107	2-Propenoic acid	1*	4	U008	D	5000 (2270)
Acrylonitrile	107131	2-Propenenitrile	100	1,2,4	U009	B	100 (45.4)
Adipic acid	124049		5000	1		D	5000 (2270)
Aldicarb	116063	Propanal, 2-methyl-2-(methylthio)-,O-((methylamino) carbonyl)oxime	1*	4	P070	X	1 (0.454)
Aldrin	309002	1,4,5,8-Dimethanonaphthalene, 1,2,3,4,10,10-10-hexachloro-1,4,4a,5,8,8a-hexahydro-, (1alpha,4alpha,4abeta,5alpha, 8alpha,8abeta)-	1	1,2,4	P004	X	1 (0.454)
Allyl alcohol	107186	2-Propen-1-ol	100	1,4	P005	B	100 (45.4)
Allyl chloride	107051		1000	1		C	1000 (454)
Aluminum phosphide	20859738		1*	4	P006	B	100 (45.4)
Aluminum sulfate	10043013		5000	1		D	5000 (2270)
5-(Aminomethyl)-3-isoxazolol	2763964	Muscimol 3(2H)-Isoxazolone, 5-(aminomethyl)-	1*	4	P007	C	1000 (454)
4-Aminopyridine	504245	4-Pyridinamine	1*	4	P008	C	1000 (454)
Amitrole	61825	1H-1,2,4-Triazol-3-amine	1*	4	U011	A	10 (4.54)
Ammonia	7664417		100	1		B	100 (45.4)
Ammonium acetate	631618		5000	1		D	5000 (2270)
Ammonium benzoate	1863634		5000	1		D	5000 (2270)
Ammonium bicarbonate	1066337		5000	1		D	5000 (2270)
Ammonium bichromate	7789095		1000	1		A	10 (4.54)
Ammonium bifluoride	1341497		5000	1		B	100 (45.4)
Ammonium bisulfite	10192300		5000	1		D	5000 (2270)
Ammonium carbamate	1111780		5000	1		D	5000 (2270)
Ammonium carbonate	506876		5000	1		D	5000 (2270)
Ammonium chloride	12125029		5000	1		D	5000 (2270)
Ammonium chromate	7788989		1000	1		A	10 (4.54)
Ammonium citrate, dibasic	3012655		5000	1		D	5000 (2270)
Ammonium fluoborate	13826830		5000	1		D	5000 (2270)
Ammonium fluoride	12125018		5000	1		B	100 (45.4)
Ammonium hydroxide	1336216		1000	1		C	1000 (454)
Ammonium oxalate	6009707		5000	1		D	5000 (2270)

| Hazardous Substance | CASRN | Regulatory Synonyms | Statutory | | | Final RQ | |
			RQ	Code†	RCRA Waste #	Category	Pounds (Kg)
	5972736		5000	1		D	5000 (2270)
	14258492		5000	1		D	5000 (2270)
Ammonium picrate	131748	Phenol, 2,4,6-trinitro-, ammonium salt	1*	4	P009	A	10 (4.54)
Ammonium silicofluoride	16919190		1000	1		C	1000 (454)
Ammonium sulfamate	7773060		5000	1		D	5000 (2270)
Ammonium sulfide	12135761		5000	1		B	100 (45.4)
Ammonium sulfite	10196040		5000	1		D	5000 (2270)
Ammonium tartrate	14307438		5000	1		D	5000 (2270)
	3164292		5000	1		D	5000 (2270)
Ammonium thiocyanate	1762954		5000	1		D	5000 (2270)
Ammonium vanadate	7803556	Vanadic acid, ammonium salt	1*	4	P119	C	1000 (454)
Amyl acetate	628637		1000	1		D	5000 (2270)
iso-Amyl acetate	123922		1000	1		D	5000 (2270)
sec-Amyl acetate	626380		1000	1		D	5000 (2270)
tert-Amyl acetate	625161		1000	1		D	5000 (2270)
Aniline	62533	Benzenamine	1000	1,4	U012	D	5000 (2270)
Anthracene	120127		1*	2		D	5000 (2270)
Antimony††	7440360		1*	2		D	5000 (2270)
ANTIMONY AND COMPOUNDS	N/A		1*	2			**
Antimony pentachloride	7647189		1000	1		C	1000 (454)
Antimony potassium tartrate	28300745		1000	1		B	100 (45.4)
Antimony tribromide	7789619		1000	1		C	1000 (454)
Antimony trichloride	10025919		1000	1		C	1000 (454)
Antimony trifluoride	7783564		1000	1		C	1000 (454)
Antimony trioxide	1309644		5000	1		C	1000 (454)
Argentate(1-), bis(cyano-C)-, potassium	506616	Potassium silver cyanide	1*	4	P099	X	1 (0.454)
Aroclor 1016	12674112	POLYCHLORINATED BIPHENYLS (PCBs)	10	1,2		X	1 (0.454)
Aroclor 1221	11104282	POLYCHLORINATED BIPHENYLS (PCBs)	10	1,2		X	1 (0.454)
Aroclor 1232	11141165	POLYCHLORINATED BIPHENYLS (PCBs)	10	1,2		X	1 (0.454)

Hazardous Substance	CASRN	Regulatory Synonyms	Statutory			Final RQ	
			RQ	Code†	RCRA Waste #	Cate-gory	Pounds (Kg)
Aroclor 1242	53469219	POLYCHLORINATED BIPHENYLS (PCBs)	10	1,2		X	1 (0.454)
Aroclor 1248	12672296	POLYCHLORINATED BIPHENYLS (PCBs)	10	1,2		X	1 (0.454)
Aroclor 1254	11097691	POLYCHLORINATED BIPHENYLS (PCBs)	10	1,2		X	1 (0.454)
Aroclor 1260	11096825	POLYCHLORINATED BIPHENYLS (PCBs)	10	1,2		X	1 (0.454)
Arsenic††	7440382		1*	2,3		X	1 (0.454)
Arsenic acid	1327522	Arsenic acid H3AsO4	1*	4	P010	X	1 (0.454)
	7778394						
Arsenic acid H3AsO4	1327522	Arsenic acid	1*	4	P010	X	1 (0.454)
	7778394		1*	4	P010	X	1(0.454)
ARSENIC AND COMPOUNDS	N/A		1*	2			**
Arsenic disulfide	1303328		5000	1		X	1 (0.454)
Arsenic oxide As2O3	1327533	Arsenic trioxide	5000	1,4	P012	X	1 (0.454)
Arsenic oxide As2O5	1303282	Arsenic pentoxide	5000	1,4	P011	X	1 (0.454)
Arsenic pentoxide	1303282	Arsenic oxide As2O5	5000	1,4	P011	X	1 (0.454)
Arsenic trichloride	7784341		5000	1		X	1 (0.454)
Arsenic trioxide	1327533	Arsenic oxide As2O3	5000	1,4	P012	X	1 (0.454)
Arsenic trisulfide	1303339		5000	1		X	1 (0.454)
Arsine, diethyl-	692422	Diethylarsine	1*	4	P038	X	1 (0.454)
Arsinic acid, dimethyl-	75605	Cacodylic acid	1*	4	U136	X	1 (0.454)
Arsonous dichloride, phenyl-	696286	Dichlorophenylarsine	1*	4	P036	X	1 (0.454)
Asbestos†††	1332214		1*	2,3		X	1 (0.454)
Auramine	492808	Benzenamine, 4,4'-carbonimidoylbis (N,N-dimethyl-	1*	4	U014	B	100 (45.4)
Azaserine	115026	L-Serine, diazoacetate (ester)	1*	4	U015	X	1 (0.454)
Aziridine	151564	Ethylenimine	1*	4	P054	X	1 (0.454)
Aziridine, 2-methyl-	75558	1,2-Propylenimine	1*	4	P067	X	1 (0.454)
Azirino[2',3':3,4]pyrrolo[1,2-a]indole-4,7-dione, 6-amino-8-[[(aminocarbonyloxy]methyl]-1,1a,2,8,8a,8b-hexahydro-8a-methoxy-5-methyl-,[1aS-(1aalpha,8beta,8aalpha,8balpha)]-	50077	Mitomycin C	1*	4	U010	A	10 (4.54)

Hazardous Substance	CASRN	Regulatory Synonyms	Statutory			Final RQ	
			RQ	Code†	RCRA Waste #	Category	Pounds (Kg)
Barium cyanide	542621		10	1,4	P013	A	10 (4.54)
Benz[j]aceanthrylene, 1,2-dihydro-3-methyl-	56495	3-Methylcholanthrene	1*	4	U157	A	10 (4.54)
Benz(c)acridine	225514		1*	4	U016	B	100 (45.4)
Benzal chloride	98873	Benzene, dichloromethyl-	1*	4	U017	D	5000 (2270)
Benzamide, 3,5-dichloro-N-(1,1-dimethyl-2-propynyl)-	23950585	Pronamide	1*	4	U192	D	5000 (2270)
Benz[a]anthracene	56553	Benzo(a)anthracene 1,2-Benzanthracene	1*	2,4	U018	A	10 (4.54)
1,2-Benzanthracene	56553	Benz[a]anthracene Benzo[a]anthracene	1*	2,4	U018	A	10 (4.54)
Benz[a]anthracene, 7,12-dimethyl-	57976	7,12-Dimethylbenz[a]anthracene	1*	4	U094	X	1 (0.454)
Benzenamine	62533	Aniline	1000	1,4	U012	D	5000 (2270)
Benzenamine, 4,4'-carbonimidoylbis (N,N-dimethyl-	492808	Auramine	1*	4	U014	B	100 (45.4)
Benzenamine, 4-chloro-	106478	p-Chloroaniline	1*	4	P024	C	1000 (454)
Benzenamine, 4-chloro-2-methyl-, hydrochloride	3165933	4-Chloro-o-toluidine, hydrochloride	1*	4	U049	B	100 (45.4)
Benzenamine, N,N-dimethyl-4(phenylazo-)	60117	p-Dimethylaminoazobenzene	1*	4	U093	A	10 (4.54)
Benzenamine, 2-methyl-	95534	o-Toluidine	1*	4	U328	B	100 (45.4)
Benzenamine, 4-methyl-	106490	p-Toluidine	1*	4	U353	B	100 (45.4)
Benzenamine, 4,4'-methylenebis(2-chloro-	101144	4,4'-Methylenebis(2-chloroaniline)	1*	4	U158	A	10 (4.54)
Benzenamine, 2-methyl-, hydrochloride	636215	o-Toluidine hydrochloride	1*	4	U222	B	100 (45.4)
Benzenamine, 2-methyl-5-nitro	99558	5-Nitro-o-toluidine	1*	4	U181	B	100 (45.4)
Benzenamine, 4-nitro-	100016	p-Nitroaniline	1*	4	P077	D	5000 (2270)
Benzene	71432		1000	1,2,3,4	U109	A	10 (4.54)
Benzeneacetic acid, 4-chloro-alpha-(4-chlorophenyl)-alpha-hydroxy-, ethyl ester	510156	Chlorobenzilate	1*	4	U038	A	10 (4.54)
Benzene, 1-bromo-4-phenoxy-	101553	4-Bromophenyl phenyl ether	1*	2,4	U030	B	100 (45.4)
Benzenebutanoic acid, 4-[bis(2-chloroethyl)amino]-	305033	Chlorambucil	1*	4	U035	A	10 (4.54)
Benzene, chloro-	108907	Chlorobenzene	100	1,2,4	U037	B	100 (45.4)
Benzene, chloromethyl-	100447	Benzyl chloride	100	1,4	P028	B	100 (45.4)

| Hazardous Substance | CASRN | Regulatory Synonyms | Statutory | | | Final RQ | |
			RQ	Code†	RCRA Waste #	Cate-gory	Pounds (Kg)
Benzenediamin, ar-methyl-	95807	Toluenediamine	1*	4	U221	A	10 (4.54)
	496720		1*	4	U221	A	10 (4.54)
	823405		1*	4	U221	A	10 (4.54)
1,2-Benzenedicarboxylic acid, dioctyl ester	117840	Di-n-octyl phthalate	1*	2,4	U107	D	5000 (2270)
1,2-Benzenedicarboxylic acid, [bis(2-ethylhexyl)]-ester	117817	Bis (2-ethylhexyl)phthalate Diethylhexyl phthalate	1*	2,4	U028	B	100 (45.4)
1,2-Benzenedicarboxylic acid, dibutyl ester	84742	Di-n-butyl phthalate Dibutyl phthalate n-Butyl phthalate	100	1,2,4	U069	A	10 (4.54)
1,2-Benzenedicarboxylic acid, diethyl ester	84662	Diethyl phthalate	1*	2,4	U088	C	1000 (454)
1,2-Benzenedicarboxylic acid, dimethyl ester	131113	Dimethyl phthalate	1*	2,4	U102	D	5000 (2270)
Benzene, 1,2-dichloro-	95501	o-Dichlorobenzene 1,2-Dichlorobenzene	100	1,2,4	U070	B	100 (45.4)
Benzene, 1,3-dichloro-	541731	m-Dichlorobenzene 1,3-Dichlorobenzene	1*	2,4	U071	B	100 (45.4)
Benzene, 1,4-dichloro-	106467	p-Dichlorobenzene 1,4-Dichlorobenzene	100	1,2,4	U072	B	100 (45.4)
Benzene, 1,1'-(2,2-dichloroethylidene)bis[4-chloro-	72548	DDD TDE 4,4' DDD	1	1,2,4	U060	X	1 (0.454)
Benzene, dichloromethyl-	98873	Benzal chloride	1*	4	U017	D	5000 (2270)
Benzene, 1,3-diisocyanatomethyl-	584849	Toluene diisocyanate	1*	4	U223	B	100 (45.4)
	91087		1*	4	U223	B	100 (45.4)
	26471625		1*	4	U223	B	100 (45.4)
Benzene, dimethyl	1330207	Xylene (mixed)	1000	1,4	U239	C	1000 (454)
m-Benzene, dimethyl	108383	m-Xylene	1000	1,4	U239	C	1000 (454)
o-Benzene, dimethyl	95476	o-Xylene	1000	1,4	U239	C	1000 (454)
p-Benzene, dimethyl	106423	p-Xylene	1000	1,4	U239	C	1000 (454)
1,3-Benzenediol	108463	Resorcinol	1000	1,4	U201	D	5000 (2270)
1,2-Benzenediol,4-[1-hydroxy-2-(methylamino)ethyl]-	51434	Epinephrine	1*	4	P042	C	1000 (454)
Benzeneethanamine, alpha,alpha-dimethyl-	122098	alpha,alpha-Dimethylphenethylamine	1*	4	P046	D	5000 (2270)
Benzene, hexachloro-	118741	Hexachlorobenzene	1*	2,4	U127	A	10 (4.54)
Benzene, hexahydro-	110827	Cyclohexane	1000	1,4	U056	C	1000 (454)

Hazardous Substance	CASRN	Regulatory Synonyms	Statutory			Final RQ	
			RQ	Code †	RCRA Waste #	Category	Pounds (Kg)
Benzene, hydroxy-	108952	Phenol	1000	1,2,4	U188	C	1000 (454)
Benzene, methyl-	108883	Toluene	1000	1,2,4	U220	C	1000 (454)
Benzene, 2-methyl-1,3-dinitro-	606202	2,6-Dinitrotoluene	1000	1,2,4	U106	B	100 (45.4)
Benzene, 1-methyl-2,4-dinitro-	121142	2,4-Dinitrotoluene	1000	1,2,4	U105	A	10 (4.54)
Benzene, 1-methylethyl-	98828	Cumene	1*	4	U055	D	5000 (2270)
Benzene, nitro-	98953	Nitrobenzene	1000	1,2,4	U169	C	1000 (454)
Benzene, pentachloro-	608935	Pentachlorobenzene	1*	4	U183	A	10 (4.54)
Benzene, pentachloronitro-	82688	Pentachloronitrobenzene (PCNB)	1*	4	U185	B	100 (45.4)
Benzenesulfonic acid chloride	98099	Benzenesulfonyl chloride	1*	4	U020	B	100 (45.4)
Benzenesulfonyl chloride	98099	Benzenesulfonic acid chloride	1*	4	U020	B	100 (45.4)
Benzene, 1,2,4,5-tetrachloro-	95943	1,2,4,5-Tetrachlorobenzene	1*	4	U207	D	5000 (2270)
Benzenethiol	108985	Thiophenol	1*	4	P014	B	100 (45.4)
Benzene, 1,1'-(2,2,2-trichloroethylidene)bis(4-chloro-	50293	DDT 4,4'DDT	1	1,2,4	U061	X	1 (0.454)
Benzene, 1,1'-(trichloroethylidene) bis(4-methoxy-	72435	Methoxychlor	1	1,4	U247	X	1 (0.454)
Benzene, (trichloromethyl)-	98077	Benzotrichloride	1*	4	U023	A	10 (4.54)
Benzene, 1,3,5-trinitro-	99354	1,3,5-Trinitrobenzene	1*	4	U234	A	10 (4.554)
Benzidine	92875	(1,1'-Biphenyl)-4,4'diamine	1*	2,4	U021	X	1 (0.454)
1,2-Benzisothiazol-3(2H)-one, 1,1-dioxide	81072	Saccharin and salts	1*	4	U202	B	100 (45.4)
Benzo[a]anthracene	56553	Benz[a]anthracene 1,2-Benzanthracene	1*	2,4	U018	A	10 (4.54)
Benzo(b)fluoranthene	205992		1*	2		X	1 (0.454)
Benzo(k)fluoranthene	207089		1*	2		D	5000 (2270)
Benzo(j,k)fluorene	206440	Fluoranthene	1*	2,4	U120	B	100 (45.4)
1,3-Benzodioxole, 5-(1-propenyl)-	120581	Isosafrole	1*	4	U141	B	100 (45.4)
1,3-Benzodioxole, 5-(2-propenyl)-	94597	Safrole	1*	4	U203	B	100 (45.4)
1,3-Benzodioxole, 5-propyl-	94586	Dihydrosafrole	1*	4	U090	A	10 (4.54)
Benzoic acid	65850		5000	1		D	5000 (2270)
Benzonitrile	100470		1000	1		D	5000 (2270)
Benzo[rst]pentaphene	189559	Dibenz(a,i)pyrene	1*	4	U064	A	10 (4.54)
Benzo(ghi)perylene	191242		1*	2		D	5000 (2270)

Hazardous Substance	CASRN	Regulatory Synonyms	Statutory			Final RQ	
			RQ	Code†	RCRA Waste #	Category	Pounds (Kg)
2H-1 Benzopyran-2-one, 4-hydroxy-3-(3-oxo-1-phenyl-butyl)- & salts, when present at concentrations greater than 0.3%	81812	Warfarin, & salts, when present at concentrations greater than 0.3%	1*	4	P001	B	100 (45.4)
Benzo[a]pyrene	50328	3,4-Benzopyrene	1*	2,4	U022	X	1 (0.454)
3,4-Benzopyrene	50328	Benzo[a]pyrene	1*	2,4	U022	X	1 (0.454)
p-Benzoquinone	106514	2,5-Cyclohexadiene-1,4-dione	1*	4	U197	A	10 (4.54)
Benzotrichloride	98077	Benzene, (trichloromethyl)-	1*	4	U023	A	10 (4.54)
Benzoyl chloride	98884		1000	1		C	1000 (454)
1,2-Benzphenanthrene	218019	Chrysene	1*	2,4	U050	B	100 (45.4)
Benzyl chloride	100447	Benzene, chloromethyl-	100	1,4	P028	B	100 (45.4)
Beryllium ††	7440417	Beryllium dust ††	1*	2,3,4	P015	A	10 (4.54)
BERYLLIUM AND COMPOUNDS	N/A		1*	2			**
Beryllium chloride	7787475		5000	1		X	1 (0.454)
Beryllium dust ††	7440417	Beryllium ††	1*	2,3,4	P015	A	10 (4.54)
Beryllium fluoride	7787497		5000	1		X	1 (0.454)
Beryllium nitrate	13597994		5000	1		X	1 (0.454)
	7787555		5000	1		X	1 (0.454)
alpha-BHC	319846		1*	2		A	10 (4.54)
beta-BHC	319857		1*	2		X	1 (0.454)
delta-BHC	319868		1*	2		X	1 (0.454)
gamma-BHC	58899	Cyclohexane, 1,2,3,4,5,6-hexachloro-,(1alpha,2alpha,3beta, 4alpha,5alpha,6 beta)- Hexachlorocyclohexane (gamma isomer) Lindane	1	1,2,4	U129	X	1 (0.454)
2,2'-Bioxirane	1464535	1,2:3,4-Diepoxybutane	1*	4	U085	A	10 (4.54)
(1,1'-Biphenyl)-4,4'diamine	92875	Benzidine	1*	2,4	U021	X	1 (0.454)
[1,1'-Biphenyl]-4,4'diamine,3,3'dichloro-	91941	3,3'-Dichlorobenzidine	1*	2,4	U073	X	1 (0.454)
[1,1'-Biphenyl]-4,4'diamine,3,3'dimethoxy-	119904	3,3'-Dimethoxybenzidine	1*	4	U091	B	100 (45.4)
[1,1'-Biphenyl]-4,4'-diamine,3,3'-dimethyl-	119937	3,3'-Dimethylbenzidine	1*	4	U095	A	10 (4.54)

Hazardous Substance	CASRN	Regulatory Synonyms	Statutory			Final RQ	
			RQ	Code †	RCRA Waste #	Category	Pounds (Kg)
Bis (2-chloroethyl) ether	111444	Dichloroethyl ether Ethane, 1,1'-oxybis(2-chloro-	1*	2,4	U025	A	10 (4.54)
Bis(2-chloroethoxy) methane	111911	Dichloromethoxy ethane Ethane, 1,1'-(methylenebis(oxy)) bis(2-chloro-	1*	2,4	U024	C	1000 (454)
Bis (2-ethylhexyl)phthalate	117817	Diethylhexyl phthalate 1,2-Benzenedicarboxylic acid, [bis(2-ethylhexyl)] ester	1*	2,4	U028	B	100 (45.4)
Bromoacetone	598312	2-Propanone, 1-bromo-	1*	4	P017	C	1000 (454)
Bromoform	75252	Methane, tribromo-	1*	2,4	U225	B	100 (45.4)
4-Bromophenyl phenyl ether	101553	Benzene, 1-bromo-4-phenoxy-	1*	2,4	U030	B	100 (45.4)
Brucine	357573	Strychnidin-10-one, 2,3- dimethoxy-	1*	4	P018	B	100 (45.4)
1,3-Butadiene, 1,1,2,3,4,4- hexachloro-	87683	Hexachlorobutadiene	1*	2,4	U128	X	1 (0.454)
1-Butanamine, N-butyl-N-nitroso-	924163	N-Nitrosodi-n-butylamine	1*	4	U172	A	10 (4.54)
1-Butanol	71363	n-Butyl alcohol	1*	4	U031	D	5000 (2270)
2-Butanone	78933	Methyl ethyl ketone (MEK)	1*	4	U159	D	5000 (2270)
2-Butanone peroxide	1338234	Methyl ethyl ketone peroxide	1*	4	U160	A	10 (4.54)
2 Butanone, 3,3-dimethyl-1- (methylthio)-, O[(methylamino) carbonyl] oxime.	39196184	Thiofanox	1*	4	P045	B	100 (45.4)
2-Butenal	123739	Crotonaldehyde	100	1,4	U053	B	100 (45.4)
	4170303						
2-Butene, 1,4-dichloro-	764410	1,4-Dichloro-2-butene	1*	4	U074	X	1 (0.454)
2-Butenoic acid, 2-methyl, 7[(2,3- dihydroxy-2-(1-methoxyethyl)-3- methyl-1-oxobutoxy]methyl]- 2,3,5,7a-tetrahydro-1H-pyrrolizin- 1-ylester, (1S-[1alpha(Z), 7(2S*,3R*),7aalpha])-	303344	Lasiocarpine	1*	4	U143	A	10 (4.54)
Butyl acetate	123864		5000	1		D	5000 (2270)
iso-Butyl acetate	110190		5000	1		D	5000 (2270)
sec-Butyl acetate	105464		5000	1		D	5000 (2270)
tert-Butyl acetate	540885		5000	1		D	5000 (2270)
n-Butyl alcohol	71363	1-Butanol	1*	4	U031	D	5000 (2270)
Butylamine	109739		1000	1		C	1000 (454)
iso-Butylamine	78819		1000	1		C	1000 (454)

Hazardous Substance	CASRN	Regulatory Synonyms	Statutory			Final RQ	
			RQ	Code†	RCRA Waste #	Category	Pounds (Kg)
sec-Butylamine	513495		1000	1		C	1000 (454)
	13952846		1000	1		C	1000 (454)
tert-Butylamine	75649		1000	1		C	1000 (454)
Butyl benzyl phthalate	85687		1*	2		B	100 (45.4)
n-Butyl phthalate	84742	Di-n-butyl phthalate Dibutyl phthalate 1,2-Benzenedicarboxylic acid, dibutyl ester	100	1,2,4	U069	A	10 (4.54)
Butyric acid	107926		5000	1		D	5000 (2270)
iso-Butyric acid	79312						
Cacodylic acid	75605	Arsinic acid, dimethyl-	1*	4	U136	X	1 (0.454)
Cadmium††	7440439		1*	2		A	10 (4.54)
Cadmium acetate	543908		100	1		A	10 (4.54)
CADMIUM AND COMPOUNDS	N/A		1*	2			**
Cadmium bromide	7789426		100	1		A	10 (4.54)
Cadmium chloride	10108642		100	1		A	10 (4.54)
Calcium arsenate	7778441		1000	1		X	1 (0.454)
Calcium arsenite	52740166		1000	1		X	1 (0.454)
Calcium carbide	75207		5000	1		A	10 (4.54)
Calcium chromate	13765190	Chromic acid H2CrO4, calcium salt	1000	1,4	U032	A	10 (4.54)
Calcium cyanide	592018	Calcium cyanide Ca(CN)2	10	1,4	P021	A	10 (4.54)
Calcium cyanide Ca(CN)2	592018	Calcium cyanide	10	1,4	P021	A	10 (4.54)
Calcium dodecylbenzenesulfonate	26264062		1000	1		C	1000 (454)
Calcium hypochlorite	7778543		100	1		A	10 (4.54)
Camphene, octachloro-	8001352	Toxaphene	1	1,2,4	P123	X	1 (0.454)
Captan	133062		10	1		A	10 (4.54)
Carbamic acid, ethyl ester	51796	Ethyl carbamate (urethane)	1*	4	U238	B	100 (45.4)
Carbamic acid, methylnitroso-, ethyl ester	615532	N-Nitroso-N-methylurethane	1*	4	U178	X	1 (0.454)
Carbamic chloride, dimethyl-	79447	Dimethylcarbamoyl chloride	1*	4	U097	X	1 (0.454)
Carbamodithioic acid, 1,2-ethanediylbis, salts & esters	111546	Ethylenebisdithiocarbamic acid, salts & esters	1*	4	U114	D	5000 (2270)
Carbamothioic acid, bis(1-methylethyl)-, S-(2,3-dich-loro-2-propenyl) ester	2303164	Diallate	1*	4	U062	B	100 (45.4)

Hazardous Substance	CASRN	Regulatory Synonyms	Statutory			Final RQ	
			RQ	Code†	RCRA Waste #	Category	Pounds (Kg)
Carbaryl	63252		100	1		B	100 (45.4)
Carbofuran	1563662		10	1		A	10 (4.54)
Carbon disulfide	75150		5000	1,4	P022	B	100 (45.4)
Carbon oxyfluoride	353504	Carbonic difluoride	1*	4	U033	C	1000 (454)
Carbon tetrachloride	56235	Methane, tetrachloro-	5000	1,2,4	U211	A	10 (4.54)
Carbonic acid, dithallium(1+) salt	653739	Thallium(I) carbonate	1*	4	U215	B	100 (45.4)
Carbonic dichloride	75445	Phosgene	5000	1,4	P095	A	10 (4.54)
Carbonic difluoride	353504	Carbon oxyfluoride	1*	4	U033	C	1000 (454)
Carbonochloridic acid, methyl ester	79221	Methyl chlorocarbonate Methyl chloroformate	1*	4	U156	C	1000 (454)
Chloral	75876	Acetaldehyde, trichloro-	1*	4	U034	D	5000 (2270)
Chlorambucil	305033	Benzenebutanoic acid, 4-[bis(2-chloroethyl)amino]-	1*	4	U035	A	10 (4.54)
Chlordane	57749	Chlordane, alpha & gamma isomers Chlordane, technical 4,7-Methano-1H-indene, 1,2,4,5,6,7,8,8-octachloro-2,3,3a,4,7,7a-hexahydro-	1	1,2,4	U036	X	1 (0.454)
CHLORDANE (TECHNICAL MIXTURE AND METABOLITES)	N/A		1*	2			**
Chlordane, alpha & gamma isomers	57749	Chlordane Chlordane, technical 4,7-Methano-1H-indene, 1,2,4,5,6,7,8,8-octachloro-2,3,3a,4,7,7a-hexahydro-	1	1,2,4	U036	X	1 (0.454)
Chlordane, technical	57749	Chlordane Chlordane, alpha & gamma isomers 4,7-Methano-1H-indene, 1,2,4,5,6,7,8,8-octachloro-2,3,3a,4,7,7a-hexahyrdo-	1	1,2,4	U036	X	1 (0.454)
CHLORINATED BENZENES	N/A		1*	2			**
CHLORINATED ETHANES	N/A		1*	2			**
CHLORINATED NAPHTHALENE	N/A		1*	2			**
CHLORINATED PHENOLS	N/A		1*	2			**
Chlorine	7782505		10	1		A	10 (4.54)
Chlornaphazine	494031	Naphthalenamine, N,N'-bis(2-chloroethyl)-	1*	4	U026	B	100 (45.4)
Chloroacetaldehyde	107200	Acetaldehyde, chloro-	1*	4	P023	C	1000 (454)

Hazardous Substance	CASRN	Regulatory Synonyms	Statutory			Final RQ	
			RQ	Code†	RCRA Waste #	Cate-gory	Pounds (Kg)
CHLOROALKYL ETHERS	N/A		1*	2			**
p-Chloroaniline	106478	Benzenamine, 4-chloro-	1*	4	P024	C	1000 (454)
Chlorobenzene	108907	Benzene, chloro-	100	1,2,4	U037	B	100 (45.4)
Chlorobenzilate	510156	Benzeneacetic acid, 4-chloro-alpha-(4-chloro-phenyl)-alpha-hydroxy-, ethyl ester	1*	4	U038	A	10 (4.54)
4-Chloro-m-cresol	59507	p-Chloro-m-cresol Phenol, 4-chloro-3-methyl	1*	2,4	U039	D	5000 (2270)
p-Chloro-m-cresol	59507	Phenol, 4-chloro-3-methyl-4-Chloro-m-cresol	1*	2,4	U039	D	5000 (2270)
Chlorodibromomethane	124481		1*	2		B	100 (45.4)
Chloroethane	75003		1*	2		B	100 (45.4)
2-Chloroethyl vinyl ether	110758	Ethane, 2-chloroethoxy-	1*	2,4	U042	C	1000 (454)
Chloroform	67663	Methane, trichloro-	5000	1,2,4	U044	A	10 (4.54)
Chloromethyl methyl ether	107302	Methane, chloromethoxy-	1*	4	U046	A	10 (4.54)
beta-Chloronaphthalene	91587	Naphthalene, 2-chloro-2-Chloronaphthalene	1*	2,4	U047	D	5000 (2270)
2-Chloronaphthalene	91587	beta-Chloronaphthalene Naphthalene, 2-chloro-	1*	2,4	U047	D	5000 (2270)
2-Chlorophenol	95578	o-Chlorophenol Phenol, 2-chloro-	1*	2,4	U048	B	100 (45.4)
o-Chlorophenol	95578	Phenol, 2-chloro-2-Chlorophenol	1*	2,4	U048	B	100 (45.4)
4-Chlorophenyl phenyl ether	7005723		1*	2		D	5000 (2270)
1-(o-Chlorophenyl)thiourea	5344821	Thiourea, (2-chlorophenyl)-	1*	4	P026	B	100 (45.4)
3-Chloropropionitrile	542767	Propanenitrile, 3-chloro-	1*	4	P027	C	1000 (454)
Chlorosulfonic acid	7790945		1000	1		C	1000 (454)
4-Chloro-o-toluidine, hydrochloride	3165933	Benzenamine, 4-chloro-2-methyl-, hydrochloride	1*	4	U049	B	100 (45.4)
Chlorpyrifos	2921882		1	1		X	1 (0.454)
Chromic acetate	1066304		1000	1		C	1000 (454)
Chromic acid	11115745		1000	1		A	10 (4.54)
	7738945		1000	1		A	10 (4.54)
Chromic acid H2CrO4, calcium salt	13765190	Calcium chromate	1000	1,4	U032	A	10 (4.54)
Chromic sulfate	10101538		1000	1		C	1000 (454)
Chromium††	7440473		1*	2		D	5000 (2270)

Hazardous Substance	CASRN	Regulatory Synonyms	Statutory			Final RQ	
			RQ	Code†	RCRA Waste #	Category	Pounds (Kg)
CHROMIUM AND COMPOUNDS	N/A		1*	2			**
Chromous chloride	10049055		1000	1		C	1000 (454)
Chrysene	218019	1,2-Benzphenanthrene	1*	2,4	U050	B	100 (45.4)
Cobaltous bromide	7789437		1000	1		C	1000 (454)
Cobaltous formate	544183		1000	1		C	1000 (454)
Cobaltous sulfamate	14017415		1000	1		C	1000 (454)
Coke Oven Emissions	N/A		1*	3		X	1 (0.454)
Copper cyanide CuCN	544923	Copper cyanide	1*	4	P029	A	10 (4.54)
Copper††	7440508		1*	2		D	5000 (2270)
COPPER AND COMPOUNDS	N/A		1*	2			**
Copper cyanide	544923	Copper cyanide CuCN	1*	4	P029	A	10 (4.54)
Coumaphos	56724		10	1		A	10 (4.54)
Creosote	8001589		1*	4	U051	X	1 (0.454)
Cresol(s)	1319773	Cresylic acid Phenol, methyl-	1000	1,4	U052	C	1000 (454)
m-Cresol	108394	m-Cresylic acid	1000	1,4	U052	C	1000 (454)
o-Cresol	95487	o-Cresylic acid	1000	1,4	U052	C	1000 (454)
p-Cresol	106445	p-Cresylic acid	1000	1,4	U052	C	1000 (454)
Cresylic acid	1319773	Cresol(s) Phenol, methyl-	1000	1,4	U052	C	1000 (454)
m-Cresol	108394	m-Cresylic acid	1000	1,4	U052	C	1000 (454)
o-Cresol	95487	o-Cresylic acid	1000	1,4	U052	C	1000 (454)
p-Cresol	106445	p-Cresylic acid	1000	1,4	U052	C	1000 (454)
Crotonaldehyde	123739	2-Butenal	100	1,4	U053	B	100 (45.4)
	4170303						
Cumene	98828	Benzene, 1-methylethyl-	1*	4	U055	D	5000 (2270)
Cupric acetate	142712		100	1		B	100 (45.4)
Cupric acetoarsenite	12002038		100	1		X	1 (0.454)
Cupric chloride	7447394		10	1		A	10 (4.54)
Cupric nitrate	3251238		100	1		B	100 (45.4)
Cupric oxalate	5893663		100	1		B	100 (45.4)
Cupric sulfate	7758987		10	1		A	10 (4.54)
Cupric sulfate, ammoniated	10380297		100	1		B	100 (45.4)

Hazardous Substance	CASRN	Regulatory Synonyms	Statutory			Final RQ	
			RQ	Code †	RCRA Waste #	Cate-gory	Pounds (Kg)
Cupric tartrate	815827		100	1		B	100 (45.4)
CYANIDES	N/A		1*	2			**
Cyanides (soluble salts and complexes) not otherwise specified	57125		1*	4	P030	A	10 (4.54)
Cyanogen	460195	Ethanedinitrile	1*	4	P031	B	100 (45.4)
Cyanogen bromide	506683	Cyanogen bromide (CN)Br	1*	4	U246	C	1000 (454)
Cyanogen bromide (CN)Br	506683	Cyanogen bromide	1*	4	U246	C	1000 (454)
Cyanogen chloride	506774	Cyanogen chloride (CN)Cl	10	1,4	P033	A	10 (4.54)
Cyanogen chloride (CN)Cl	506774	Cyanogen chloride	10	1,4	P033	A	10 (4.54)
2,5-Cyclohexadiene-1,4-dione	106514	p-Benzoquinone	1*	4	U197	A	10 (4.54)
Cyclohexane	110827	Benzene, hexahydro-	1000	1,4	U056	C	1000 (454)
Cyclohexane, 1,2,3,4,5,6-hexachloro-,(1alpha, 2alpha, 3beta,4alpha,5alpha,6,beta)-	58899	gamma—BHC	1	1,2,4	U129	X	1 (0.454)
Cyclohexanone	108941		1*	4	U057	D	5000 (2270)
2-Cyclohexyl-4,6-dinitrophenol	131895	Phenol, 2-cyclohexyl-4,6-dinitro-	1*	4	P034	B	100 (45.4)
1,3-Cyclopentadiene, 1,2,3,4,5,5-hexachloro-	77474	Hexachlorocyclopentadiene	1	1,2,4	U130	A	10 (4.54)
Cyclophosphamide	50180	2H-1,3,2-Oxazaphosphorin-2-amine, N,N-bis(2-chloroethyl) tetrahydro-,2-oxide	1*	4	U058	A	10 (4.54)
2,4-D Acid	94757	Acetic acid (2,4-dichlorophenoxy)-2,4-D, salts and esters	100	1,4	U240	B	100 (45.4)
2,4-D Ester	94111		100	1		B	100 (45.4)
	94791		100	1		B	100 (45.4)
	94804		100	1		B	100 (45.4)
	1320189		100	1		B	100 (45.4)
	1928387		100	1		B	100 (45.4)
	1928616		100	1		B	100 (45.4)
	1929733		100	1		B	100 (45.4)
	2971382		100	1		B	100 (45.4)
	25168267		100	1		B	100 (45.4)
	53467111		100	1		B	100 (45.4)
2,4-D, salts and esters	94757	Acetic acid (2,4-dichlorophenoxy)-2,4-D Acid	100	1,4	U240	B	100 (45.4)

Hazardous Substance	CASRN	Regulatory Synonyms	Statutory			Final RQ	
			RQ	Code †	RCRA Waste #	Category	Pounds (Kg)
Daunomycin	20830813	5,12-Naphthacenedione, 8-acetyl-10-[3-amino-2,3,6- trideoxy-alpha-L-lyxo-hexo-pyranosyl)oxyl-7,8,9,10-tetrahydro-6,8,11-trihydroxy-1-methoxy-, (8S-cis)-	1*	4	U059	A	10 (4.54)
DDD	72548	Benzene, 1,1'-(2,2-dichloroethylidene)bis(4-chloro-TDE 4,4' DDD	1	1,2,4	U060	X	1 (0.454)
4,4' DDD	72548	Benzene, 1,1'-(2,2-dichloroethylidene)bis(4-chloro-DDD TDE	1	1,2,4	U060	X	1 (0.454)
DDE	72559	4,4' DDE	1*	2		X	1 (0.454)
4,4' DDE	72559	DDE	1*	2		X	1 (0.454)
DDT	50293	Benzene, 1,1'-(2,2,2-trichloroethylidene)bis(4-chloro-4,4'DDT	1	1,2,4	U061	X	1 (0.454)
4,4' DDT	50293	Benzene, 1,1'-(2,2,2-trichloroethylidene)bis(4-chloro-DDT	1	1,2,4	U061	X	1 (0.454)
DDT AND METABOLITES	N/A		1*	2			**
Diallate	2303164	Carbamothioic acid, bis(1-methylethyl)-, S-(2,3,-dich-loro-2-propenyl) ester	1*	4	U062	B	100 (45.4)
Diazinon	333415		1	1		X	1 (0.454)
Dibenz[a,h]anthracene	53703	Dibenzo[a,h]anthracene 1,2:5,6-Dibenzanthracene	1*	2,4	U063	X	1 (0.454)
1,2:5,6-Dibenzanthracene	53703	Dibenz[a,h]anthracene Dibenzo[a,h]anthracene	1*	2,4	U063	X	1 (0.454)
Dibenzo[a,h]anthracene	53703	Dibenz[a,h]anthracene 1,2:5,6-Dibenzanthracene	1*	2,4	U063	X	1 (0.454)
Dibenz[a,i]pyrene	189559	Benzo[rst]pentaphene	1*	4	U064	A	10 (4.54)
1,2-Dibromo-3-chloropropane	96128	Propane, 1,2-dibromo-3-chloro-	1*	4	U066	X	1 (0.454)
Dibutyl phthalate	84742	Dibutyl phthalate n-Butyl phthalate 1,2-Benzenedicarboxylic acid, dibutyl ester	100	1,2,4	U069	A	10 (4.54)
Di-n-butyl phthalate	84742	Dibutyl phthalate n-Butyl phthalate 1,2-Benzenedicarboxylic acid, dibutyl ester	100	1,2,4	U069	A	10 (4.54)
Dicamba	1918009		1000	1		C	1000 (454)

Hazardous Substance	CASRN	Regulatory Synonyms	Statutory			Final RQ	
			RQ	Code†	RCRA Waste #	Category	Pounds (Kg)
Dichlobenil	1194656		1000	1		B	100 (45.4)
Dichlone	117806		1	1		X	1 (0.454)
Dichlorobenzene	25321226		100	1		B	100 (45.4)
1,2-Dichlorobenzene	95501	Benzene, 1,2-dichloro- o-Dichlorobenzene	100	1,2,4	U070	B	100 (45.4)
1,3-Dichlorobenzene	541731	Benzene, 1,3-dichloro m-Dichlorobenzene	1*	2,4	U071	B	100 (45.4)
1,4-Dichlorobenzene	106467	Benzene, 1,4-dichloro p-Dichlorobenzene	100	1,2,4	U072	B	100 (45.4)
m-Dichlorobenzene	541731	Benzene, 1,3-dichloro 1,3-Dichlorobenzene	1*	2,4	U071	B	100 (45.4)
o-Dichlorobenzene	95501	Benzene, 1,2-dichloro 1,2-Dichlorobenzene	100	1,2,4	U070	B	100 (45.4)
p-Dichlorobenzene	106467	Benzene, 1,4-dichloro 1,4-Dichlorobenzene	100	1,2,4	U072	B	100 (45.4)
DICHLOROBENZIDINE	N/A		1*	2			**
3,3'-Dichlorobenzidine	91941	[1,1'-Biphenyl]-4,4'diamine,3,3'dichloro-	1*	2,4	U073	X	1 (0.454)
Dichlorobromomethane	75274		1*	2		D	5000 (2270)
1,4-Dichloro-2-butene	764410	2-Butene, 1,4-dichloro-	1*	4	U074	X	1 (0.454)
Dichlorodifluoromethane	75718	Methane, dichlorodifluoro-	1*	4	U075	D	5000 (2270)
1,1-Dichloroethane	75343	Ethane, 1,1-dichloro-Ethylidene dichloride	1*	2,4	U076	C	1000 (454)
1,2-Dichloroethane	107062	Ethane, 1,2-dichloro-Ethylene dichloride	5000	1,2,4	U077	B	100 (45.4)
1,1-Dichloroethylene	75354	Ethene, 1,1-dichloro-Vinylidene chloride	5000	1,2,4	U078	B	100 (45.4)
1,2-Dichloroethylene	156605	Ethene 1,2-dichloro- (E)	1*	2,4	U079	C	1000 (454)
Dichloroethyl ether	111444	Bis (2-chloroethyl) ether Ethane, 1,1'-oxybis[2-chloro-	1*	2,4	U025	A	10 (4.54)
Dichloroisopropyl ether	108601	Propane, 2,2'-oxybis[2-chloro-	1*	2,4	U027	C	1000 (454)
Dichloromethoxy ethane	111911	Bis(2-chloroethoxy) methane Ethane, 1,1'-[methylenebis(oxy)]bis(2-chloro-	1*	2,4	U024	C	1000 (454)
Dichloromethyl ether	542881	Methane, oxybis(chloro-	1*	4	P016	A	10 (4.54)
2,4-Dichlorophenol	120832	Phenol, 2,4-dichloro-	1*	2,4	U081	B	100 (45.4)
2,6-Dichlorophenol	87650	Phenol, 2,6-dichloro-	1*	4	U082	B	100 (45.4)
Dichlorophenylarsine	696286	Arsonous dichloride, phenyl-	1*	4	P036	X	1 (0.454)

Hazardous Substance	CASRN	Regulatory Synonyms	Statutory				Final RQ	
			RQ	Code †	RCRA Waste #	Category	Pounds (Kg)	
Dichloropropane	26638197		5000	1		C	1000 (454)
1,1-Dichloropropane	78999		5000	1		C	1000 (454)
1,3-Dichloropropane	142289		5000	1		C	1000 (454)
1,2-Dichloropropane	78875	Propane, 1,2-dichloro- Propylene dichloride	5000	1,2,4	U083	C	1000 (454)
Dichloropropane — Dichloropropene (mixture)	8003198		5000	1		B	100 (45.4)
Dichloropropene	26952238		5000	1		B	100 (45.4)
2,3-Dichloropropene	78886		5000	1		B	100 (45.4)
1,3-Dichloropropene	542756	1-Propene, 1,3-dichloro-	5000	1,2,4	U084	B	100 (45.4)
2,2-Dichloropropionic acid	75990		5000	1		D	5000 (2270)
Dichlorvos	627737		10	1		A	10 (4.54)
Dicofol	115322		5000	1		A	5000 (2270)
Dieldrin	60571	2,7:3,6-Dimethanonaphth(2,3-b]oxirene,3,4,5,6,9,9-hexachloro-1a,2,2a,3,6,6a,7,7a-octahydro-,(1aalpha,2beta,2aalpha,3beta,6beta,6aalpha,7beta, 7aalpha)-	1	1,2,4	P037	X	1 (0.454)
1,2:3,4-Diepoxybutane	1464535	2,2'-Bioxirane	1 *	4	U085	A	10 (4.54)
Diethylamine	109897		1000	1		B	100 (454.4)
Diethylarsine	692422	Arsine, diethyl-	1 *	4	P038	X	1 (0.454)
1,4-Diethylenedioxide	123911	1,4-Dioxane	1 *	4	U108	B	100 (45.4)
Diethylhexyl phthalate	117817	Bis (2-ethylhexyl)phthalate 1,2,-Benzenedicarboxylic acid, [bis(2-ethylhexyl)] ester	1 *	2,4	U028	B	100 (45.4)
N,N-'Diethylhydrazine	1615801	Hydrazine, 1,2-diethyl-	1 *	4	U086	A	10 (4.54)
O,O-Diethyl S-methyl dithiophosphate	3288582	Phosphorodithioic acid, O,O-diethyl S-methyl ester	1 *	4	U087	D	5000 (2270)
Diethyl-p-nitrophenyl phosphate	311455	Phosphoric acid, diethyl 4-nitrophenyl ester	1 *	4	P041	B	100 (45.4)
Diethyl phthalate	84662	1,2-Benzenedicarboxylic acid, diethyl ester	1 *	2,4	U088	C	1000 (454)
O,O-Diethyl O-pyrazinyl phosphorothioate	297972	Phosphorothioic acid, O,O-diethyl O-pyrazinyl ester	1 *	4	P040	B	100 (45.4)
Diethylstilbestrol	56531	Phenol, 4,4'-(1,2-diethyl-1,2-ethenediyl)bis-, (E)	1 *	4	U089	X	1 (0.454)
Dihydrosafrole	94586	1,3-Benzodioxole, 5-propyl-	1 *	4	U090	A	10 (4.54)

| Hazardous Substance | CASRN | Regulatory Synonyms | Statutory | | | Final RQ | |
			RQ	Code †	RCRA Waste #	Cate-gory	Pounds (Kg)
Diisopropylfluorophosphate	55914	Phosphorofluoridic acid, bis(1-methylethyl) ester	1*	4	P043	B	100 (45.4)
1,4,5,8-Dimethanonaphthalene, 1,2,3,4,10,10-,10-hexachloro-1,4,4a,5,8,8a-hexahydro-, (1alpha,4alpha,4abeta,5alpha,' 8alpha,	309002	Aldrin	1	1,2,4	P004	X	1 (0.454)
8abeta)-1,4,5,8-Dimenthanonaphthalene, 1,2,3,4,10,10-hexachloro-1,4,4a,5,8,8a-hexahydro,(1alpha,4alpha,4abeta, 5abeta,8beta,	465736	Isodrin	1*	4	P060	X	1 (0.454)
8abeta)-2,7:3,6-Dimethanonaphth(2,3-b]oxirene, 3,4,5,6,9,9-hexachloro-1a,2,2a,3, 6,6a,7,7a-octahydro-, (1aalpha, 2beta,2aalpha,3beta,6beta,	60571	Dieldrin	1	1,2,4	P037	X	1 (0.454)
6aalpha,7beta,7aalpha)-2,7:3,6-Dimethanonaphth(2,3-b]oxirene, 3,4,5,6,9,9-hexachloro-1a,2,2a,3, 6,6a,7,7a-octa-hydro-,(1aalpha, 2beta,2abeta,3alpha,6alpha,	72208	Endrin Endrin & metabolites	1	1,2,4	P051	X	1 (0.454)
6abeta,7beta,7aalpha)-Dimethoate	60515	Phosphorodithioic acid, O,O-dimethyl S-[2(methyla-mino)-2-oxoethyl] ester	1*	4	P044	A	10 (4.54)
3,3'-Dimethoxybenzidine	119904	[1,1'-Biphenyl)-4,4'diamine,3, 3'dimethoxy-	1*	4	U091	B	100 (45.4)
Dimethylamine	124403	methanamine, N-methyl	1000	1,4	U092	C	1000 (454)
p-Dimethylaminoazobenzene	60117	Benzenamine, N,N-dimethyl-4-(phenylazo-)	1*	4	U093	A	10 (4.54)
7,12-Dimethylbenz[a]anthracene	57976	Benz[a]anthracene, 7,12-dimethyl-	1*	4	U094	X	1 (0.454)
3,3'-Dimethylbenzidine	119937	[1,1'Biphynyl]-4,4'diamine,3,3'-dimethyl-	1*	4	U095	A	10 (4.54)
alpha,alpha-Dimethylbenzylhydroperoxide	80159	Hydroperoxide, 1-mehtyl-1-phenylethyl-	1*	4	U096	A	10 (4.54)
Dimethylcarbamoyl chloride	79447	Carbamic chloride, dimethyl-	1*	4	U097	X	1 (0.454)
1,1-Dimethylhydrazine	57147	Hydrazine, 1,1-dimethyl-	1*	4	U098	A	10 (4.54)
1,2-Dimethylhydrazine	540738	Hydrazine, 1,2-dimethyl-	1*	4	U099	X	1 (0.454)
alpha,alpha-Dimethylphenethylamine	122098	Benzeneethanamine, alpha,alpha-dimethyl-	1*	4	P046	D	5000 (2270)
2,4-Dimethylphenol	105679	Phenol, 2,4-dimethyl-	1*	2,4	U101	B	100 (45.4)

Hazardous Substance	CASRN	Regulatory Synonyms	Statutory			Final RQ	
			RQ	Code†	RCRA Waste #	Cate-gory	Pounds (Kg)
Dimethyl phthalate	131113	1,2-Benzenedicarboxylic acid, dimethyl ester	1*	2,4	U102	D	5000 (2270)
Dimethyl sulfate	77781	Sulfuric acid, dimethyl ester	1*	4	U103	B	100 (45.4)
Dinitrobenzene (mixed)	25154545		1000	1		B	100 (45.4)
m-Dinitrobenzene	99650		1000	1		B	100 (45.4)
o-Dinitrobenzene	528290		1000	1		B	100 (45.4)
p-Dinitrobenzene	100254		1000	1		B	100 (45.4)
4,6-Dinitro-o-cresol and salts	534521	Phenol, 2-methyl-4,6-dinitro-	1*	2,4	P047	A	10 (4.54)
Dinitrophenol	25550587		1000	1		A	10 (4.54)
2,5-Dinitrophenol	329715		1000	1		A	10 (4.54)
2,6-Dinitrophenol	573568		1000	1		A	10 (4.54)
2,4-Dinitrophenol	51285	Phenol, 2,4-dinitro-	1000	1,2,4	P048	A	10 (4.54)
Dinitrotoluene	25321146		1000	1,2		A	10 (4.54)
3,4-Dinitrotoluene	610399						
2,4-Dinitrotoluene	121142	Benzene, 1-methyl-2,4-dinitro-	1000	1,2,4	U105	A	10 (4.54)
2,6-Dinitrotoluene	606202	Benzene, 2-methyl-1,3-dinitro-	1000	1,2,4	U106	B	100 (45.4)
Dinoseb	88857	Phenol, 2-(1-methylpropyl)-4,6-dinitro	1*	4	P020	C	1000 (454)
Di-n-octyl phthalate	117840	1,2-Benzenedicarboxylic acid, dioctyl ester	1*	2,4	U107	D	5000 (2270)
1,4-Dioxane	123911	1,4-Diethylenedioxide	1*	4	U108	B	100 (45.4)
DIPHENYLHYDRAZINE	N/A		1*	2			**
1,2-Diphenylhydrazine	122667	Hydrazine, 1,2-diphenyl	1*	2,4	U109	A	10 (4.54)
Diphosphoramide, octamethyl-	152169	Octamethylpyrophosphoramide	1*	4	P085	B	100 (45.4)
Diphosphoric acid, tetraethyl ester	107493	Tetraethyl pyrophosphate	100	1,4	P111	A	10 (4.54)
Dipropylamine	142847	1-Propanamine, N-propyl-	1*	4	U110	D	5000 (2270)
Di-n-propylnitrosamine	621647	1-Propanamine, N-nitroso-N-propyl-	1*	2,4	U111	A	10 (4.54)
Diquat	85007		1000	1		C	1000 (454)
	2764729		1000	1		C	1000 (454)
Disulfoton	298044	Phosphorodithioic acid, o,o-diethyl S-[2-(ethylthio)ethyl]ester	1	1,4	P039	X	1 (0.454)
Dithiobiuret	541537	Thiomidodicarbonic diamide [(H2N)C(S)]2NH	1*	4	P049	B	100 (45.4)
Diuron	330541		100	1		B	100 (45.4)

Hazardous Substance	CASRN	Regulatory Synonyms	Statutory			Final RQ	
			RQ	Code †	RCRA Waste #	Cate-gory	Pounds (Kg)
Dodecylbenzenesulfonic acid	27176870		1000	1		C	1000 (454)
Endosulfan	115297	6,9-Methano-2,4,3-benzodioxathiepin, 6,7,8,9,10,10-hexachloro-1,5,5a,6,9,9a-hexahydro-, 3-oxide	1	1,2,4	P050	X	1 (0.454)
alpha - Endosulfan	959988		1*	2		X	1 (0.454)
beta - Endosulfan	33213659		1*	2		X	1 (0.454)
ENDOSALFAN AND METABOLITES	N/A		1*	2			**
Endosulfan sulfate	1031078		1*	2		X	1 (0.454)
Endothall	145733	7-Oxabicyclo[2.2.1]heptane-2,3-dicarboxylic acid	1*	4	P088	C	1000 (454)
Endrin	72208	Endrin, & metabolites 2,7:3,6-Dimethanonaphth[2,3-b]oxirene,3,4,5,6,9,9 -hexachloro-1a,2,2a,3,6,6a,7,7a-octa-hydro-, (1 aalpha, 2beta,2abeta,3alpha, 6alpha,6abeta,7beta, 7aalpha)-	1	1,2,4	P051	X	1 (0.454)
Endrin aldehyde	7421934		1*	2		X	1 (0.454)
ENDRIN AND METABOLITES	N/A		1*	2			**
Endrin, & metablites	72208	Endrin 2,7:3,6-Dimethanonaphth[2,3-b] oxirene, 3,4,5,6,9,9-hexachloro-1a,2,2a,3,6,6a,7,7a-octa-hydro-, (1 aalpha,2beta, 2abeta,3alpha, 6alpha, ... 6abeta,7beta, 7aalpha)-	1	1,2,4	P051	X	1 (0.454)
Epichlorohydrin	106898	Oxirane, (chloromethyl)-	1000	1,4	U041	B	100 (45.4)
Epinephrine	51434	1,2-Benzenediol,4-[1-hydroxy-2-(methylamino)ethyl]-	1*	4	P042	C	1000 (454)
Ethanal	75070	Acetaldehyde	1000	1,4	U001	C	1000 (454)
Ethanamine, N-ethyl-N-nitroso-	55185	N-Nitrosodiethylamine	1*	4	U174	X	1 (0.454)
1,2-Ethanediamine, N,N-dimethyl-N'-2-pyridinyl-N'-(2-thienylmethyl)-	91805	Methapyrilene	1*	4	U155	D	5000 (2270)
Ethane, 1,2-dibromo-	106934	Ethylene dibromide	1000	1,4	U067	X	1 (0.454)
Ethane, 1,1-dichloro-	75343	Ethylidene dichloride 1,1-Dichloroethane	1*	2,4	U076	C	1000 (454)
Ethane, 1,2-dichloro-	107062	Ethylene dichloride 1,2-Dichlorethane	5000	1,2,4	U077	B	100 (45.4)
Ethanedinitrile	460195	Cyanogen	1*	4	P031	B	100 (45.4)

| Hazardous Substance | CASRN | Regulatory Synonyms | Statutory | | | Final RQ | |
			RQ	Code†	RCRA Waste #	Category	Pounds (Kg)
Ethane, hexachloro-	67721	Hexachloroethane	1*	2,4	U131	B	100 (45.4)
Ethane, 1,1'-[methylenebis(oxy)]bis(2-chloro-	111911	Bis(2-chloroethoxy) methane Dichloromethoxy ethane	1*	2,4	U024	C	1000 (454)
Ethane, 1,1'-oxybis-	60297	Ethyl ether	1*	4	U117	B	100 (45.4)
Ethane, 1,1'-oxybis(2-chloro-	111444	Bis (2-chloroethyl) ether Dichloroethyl ether	1*	2,4	U025	A	10 (4.54)
Ethane, pentachloro-	76017	Pentachloroethane	1*	4	U184	A	10 (4.54)
Ethane, 1,1,1,2-tetrachloro	630206	1,1,1,2-Tetrachloroethane	1*	4	U208	B	100 (45.4)
Ethane, 1,1,2,2-tetrachloro	79345	1,1,2,2-Tetrachloroethane	1*	2,4	U209	B	100 (45.4)
Ethanethioamide	62555	Thioacetamide	1*	4	U218	A	10 (4.54)
Ethane, 1,1,1-trichloro	71556	Methyl chloroform 1,1,1-Trichloroethane	1*	2,4	U226	C	1000 (454)
Ethane, 1,1,2-trichloro-	79005	1,1,2-Trichloroethane	1*	2,4	U227	B	100 (45.4)
Ethanimidothioic acid, N-[[(methyl-amino)carbonyl]oxy]-, methyl ester	16752775	Methomyl	1*	4	P066	B	100 (45.4)
Ethanol, 2-ethoxy-	110805	Ethylene glycol monoethyl ether	1*	4	U359	C	1000 (454)
Ethanol, 2,2'-(nitrosoimino)bis-	1116547	N-Nitrosodiethanolamine	1*	4	U173	X	1 (0.454)
Ethanone, 1-phenyl-	98862	Acetophenone	1*	4	U004	D	5000 (2270)
Ethene, chloro-	75014	Vinyl chloride	1*	2,3,4	U043	X	1 (0.454)
Ethene, 2-Cloroethoxy-	110758	2-Chloroethyl vinyl ether	1*	2,4	U042	C	1000 (454)
Ethene, 1,1-dichloro-	75354	Vinylidene chloride 1,1-Dichloroethylene	5000	1,2,4	U078	B	100 (45.4)
Ethene, 1,2-dichloro-	156605	1,2-Dichloroethylene	1*	2,4	U079	C	1000 (45.4)
Ethene, tetrachloro-	127184	Perchloroethylene Tetrachlorethene Tetrachloroethylene	1*	2,4	U210	B	100 (45.4)
Ethene, trichloro-	79016	Trichloroethene Trichloroethylene	1000	1,2,4	U228	B	100 (45.4)
Ethion	563122		10	1		A	10 (4.54)
Ethyl acetate	141786	Acetic acid, ethyl ester	1*	4	U112	D	5000 (2270)
Ethyl acrylate	140885	2-Propenoic acid, ethyl ester	1*	4	U113	C	1000 (454)
Ethylbenzene	100414		1000	1,2		C	1000 (454)
Ethyl carbamate (urethane)	51796	Carbamic acid, ethyl ester	1*	4	U238	B	100 (45.4)
Ethyl cyanide	107120	Propanenitril	1*	4	P101	A	10 (4.54)

Hazardous Substance	CASRN	Regulatory Synonyms	Statutory			Final RQ	
			RQ	Code†	RCRA Waste #	Cate-gory	Pounds (Kg)
Ethylenebisdithiocarbamic acid, salts & esters	111546	Carbamodithioic acid, 1,2-ethanediylbis, salts & esters	1*	4	U114	D	5000 (2270)
Ethylenediamine	107153		1000	1		D	5000 (2270)
Ethylenediamine-tetraacetic acid (EDTA)	60004		5000	1		D	5000 (2270)
Ethylene dibromide	106934	Ethane, 1,2-dibromo-	1000	1,4	U067	X	1 (0.454)
Ethylene dichloride	107062	Ethane, 1,2-dichloro-1,2-Dichloroethane	5000	1,2,4	U077	B	100 (45.4)
Ethyllene glycol monoethy ether	110805	Ethanol, 2-ethoxy-	1*	4	U359	C	1000 (454)
Ethylene oxide	75218	Oxirane	1*	4	U115	A	10 (4.54)
Ethylenethiourea	96457	2-Imidazolidinethione	1*	4	U116	A	10 (4.54)
Ethylenimine	151564	Aziridine	1*	4	P054	X	1 (0.454)
Ethyl ether	60297	Ethane, 1,1'-oxybis	1*	4	U117	B	100 (45.4)
Ethylidene dichloride	75343	Ethane, 1,1'-dichloro-1,1-Dichloroethane	1*	2,4	U076	C	1000 (454)
Ethyl methacrylate	97632	2-Propenoic acid, 2-methyl-, ethyl ester	1*	4	U118	C	1000 (454)
Ethyl methanesulfonate	62500	Methanesulfonic acid, ethyl ester	1*	4	U119	X	1 (0.454)
Famphur	52857	Phosphorothioic acid, O,[4-[(di-methylamino) sulfonyl] phenyl] O,O-dimethyl ester	1*	4	P097	C	1000 (454)
Ferric ammonium citrate	1185575		1000	1		C	1000 (454)
Ferric ammonium oxalate	2944674		1000	1		C	1000 (454)
	55488874		1000	1		C	1000 (454)
Ferric chloride	7705080		1000	1		C	1000 (454)
Ferric flouride	7783508		100	1		B	100 (45.4)
Ferric nitrate	10421484		1000	1		C	1000 (454)
Ferric sulfate	10028225		1000	1		C	1000 (454)
Ferrous ammonium sulfate	10045893		1000	1		C	1000 (454)
Ferrous chloride	7758943		100	1		B	100 (45.4)
Ferrous sulfate	7720787		1000	1		C	1000 (454)
	7782630		1000	1		C	1000 (454)
Flouranthene	206440	Benzo[j,k]flourene	1*	2,4	U120	B	100 (45.4)
Flourene	86737		1*	2		D	5000 (2270)
Flourine	7782414		1*	4	P056	A	10 (4.54)

Hazardous Substance	CASRN	Regulatory Synonyms	Statutory			Final RQ	
			RQ	Code †	RCRA Waste #	Category	Pounds (Kg)
Flouroacetamide	640197	Acetamide, 2-fluoro-	1*	4	P057	B	100 (45.4)
Flouracetic acid, sodium salt	62748	Acetic acid, fluoro-, sodium salt	1*	4	P058	A	10 (4.54)
Formaldehyde	50000		1000	1,4	U122	B	100 (45.4)
Formic acid	64186		5000	1,4	U123	D	5000 (2270)
Fulminic acid, mercury(2+)salt	628864	Mercury fulminate	1*	4	P065	A	10 (4.54)
Fumaric acid	110178		5000	1		D	5000 (2270)
Furan	110009	Furfuran	1*	4	U124	B	100 (45.4)
Furan, tetrahydro-	109999	Tetrahydrofuran	1*	4	U213	C	1000 (454)
2-Furancarboxaldehyde	98011	Furfural	1000	1,4	U125	D	5000 (2270)
2,5-Furandione	108316	Maleic anhydride	5000	1,4	U147	D	5000 (2270)
Furfural	98011	2-Furancarboxaldehyde	1000	1,4	U125	D	5000 (2270)
Furfuran	110009	Furan	1*	4	U124	B	100 (45.4)
Glucopyranose, 2-deoxy-2-(3-methyl-3-nitrosoureido)-	18883664	D-Glucose, 2-deoxy-2-[[(methylnitrosoamino)-carbonyl]amino] Streptozotocin	1*	4	U206	X	1 (0.454)
D-Glucose, 2-deoxy-2-[[(methylnitrosoamino)-carbonyl]amino]-	18883664	Glucopyranose, 2-deoxy-2-(3-methyl-3-nitrosoureido)-	1*	4	U206	X	1 (0.45)
Glycidylaldehyde	765344	Oxiranecarboxaldehyde	1*	4	U126	A	10 (4.54)
Guanidien, N-methyl-N'-nitro-N-nitroso-	70257	MNNG	1*	4	U163	A	10 (4.54)
Guthion	865500		1	1		X	1 (0.454)
HALOETHERS	N/A		1*	2			**
HALOMETHANES	N/A		1*	2			**
Heptachlor	76448	4,7-Methano-1H-indene, 1,4,5,6,7,8,8-heptachloro-3a,4,7,7a-tetrahydro-	1	1,2,4	P059	X	1 (0.454)
HEPTACHLOR AND METABOLITES	N/A		1*	2			**
Heptachlor epoxide	1024573		1*	2		X	1 (0.454)
Hexachlorobenzene	118741	Benzene, hexachloro-	1*	2,4	U127	A	10 (4.54)
Hexachlorobutadiene	87683	1,3-Butadiene, 1,1,2,3,4,5-hexachloro-	1*	2,4	U128	X	1 (0.454)
HEXACHLOROCYCLOHEXANE (all isomers)	608731		1*	2			**

Hazardous Substance	CASRN	Regulatory Synonyms	Statutory				Final RQ	
			RQ	Code†	RCRA Waste #	Category	Pounds (Kg)	
Hexachlorocyclohexane (gammer isomer)	58899	Cyclohexane, 1,2,3,4,5,6-hexachloro-,(1alpha,2alpha,3beta,4alpha,5alpha,6beta)-gamma-BHC Lindane	1	1,2,4	U129	X	1 (0.454)	
Hexachlorocyclopentadiene	77474	1,3-Cyclopentadiene,1,2,3,4,5,5-hexachloro-	1	1,2,4	U130	A	10 (4.54)	
Hexachloroethane	67721	Ethane, hexachloro-	1*	2,4	U131	B	100 (45.4)	
Hexachlorophene	70304	Phenol, 2,2'-methylenebis[3,4,5-trichloro-	1*	4	U132	B	100 (45.4)	
Hexachloropropene	1888717	1-Propene, 1,1,2,3,3,3-hexachloro-	1*	4	U243	C	1000 (454)	
Hexaethyl tetraphosphate	757584	Tetraphosphoric acid, hexaethyl ester	1*	4	P062	B	100 (45.4)	
Hydrazine	302012		1*	4	U133	X	1 (0.454)	
Hydrazine, 1,2-diethyl-	1615801	N,N'-Diethylhydrazine	1*	4	U086	A	10 (4.54)	
Hydrazine, 1,1-dimethyl-	57147	1,1-Dimethylhydrazine	1*	4	U098	A	10 (4.54)	
Hydrazine, 1,2-dimethyl-	540738	1,2-Dimethylhydrazine	1*	4	U099	X	1 (0.454)	
Hydrazine, 1,2-diphenyl-	122667	1,2-Diphenylhydrazine	1*	2,4	U109	A	10 (4.54)	
Hydrazine, methyl-	60344	Methyl hydrazine	1*	4	P068	A	10 (4.54)	
Hydrazinecarbothioamide	79196	Thiosemicarbazide	1*	4	P116	B	100 (45.4)	
Hydrochloric acid	7647010	Hydrogen chloride	5000	1		D	5000 (2270)	
Hydrocyanic acid	74908	Hydrogen cyanide	10	1,4	P063	A	10 (4.54)	
Hydrofluoric acid	7664393	Hydrogen flouride	5000	1,4	U134	B	100 (45.4)	
Hydrogen chloride	7647010	Hydrochloric acid	5000	1		D	5000 (2270)	
Hydrogen cyanide	74908	Hydrocyanic acid	10	1,4	P063	A	10 (4.54)	
Hydrogen fluoride	7664393	Hydrofluoric acid	5000	1,4	U134	B	100 (45.4)	
Hydrogen sulfide	7783064	Hydrogen sulfide H2S	100	1,4	U135	B	100 (45.4)	
Hydrogen sulfide H2S	7783064	Hydrogen sulfide	100	1,4	U135	B	100 (45.4)	
Hydroperoxide, 1-methyl-1-phenylethyl-	80159	alpha,alpha-Dimethylbenzylhydroperoxide	1*	4	U096	A	10 (4.54)	
2-Imidazolidinethione	96457	Ethylenethiourea	1*	4	U116	A	10 (4.54)	
Indeno(1,2,3-cd)pyrene	193395	1,10-(1,2-Phenylene)pyrene	1*	2,4	U137	B	100 (45.4)	
1,3-Isobenzofurandione	85449	Phthalic anhydride	1*	4	U190	D	5000 (2270)	
Isobutyl alcohol	78831	1-Propanol, 2-methyl-	1*	4	U140	D	5000 (2270)	

Hazardous Substance	CASRN	Regulatory Synonyms	Statutory			Final RQ	
			RQ	Code†	RCRA Waste #	Cate-gory	Pounds (Kg)
Isodrin	465736	1,4,5,8-Dimethanonaphthalene, 1,2,3,4,10,10-hexachloro-1,4,4a, 5,8,8a-hexahydro,(1alpha,4alpha, 4abeta,5beta,8beta,8abeta)-	1*	4	P060	X	1 (0.454)
Isophorone	78591		1*	2		D	5000 (2270)
Isoprene	78795		1000	1		B	100 (45.4)
Isopropanolamine dodecylbenzenesulfonate	42504461		1000	1		C	1000 (454)
Isosafrole	120581	1,3-Benzodioxole,5-)1-propenly)-	1*	4	U141	B	100 (45.4)
3(2H)-Isoxazolone, 5-(aminomethyl)-	2763964	Muscimol 5-(Aminomethyl)-3-isoxazolol	1*	4	P007	C	1000 (454)
Kepone	143500	1,2,4-Metheno-2H-cyclobuta[cd] pentalen-2-one,1,1a,3,3a,4,5,5, 5a,5b,6-decachlorooctahydro-	1	1,4	U142	X	1 (0.454)
Lasiocarpine	303344	2-Butenoic acid, 2-methyl-, 7[(2,3-dihydroxy-2-(1-methoxyethyl)-3-methyl-1-oxobutoxy]methyl]-2,3,5, 7a-tetrahydro-1H-pyrrolizin-1-yl ester, [1S-(1alpha(Z), 7(2S*,3R*), 7aalpha)]-	1*	4	U143	A	10 (4.54)
Lead††	7439921		1*	2	U143	A	10 (4.54)
Lead acetate	301042	Acetic acid, lead(2+) salt	5000	1,4	U144		#
LEAD AND COMPOUNDS	N/A		1*	2			**
Lead arsenate	7784409		5000	1		X	1 (0.454)
	7645252		5000	1		X	1 (0.454)
	10102484		5000	1		X	1 (0.454)
Lead, bis(acetato-O)tetrahydroxytri	1335326	Lead subacetate	1*	4	U146	B	100 (45.4)
Lead chloride	7758954		5000	1		B	100 (45.4)
Lead fluoborate	13814965		5000	1		B	100 (45.4)
Lead fluoride	7783462		1000	1		B	100 (45.4)
Lead iodide	10101630		5000	1		B	100 (45.4)
Lead nitrate	10099748		5000	1		B	100 (45.4)
Lead phosphate	7446277	Phosphoric acid, lead(2+) salt (2:3)	1*	4	U145		#
Lead stearate	7428480		5000	1		D	# 5000 (2270)
	1072351		5000	1		D	# 5000 (2270)

Hazardous Substance	CASRN	Regulatory Synonyms	Statutory			Final RQ	
			RQ	Code †	RCRA Waste #	Category	Pounds (Kg)
	52652592		5000	1		D	# 5000 (2270)
	56189094		5000	1		D	# 5000 (2270)
Lead subacetate	1335326	Lead, bis(acetato-O)tetrahydroxytri	1*	4	U146	B	100 (45.4)
Lead sulfate	15739807		5000	1		B	100 (45.4)
	7446142		5000	1		B	100 (45.4)
			5000	1		B	100 (45.4)
Lead sulfide	1314870		5000	1		D	# 5000 (2270)
Lead thiocyanate	592870		5000	1		B	100 (45.4)
Lindane	58899	Cyclohexane, 1,2,3,4,5,6-hexachloro-,(1alpha,2alpha,3beta,4alpha,5alpha,6beta)-gamma-BHC Hexachlorocyclohexane (gamma isomer)	1	1,2,4	U129	X	1 (0.454)
Lithium Chromate	14307358		1000	1		A	10 (4.54)
Malathion	121755		10	1		B	100 (45.4)
Maleic acid	110167		5000	1		D	5000 (2270)
Maleic anhydride	108316	2,5-Furandione	5000	1,4	U147	D	5000 (2270)
Maleic hydrazide	123331	3,6-Pyridazinedione, 1,2-dihydro-	1*	4	U148	D	5000 (2270)
Malononitrile	109773	Propanedinitrile	1*	4	U149	C	1000 (454)
Melphalan	148823	L-Phenylalanine, 4-[bis(2-chloroethyl)amino]	1*	4	U150	X	1 (0.454)
Mercaptodimethur	2032657		100	1		A	10 (4.54)
Mercuric cyanide	592041		1	1		X	1 (0.454)
Mercuric nitrate	10045940		10	1		A	10 (4.54)
Mercuric sulfate	7783359		10	1		A	10 (4.54)
Mercuric thiocyanate	592858		10	1		A	10 (4.54)
Mercurous nitrate	10415755		10	1		A	10 (4.54)
	7782867		10	1		A	10 (4.54)
Mercury	7439976		1*	2,3,4	U151	X	1 (0.454)
MERCURY AND COMPOUNDS	N/A		1*	2			**
Mercury, (acetate-O)phenyl	62384	Phenylmercury acetate	1*	4	P092	B	100 (45.4)
Mercury fulminate	628864	Fulminic acid, mercury(2+)salt	1*	4	P065	A	10 (4.54)

Hazardous Substance	CASRN	Regulatory Synonyms	Statutory			Final RQ	
			RQ	Code†	RCRA Waste #	Category	Pounds (Kg)
Methacrylonitrile	126987	2-Propenenitrile,2-methyl-	1*	4	U152	C	1000 (454)
Methanamine, N-methyl-	124403	Dimethylamine	1000	1,4	U092	C	1000 (454)
Methanamine, N-methyl-N-nitroso-	62759	N-Nitrosodimethylamine	1*	2,4	P082	A	10 (4.54)
Methane, bromo-	74839	Methyl bromide	1*	2,4	U029	C	1000 (454)
Methane, chloro-	74873	Methyl chloride	1*	2,4	U045	B	100 (45.4)
Methane, chloromethoxy-	107302	Chloromethyl methyl ether	1*	4	U046	A	10 (4.54)
Methane, dibromo-	74953	Methylene bromide	1*	4	U068	C	1000 (454)
Methane, dichloro-	75092	Methylene chloride	1*	2,4	U080	C	1000 (454)
Methane, dichlorodifluoro-	75718	Dichlorodifluoromethane	1*	4	U075	D	5000 (2270)
Methane, iodo-	74884	Methyl iodide	1*	4	U138	B	100 (45.4)
Methane, isocyanato-	624839	Methyl isocyanate	1*	4	P064		##
Methane, oxybis(chloro-	542881	Dichloromethyl ether	1*	4	P016	A	10 (4.54)
Methanesulfenyl chloride, trichloro-	594423	Trichloromethanesulfenyl chloride	1*	4	P118	B	100 (45.4)
Methanesulfonic acid, ethyl ester	62500	Ethyl methanesulfonate	1*	4	U119	X	1 (0.454)
Methane, tetrachloro-	56235	Carbon tetrachloride	5000	1,2,4	U211	A	10 (4.54)
Methane, tetranitro	509148	Tetranitromethane	. 1*	4	P112	A	10 (4.54)
Methane, tribromo-	75252	Bromoform	1*	2,4	U225	B	100 (45.4)
Methane, trichloro-	67663	Chloroform	5000	1,2,4	U044	A	10 (4.54)
Methane, trichlorofluoro	75694	Trichloromonofluoromethane	1*	4	U121	D	5000 (2270)
Methanethiol	74931	Methylmercaptan Thiomethanol	100	1,4	U153	B	100 (45.4)
6,9-Methano-2,4,3-benzodioxathiepin,6,7,8,9,10,10-hexachloro-1,5,5a,6,9,9a-hexahydro-, 3-oxide	115297	Endosulfan	1	1,2,4	P050	X	1 (0.454)
1,3,4-Metheno-2H-cyclobuta[cd]pentalen-2-one, 1,1a,3,3a,4,5,5,5a,5b,6-decachlorooctahydro-	143500	Kepone	1	1,4	U142	X	1 (0.454)
4,7-Methano-1H-indene, 1,4,5,6,7,8,8-heptachloro-3a,4,7,7a-tetrahydro-	76448	Heptachlor	1	1,2,4	P059	X	1 (0.454)
4,7-Methano-1H-indene, 1,2,3,4,5,6,8,8-octachloro-2,3,3a,4,5,5a-hexahydro-	57749	Chlordane Chlordane, alpha & gamma isomers Chlordane, technical	1	1,2,4	U036	X	1 (0.454)
Methanol	67561	Methyl alcohol	1*	4	U154	D	5000 (2270)

Hazardous Substance	CASRN	Regulatory Synonyms	Statutory			Final RQ	
			RQ	Code †	RCRA Waste #	Category	Pounds (Kg)
Methapyrilene	91805	1,2-Ethanediamine, N,N-dimethyl-N'-2-pyridinyl-N'-(2-thienylmethyl)-	1 *	4	U155	D	5000 (2270)
Methomyl	16752775	Ethanimidothioic acid, N-[[(methyl-amino)carbonyl]oxy]-, methyl ester	1 *	4	P066	B	100 (45.4)
Methoxychlor	72435	Benzene, 1,1'-(2,2,2-trichloroethylidene)bis[4-methoxy-	1	1,4	U247	X	1 (0.454)
Methyl alchohol	67561	Methanol	1 *	4	U154	D	5000 (2270)
Methyl bromide	74839	Methane, bromo-	1 *	2,4	U029	C	1000 (454)
1-Methylbutadiene	504609	1,3-Pentadiene	1 *	4	U186	B	100 (45.4)
Methyl chloride	74873	Methane, chloro-	1 *	2,4	U045	B	100 (45.4)
Methyl chlorocarbonate	79221	Carbonochloridic acid, methyl ester Methyl chloroformate	1 *	4	U156	C	1000 (454)
Methyl chloroform	71556	Ethane, 1,1,1-trichloro-1,1,1-Trichloroethane	1 *	2,4	U226	C	1000 (454)
Methyl chloroformate	79221	Carbonochloridic acid, methyl ester Methyl chlorocarbonate	1 *	4	U156	C	1000 (454)
3-Methylcholanthrene	56495	Benz[j]aceanthrylene, 1,2-dihydro-3-methyl-	1 *	4	U157	A	10 (4.54)
4,4'-Methylenebis(2-chloroaniline)	101144	Benzenamine, 4,4'-methylenebis(2-chloro-	1 *	4	U158	A	10 (4.54)
Methylene bromide	74953	Methane, dibromo-	1 *	4	U068	C	1000 (454)
Methylene chloride	75092	Methane, dichloro-	1 *	2,4	U080	C	1000 (454)
Methyl ethyl ketone (MEK)	78933	2-Butanone	1 *	4	U159	D	5000 (2270)
Methyl ethyl ketone peroxide	1338234	2-Butanone peroxide	1 *	4	U160	A	10 (4.54)
Methyl hydrazine	60344	Hydrazine, methyl-	1 *	4	P068	A	10 (4.54)
Methyl iodide	74884	Methane, iodo-	1 *	4	U138	B	100 (45.4)
Methly isobutyl ketone	108101	4-Methyl-2-pentanone	1 *	4	U161	D	5000 (2270)
Methyl isocyanate	624839	Methane, isocyanato-	1 *	4	P064		##
2-Methyllactonitrile	75865	Acetone cyanohydrin Propanenitrile, 2-hydroxy-2-methyl-	10	1,4	P069	A	10 (4.54)
Methylmercaptan	74931	Methanethiol Thiomethanol	100	1,4	U153	B	100 (45.4)
Methyl methacrylate	80626	2-Propenoic acid, 2-methyl, methyl ester	5000	1,4	U162	C	1000 (454)
Methyl parathion	298000	Phosphorotioic acid,),)-dimethyl O-(4-nitro-phenyl) ester	100	1,4	P071	B	100 (45.4)
4-Methyl-2-pentanone	108101	Methyl isobutyl ketone	1 *	4	U161	D	5000 (2270)

Hazardous Substance	CASRN	Regulatory Synonyms	Statutory			Final RQ	
			RQ	Code †	RCRA Waste #	Cate-gory	Pounds (Kg)
Methylthiouracil	56042	4(1H)-Pyrimidinone, 2,3-dihydro-6-methyl-2-thioxo-	1 *	4	U164	A	10 (4.54)
Mevinphos	7786347		1	1		A	10 (4.54)
Mexacarbate	315184		1000	1		C	1000 (454)
Mitomycin C	50077	Azirino[2',3':3,4]pyrrolo[1,2-a]indole-4,7-dione,6-amino-8-[[(aminocarbonyl)oxy]methyl]-1,1a,2,8,8a,8b-hexahydro-8a-methoxy-5-methyl, [1aS-(1aalpha, 8beta,8aalpha, 8balpha)]-	1 *	4	U010	A	10 (4.54)
MNNG	70257	Guanidine, N-methyl-N'-nitro-N-nitroso-	1 *	4	U163	A	10 (4.54)
Monoethylamine	75047		1000	1		B	100 (45.4)
Monomethylamine	74895		1000	1		B	100 (45.4)
Multi Source Leachate			1 *	4	F039	X	1 (0.454)
Muscimol	2763964	3(2H)-Isoxazolone, 5-(aminomethyl)- 5-(Amino-methyl)-3-isoxazolol	1 *	4	P007	C	1000 (454)
Naled	300765		10	1		A	10 (4.54)
5,12-Naphthacenedione, 8-acetyl-10-[3-amino-2,3,6-trideoxy-alpha-L-lyxo-hexopyranosyl)oxy]-7,8,9,10-tetrahydro-6,8,11-trihydroxy-1-methoxy, (8S-cis)-	20830813	Daunomycin	1 *	4	U059	A	10 (4.54)
1-Naphthalenamine	134327	alpha-Naphthylamine	1 *	4	U167	B	100 (45.4)
2-Naphthalenamine	91598	beta-Naphthylamine	1 *	4	U168	A	10 (4.54)
Naphthalenamine,N,N'-bis(2-chloroethyl)-	494031	Chlornaphazine	1 *	4	U026	B	100 (45.4)
Naphthalene	91203		5000	1,2,4	U165	B	100 (45.4)
Naphthalene, 2-chloro-	91587	beta-Chloronaphthalene 2-Chloronaphthalene	1 *	2,4	U047	D	5000 (2270)
1,4-Naphthalenedione	130154	1,4-Naphthoquinone	1 *	4	U166	D	5000 (2270)
2,7-Naphthalenedisulfonic acid, 3,3'-[(3,3'-dimethyl-(1,1'-byphenyl)-4,4'-diyl)-bis(azo)]bis(5-amino-4-hydroxy)tetrasodium salt	72571	Trypan blue	1 *	4	U236	A	10 (4.54)
Naphthenic acid	1338245		100	1		B	100 (45.4)
1,4-Naphthoquinone	130154	1,4-Naphthalenedione	1 *	4	U166	D	5000 (2270)
alpha-Naphthylamine	134327	1,-Naphthalenamine	1 *	4	U167	B	100 (45.4)
beta-Naphthylamine	91598	2,-Naphthalenamine	1 *	4	U168	A	10 (4.54)

Hazardous Substance	CASRN	Regulatory Synonyms	Statutory			Final RQ	
			RQ	Code†	RCRA Waste #	Category	Pounds (Kg)
alpha-Naphthylthiourea	86884	Thiourea, 1-naphthalenyl-	1*	4	P072	B	100 (45.4)
Nickel††	7440020		1*	2		B	100 (45.4)
Nickel ammonium sulfate	15699180		5000	1		B	100 (45.4)
NICKEL AND COMPOUNDS	N/A		1*	2			**
Nickel carbonyl	13463393	Nickel carbonyl Ni(CO)4, (T-4)-	1*	4	P073	A .	10 (4.54)
Nickel carbonyl Ni(CO)4, (T-4)-	13463393	Nickel carbonyl	1*	4	P073	A	10 (4.54)
Nickel chloride	7718549		5000	1		B	100 (45.4)
	37211055		5000	1		B	100 (45.4)
Nickel cyanide	557197	Nickel cyanide Ni(CN)2	1*	4	P074	A	10 (4.54)
Nickel cyanide Ni(CN)2	557197	Nickel cyanide	1*	4	P074	A	10 (4.54)
Nickel hydroxide	12054487		1000	1		A	10 (4.54)
Nickel nitrate	14216752		5000	1		B	100 (45.4)
Nickel sulfate	7786814		5000	1		B	100 (45.4)
Nicotine, & salts	54115	Pyridine, 3-(1-methyl-2-pyrrolidinyl)-, (S)-	1*	4	P075	B	100 (45.4)
Nitric acid	7697372		1000	1		C	1000 (454)
Nitric acid, thalium (1 +) salt	10102451	Thallium (I) nitrate	1*	4	U217	B	100 (45.4)
Nickel oxide	10102439	Nittrogen oxide NO	1*	4	P076	A	10 (4.54)
p-Nitroaniline	100016	Benzenamine, 4-nitro-	1*	4	P077	D	5000 (2270)
Nitrobenzene	98953	Benzene, nitro-	1000	1,2,4	U169	C	1000 (454)
Nitrogen dioxide	10102440	Nitrogen oxide NO2	1000	1,4	P078	A	10 (4.54)
	10544726		1000	1,4	P078	A	10 (4.54)
Nitrogen oxide NO	10102439	Nitric oxide	1*	4	P076	A	10 (4.54)
Nitrogen oxide NO2	10102440	Nitrogen dioxide	1000	1,4	P078	A	10 (4.54)
	10544726						
Nitroglycerine	55630	1,2,3-Propanetriol, trinitrate-	1*	4	P081	A	10 (4.54)
Nitrophenol (mixed)	25154556		1000	1		B	100 (45.4)
m-Nitrophenol	554847					B	100 (45.4)
o-Nitrophenol	88755	2-Nitrophenol					
p-Nitrophenol	100027	Phenol, 4-nitro- 4-Nitrophenol					
o-Nitrophenol	88755	2-Nitrophenol	1000	1,2		B	100 (45.4)
p-Nitrophenol	100027	Phenol, 4-nitro- 4-Nitrophenol	1000	1,2,4	U170	B	100 (45.4)

Hazardous Substance	CASRN	Regulatory Synonyms	Statutory			Final RQ	
			RQ	Code†	RCRA Waste #	Category	Pounds (Kg)
2-Nitrophenol	88755	o-Nitrophenol	1000	1,2		B	100 (45.4)
4-Nitrophenol	100027	p-Nitrophenol Phenol, 4-nitro-	1000	1,2,4	U170	B	100 (45.4)
NITROPHENOLS	N/A		1*	2			**
2-Nitropropane	79469	Propane, 2-nitro-	1*	4	U171	A	10 (4.54)
NITROSAMINES	N/A		1*	2			**
N-Nitrosodi-n-butylamine	924163	1-Butanamine, N-butyl-N-nitroso-	1*	4	U172	A	10 (4.54)
N-Nitrosodiethanolamine	1116547	Ethanol, 2,2'-(nitrosoimino)bis-	1*	4	U173	X	1 (0.454)
N-Nitrosodiethylamine	55185	Ethanamine, N-ethyl-N-nitroso-	1*	4	U174	X	1 (0.454)
N-Nitrosodimethylamine	62759	Methanamine, N-methyl-N-nitroso-	1*	2,4	P082	A	10 (4.54)
N-Nitrosodiphenylamine	86306		1*	2		B	100 (45.4)
N-Nitroso-N-ethylurea	759739	Urea, N-ethyl-N-nitroso-	1*	4	U176	X	1 (0.454)
N-Nitroso-N-methylurea	684935	Urea, N-methyl-N-nitroso	1*	4	U177	X	1 (0.454)
N-Nitroso-N-methylurethane	615532	Carbamic acid, methylnitroso-, ethyl ester	1*	4	U178	X	1 (0.454)
N-Nitrosomethylvinylamine	4549400	Vinlyamine, N-methyl-N-nitroso-	1*	4	P084	A	10 (4.54)
N-Nitrosopiperidine	100754	Piperidine, 1-nitroso-	1*	4	U179	A	10 (4.54)
N-Nitrosopyrrolidine	930552	Pyrrolidine, 1-nitroso-	1*	4	U180	X	1 (0.454)
Nitrotoluene	1321126		1000	1		C	1000 (454)
m-Nitrotoluene	99081						
o-Nitrotoluene	88722						
p-Nitrotoluene	99990						
5-Nitro-o-toluidine	99558	Benzenamine, 2-methyl-5-nitro-	1*	4	U181	B	100 (45.4)
Octamethylpyrophosphoramide	152169	Diphosphoramide, octamethyl-	1*	4	P085	B	100 (45.4)
Osmium oxide OsO4 (T-4)-	20816120	Osmium tetroxide	1*	4	P087	C	1000 (454)
Osmium tetroxide	20816120	Osmium oxide OsO4 (T-4)-	1*	4	P087	C	1000 (454)
7-Oxabicyclo[2.2.1]heptane-2,3-dicarboxylic acid	145733	Endothall	1*	4	P088	C	1000 (454)
1,2-Oxathiolane, 2,2-dioxide	1120714	1,3-Propane sultone	1*	4	U193	A	10 (4.54)
2H-1,3,2-Oxazaphosphorin-2-amine, N,N-bis(2-chloroethyl) tetrahydro-, 2-oxide	50180	Cyclophosphamide	1*	4	U058	A	10 (4.54)
Oxirane	75218	Ethylene oxide	1*	4	U115	A	10 (4.54)
Oxiranecarboxyaldehyde	765344	Glycidylaldehyde	1*	4	U126	A	10 (4.54)

Hazardous Substance	CASRN	Regulatory Synonyms	Statutory			Final RQ	
			RQ	Code †	RCRA Waste #	Cate-gory	Pounds (Kg)
Oxirane, (chloromethyl)-	106898	Epichlorohydrin	1000	1,4	U041	B	100 (45.4)
Paraformaldehyde	30525894		1000	1		C	1000 (454)
Paraldehyde	123637	1,3,5-Trioxane, 2,4,6-trimethyl-	1*	4	U182	C	1000 (454)
Parathion	56382	Phosphorothioic acid, O,O-diethyl O-(4-nitrophenyl) ester	1	1,4	P089	A	10 (4.54)
Pentachlorobenzene	608935	Benzene, pentachloro-	1*	4	U183	A	10 (4.54)
Pentachloroethane	76017	Ethane, pentachloro-	1*	4	U184	A	10 (4.54)
Pentachloronitrobenzene (PCNB)	82688	Benzene, pentachloronitro-	1*	4	U185	B	100 (45.4)
Pentachlorophenol	87865	Phenol, pentachloro-	10	1,2,4	U242	A	10 (4.54)
1,3-Pentadiene	504609	1-Methylbutadiene	1*	4	U186	B	100 (45.4)
Perchloroethylene	127184	Ethene, tetrachloro- Tetrachloro-ethene Tetrachlor-oethylene	1*	2,4	U210	B	100 (45.4)
Phenacetin	62442	Acetamide, N-(4-ethoxyphenyl)-	1*	4	U187	B	100 (45.4)
Phenanthrene	85018		1*	2		D	5000 (2270)
Phenol	108952	Benzene, hydroxy-	1000	1,2,4	U188	C	1000 (454)
Phenol, 2-chloro-	95578	o-Chlorophenol 2-Chlorophenol	1*	2,4	U048	B	100 (45.4)
Phenol, 4-chloro-3-methyl-	59507	p-Chloro-m-cresol 4-Chloro-m-cresol	1*	2,4	U039	D	5000 (2270)
Phenol, 2-cyclohexyl-4,6-dinitro-	131895	2-Cyclohexyl-4,6-dinitrophenol	1*	4	P034	B	100 (45.4)
Phenol, 2,4-dichloro-	120832	2,4-Dichlorophenol	1*	2,4	U081	B	100 (45.4)
Phenol, 2,6-dichloro	87650	2,6-Dichlorophenol	1*	4	U082	B	100 (45.4)
Phenol, 4,4'-(1,2-diethyl-1,2-ethenediyl)bis-, (E)	56531	Diethylstilbestrol	1*	4	U089	X	1 (0.454)
Phenol, 2,4-dimethyl-	105679	2,4-Dimethylphenol	1*	2,4	U101	B	100 (45.4)
Phenol, 2,4-dinitro-	51285	2,4-Dinitrophenol	1000	1,2,4	P048	A	10 (4.54)
Phenol, methyl-	1319773	Cresol(s) Cresylic acid	1000	1,4	U052	C	1000 (454)
m-Cresol	108394	m-Cresylic acid	1000	1,4	U052	C	1000 (454)
o-Cresol	95487	o-Cresylic acid	1000	1,4	U052	C	1000 (454)
p-Cresol	106445	p-Cresylic acid	1000	1,4	U052	C	1000 (454)
Phenol, 2-methyl-4,6-dinitro-	534521	4,6-Dinitro-o-cresol and salts	1*	2,4	P047	A	10 (4.54)
Phenol, 2,2'-methylenebis(3,4,6-trichloro-	70304	Hexachlorophene	1*	4	U132	B	100 (45.4)
Phenol, 2-(1-methylpropyl)-4,6-dinitro	88857	Dinoseb	1*	4	P020	C	1000 (454)

Hazardous Substance	CASRN	Regulatory Synonyms	Statutory			Final RQ	
			RQ	Code†	RCRA Waste #	Cate-gory	Pounds (Kg)
Phenol, 4-nitro-	100027	p-Nitrophenol 4-Nitrophenol	1000	1,2,4	U170	B	100 (45.4)
Phenol, pentachloro-	87865	Pentachlorophenol	10	1,2,4	U242	A	10 (4.54)
Phenol, 2,3,4,6-tetrachloro-	58902	2,3,4,6-Tetrachlorophenol	1*	4	U212	A	10 (4.54)
Phenol, 2,4,5-trichloro-	95954	2,4,5-Trichlorophenol	10	1,4	U230	A	10 (4.54)
Phenol, 2,4,6-trichloro-	88062	2,4,6-Trichlorophenol	10	1,2,4	U231	A	10 (4.54)
Phenol, 2,4,6-trinitro-, ammonium salt	131748	Ammonium picrate	1*	4	P009	A	10 (4.54)
L-Phenylalanine, 4-[bis(2-chloroethyl) aminol]	148823	Melphalan	1*	4	U150	X	1 (0.454)
1,10-(1,2-Phenylene)pyrene	193395	Indeno(1,2,3-cd)pyrene	1*	2,4	U137	B	100 (45.4)
Phenylmercury acetate	62384	Mercury, (acetato-O)phenyl-	1*	4	P092	B	100 (45.4)
Phenylthiourea	103855	Thiourea, phenyl-	1*	4	P093	B	100 (45.4)
Phorate	298022	Phosphorodithioic acid, O,O-diethyl S-(ethylthio), methyl ester	1*	4	P094	A	10 (4.54)
Phosgene	75445	Carbonic dichloride	5000	1,4	P095	A	10 (4.54)
Phosphine	7803512		1*	4	P096	B	100 (45.4)
Phosphoric acid	7664382		5000	1		D	5000 (2270)
Phosphoric acid, diethyl 4-nitrophenyl ester	311455	Diethyl-p-nitrophenyl phosphate	1*	4	P041	B	100 (45.4)
Phosphoric acid, lead(2+) salt (2:3)	7446277	Lead phosphate	1*	4	U145		#
Phosphorodithioic acid, O,O-diethyl S-[2-(ethylthio)ethyl]ester	298044	Disulfoton	1	1,4	P039	X	1 (0.454)
Phosphorodithioic acid, O,O-diethyl S-(ethylthio), methyl ester	298022	Phorate	1*	4	P094	A	10 (4.54)
Phosphorodithioic acid, O,O-diethyl S-methyl ester	3288582	O,O-Diethyl S-methyl dithiophosphate	1*	4	U087	D	5000 (2270)
Phosphorodithioic acid, O,O-dimethyl S-[2(methylamino)-2-oxoethyl] ester	60515	Dimethoate	1*	4	P044	A	10 (4.54)
Phosphorofluoridic acid, bis(1-methylethyl) ester	55914	Diisopropylfluorophosphate	1*	4	P043	B	100 (45.4)
Phosphorothioic acid, O,O-diethyl O-(4-nitrophenyl) ester	56382	Parathion	1	1,4	P089	A	10 (4.54)
Phosphorothioic acid, O,[4-[(dimethylamino)sulfonyl] phenyl]O,O-dimethyl ester	52857	Famphur	1*	4	P097	C	1000 (454)

Hazardous Substance	CASRN	Regulatory Synonyms	Statutory			Final RQ	
			RQ	Code†	RCRA Waste #	Cate-gory	Pounds (Kg)
Phosphorothioic acid, O,O-dimethyl O-(4-nitrophenyl) ester	298000	Methyl parathion	100	1,4	P071	B	100 (45.4)
Phosphorothioic acid, O,O-diethyl O-pyrazinyl ester	297972	O,O-Diethyl O-pyrazinyl phosphorothioate	1*	4	P040	B	100 (45.4)
Phosphorus	7723140		1	1		X	1 (0.454)
Phosphorus oxycloride	10025873		5000	1		C	1000 (454)
Phosphorus pentasulfide	1314803	Phosphorus sulfide Sulfur phosphide	100	1,4	U189	B	100 (45.4)
Phosphorus sulfide	1314803	Phosphorus pentasulfide Sulfur phosphide	100	1,4	U189	B	100 (45.4)
Phosophorus trichloride	7719122		5000	1		C	1000 (454)
PHTHALATE ESTERS	N/A		1*	2			**
Phthalic anhydride	85449	1,3-Isobenzofurandione	1*	4	U190	D	5000 (2270)
2-Picoline	109068	Pyridine, 2-methyl-	1*	4	U191	D	5000 (2270)
Piperidine, 1-nitroso-	100754	N-Nitrosopiperidine	1*	4	U179	A	10 (4.54)
Plumbane, tetraethyl-	78002	Tetraethyl lead	100	1,4	P110	A	10 (4.54)
POLYCHLORINATED BIPHENYLS (PCBs)	1336363		10	1,2		X	1 (0.454)
Aroclor 1016	12674112	POLYCHLORINATED BIPHENYLS (PCBs)					
Aroclor 1221	11104282	POLYCHLORINATED BIPHENYLS (PCBs)					
Aroclor 1232	11141165	POLYCHLORINATED BIPHENYLS (PCBs)					
Aroclor 1242	53469219	POLYCHLORINATED BIPHENYLS (PCBs)					
Aroclor 1248	12672296	POLYCHLORINATED BIPHENYLS (PCBs)					
Aroclor 1254	11097691	POLYCHLORINATED BIPHENYLS (PCBs)					
Aroclor 1260	11096825	POLYCHLORINATED BIPHENYLS (PCBs)					
POLYNUCLEAR AROMATIC HYDROCARBONS	N/A		1*	2			**
Potassium arsenate	7784410		1000	1		X	1 (0.454)
Potassium arsenite	10124502		1000	1		X	1 (0.454)
Potassium bichromate	7778509		1000	1		A	10 (4.54)
Potassium chromate	7789006		1000	1		A	10 (4.54)

Hazardous Substance	CASRN	Regulatory Synonyms	Statutory			Final RQ	
			RQ	Code†	RCRA Waste #	Category	Pounds (Kg)
Potassium cyanide	151508	Potassium cyanide K (CN)	10	1,4	P098	A	10 (4.54)
Potassium cyanide K(CN)	151508	Potassium cyanide	10	1,4	P098	A	10 (4.54)
Potassium hydroxide	1310583		1000	1		C	1000 (454)
Potassium permanganate	7722647		100	1		B	100 (45.4)
Potassium silver cyanide	506616	Argentate (1-), bis(cyano-C)-, potassium	1*	4	P099	X	1 (0.454)
Pronamide	23950585	Benzamide, 3.5-dichloro-N-(1,1-dimethyl-2-propynyl)-	1*	4	U192	D	5000 (2270)
Propanal, 2-methyl-2-(methylthio)-, O-[(methylamino)carbonyl]oxime	116063	Aldicarb	1*	4	P070	X	1 (0.454)
1-Propanamine	107108	n-Propylamine	1*	4	U194	D	5000 (2270)
1-Propanamine, N-propyl-	142847	Dipropylamine	1*	4	U110	D	5000 (2270)
1-Propanamine, N-nitroso-N-propyl-	621647	Di-n-propylnitrosamine	1*	2,4	U111	A	10 (4.54)
Propane, 1,2-dibromo-3-chloro-	96128	1,2-Dibromo-3-chloropropane	1*	4	U066	X	1 (0.454)
Propane, 2-nitro-	79469	2-Nitropropane	1*	4	U171	A	10 (4.54)
1,3-Propane sultone	1120714	1,2-Oxathiolane, 2,2-dioxide	1*	4	U193	A	10 (4.54)
Propane, 1,2-dichloro-	78875	Propylene dichloride 1,2-Dichloropropane	5000	1,2,4	U083	C	1000 (454)
Propanedinitrile	109773	Malononitrile	1*	4	U149	C	1000 (454)
Propanenitrile	107120	Ethyl cynide	1*	4	P101	A	10 (4.54)
Propanenitrile, 3-chloro-	542767	3-Chloropropionitrile	1*	4	P027	C	1000 (454)
Propanenitrile, 2-hydroxy-2-methyl-	75865	Acetone cyanohydrin 2-Methyllactonitrile	10	1,4	P069	A	10 (4.54)
Propane, 2,2'-oxybis[2-chloro-	108601	Dichloroisopropyl ether	1*	2,4	U027	C	1000 (454)
1,2,3-Propanetriol, trinitrate-	55630	Nitroglycerine	1*	4	P081	A	10 (4.54)
1-Propanol, 2,3-dibromo-, phosphate (3:1)	126727	Tris(2,3-dibromopropyl) phosphate	1*	4	U235	A	10 (4.54)
1-Propanol, 2-methyl-	78831	Isobutyl alcohol	1*	4	U140	D	5000 (2270)
2-Propanone	67641	Acetone	1*	4	U002	D	5000 (2270)
2-Propanone, 1-bromo-	598312	Bromoacetone	1*	4	P017	C	1000 (454)
Propargite	2312358		10	1		A	10 (4.54)
Propargyl alcohol	107197	2-Propyn-1-ol	1*	4	P102	C	1000 (454)
2-Propenal	107028	Acrolein	1	1,2,4	P003	X	1 (0.454)
2-Propenamide	79061	Acrylamide	1*	4	U007	D	5000 (2270)

Hazardous Substance	CASRN	Regulatory Synonyms	Statutory RQ	Code†	RCRA Waste #	Final RQ Category	Final RQ Pounds (Kg)
1-Propene, 1,1,2,3,3,3-hexachloro-	1888717	Hexachloropropene	1*	4	U243	C	1000 (454)
1-Propene, 1,3-dichloro-	542756	1,3-Dichloropropene	5000	1,2,4	U084	B	100 (45.4)
2-Propenenitrile	107131	Acrylonitrile	100	1,2,4	U009	B	100 (45.4)
2-Propenenitrile, 2-methyl-	126987	Methacrylonitrile	1*	4	U152	C	1000 (454)
2-Propenoic acid	79107	Acrylic acid	1*	4	U008	D	5000 (2270)
2-Propenoic acid, ethyl ester	140885	Ethyl acrylate	1*	4	U113	C	1000 (454)
2-Propenoic acid, 2-methyl-, ethyl ester	97632	Ethyl methacrylate	1*	4	U118	C	1000 (454)
2-Propenoic acid, 2-methyl-, methyl ester	80626	Methyl methacrylate	5000	1,4	U162	C	1000 (454)
2-Propen-1-ol	107186	Allyl alcohol	100	1,4	P005	B	100 (45.4)
Propionic acid	79094		5000	1		D	5000 (2270)
Propionic acid, 2-(2,4,5-trichlorophenoxy)-	93721	Silvex (2,4,5-TP) 2,4,5-TP acid	100	1,4	U233	B	100 (45.4)
Propionic anhydride	123626		5000	1		D	5000 (2270)
n-Propylamine	107108	1-Propanamine	1*	4	U194	D	5000 (2270)
Propylene dichloride	78875	Propane, 1,2-dichloro- 1,2-Dichloropropane	5000	1,2,4	U083	C	1000 (454)
Propylene oxide	75569		5000	1		B	100 (45.4)
1,2-Propylenimine	75558	Aziridine, 2-methyl-	1*	4	P067	X	1 (0.454)
2-Propyn-1-ol	107197	Propargyl alcohol	1*	4	P102	C	1000 (454)
Pyrene	129000		1*	2		D	5000 (2270)
Pyrethrins	121299		1000	1		X	1 (0.454)
	121211		1000	1		X	1 (0.454)
	8003347		1000	1		X	1 (0.454)
3,6-Pyridazinedione, 1,2-dihydro-	123331	Maleic hydrazide	1*	4	U148	D	5000 (2270)
4-Pyridinamine	504245	4-Aminopyridine	1*	4	P008	C	1000 (454)
Pyridine	110861		1*	4	U196	C	1000 (454)
Pyridine, 2-methyl-	109068	2-Picoline	1*	4	U191	D	5000 (2270)
Pyridine, 3-(1-methyl-2-pyrrolidinyl)-, (S)	54115	Nicotine, & salts	1*	4	P075	B	100 (45.4)
2,4-(1H,3H)-Pyrimidinedione, 5-[bis(2-chloroethyl)amino]-	66751	Uracil mustard	1*	4	U237	A	10 (4.54)
4(1H)-Pyrimidinone, 2,3-dihydro-6-methyl-2-thioxo-	56042	Methylthiouracil	1*	4	U164	A	10 (4.54)

Hazardous Substance	CASRN	Regulatory Synonyms	Statutory			Final RQ	
			RQ	Code†	RCRA Waste #	Cate-gory	Pounds (Kg)
Pyrrolidine, 1-nitroso-	930552	N-Nitrosopyrrolidine	1*	4	U180	X	1 (0.454)
Quinoline	91225		1000	1		D	5000 (2270)
RADIONUCLIDES	N/A		1*	3			§
Reserpine	50555	Yohimban-16-carboxylic acid, 11,17-dimethoxy-18-[(3,4,5-trimethoxybenzoyl)oxy-, methyl ester (3beta, 16beta, 17alpha, 18beta, 20alpha)-	1*	4	U200	D	5000 (2270)
Resorcinol	108463	1,3-Benzenediol	1000	1,4	U201	D	5000 (2270)
Saccharin and salts	81072	1,2-Benzisothiazol-3(2H)-one, 1,1-dioxide	1*	4	U202	B	100 (45.4)
Safrole	94597	1,3-Benzodioxole, 5-(2-propenyl)-	1*	4	U203	B	100 (45.4)
Selenious acid	7783008		1*	4	U204	A	10 (4.54)
Selenious acid, dithallium (1 +) salt	12039520	Thallium selenite	1*	4	P114	C	1000 (454)
Selenium††	7782492		1*	2		B	100 (45.4)
SELENIUM AND COMPOUNDS	N/A		1*	2			**
Selenium dioxide	7446084	Selenium oxide	1000	1,4	U204	A	10 (4.54)
Selenium oxide	7446084	Selenium dioxide	1000	1,4	U204	A	10 (4.54)
Selenium sulfide	7488564	Selenium sulfide SeS2	1*	4	U205	A	10 (4.54)
Selenium sulfide SeS2	7488564	Selenium sulfide	1*	4	U205	A	10 (4.54)
Selenourea	630104		1*	4	P103	C	1000 (454)
L-Serine, diazoacetate (ester)	115026	Azaserine	1*	4	U015	X	1 (0.454)
Silver††	7440224		1*	2		C	1000 (454)
SILVER AND COMPOUNDS	N/A		1*	2			**
Silver cyanide	506649	Silver cyanide Ag(CN)	1*	4	P104	X	1 (0.454)
Silver cyanide Ag (CN)	506649	Silver cyanide	1*	4	P104	X	1 (0.454)
Silver nitrate	7761888		1	1		X	1 (0.454)
Silvex (2,4,5-TP)	93721	Propionic acid, 2-(2,4,5-trichlorophenoxy)-2,4,5-TP acid	100	1,4	U233	B	100 (45.4)
Sodium	7440235		1000	1		A	10 (4.54)
Sodium arsenate	7631892		1000	1		X	1 (0.454)
Sodium arsenite	7784465		1000	1		X	1 (0.454)
Sodium azide	26628228		1*	4	P105	C	1000 (454)
Sodium bichromate	10588019		1000	1		A	10 (4.54)

Hazardous Substance	CASRN	Regulatory Synonyms	Statutory			Final RQ	
			RQ	Code †	RCRA Waste #	Cate-gory	Pounds (Kg)
Sodium bifluoride	1333831		5000	1		B	100 (45.4)
Sodium bisulfite	7631905		5000	1		D	5000 (2270)
Sodium chromate	7775113		1000	1		A	10 (4.54)
Sodium cyanide	143339	Sodium cyanide Na(CN)	10	1,4	P106	A	10 (4.54)
Sodium cyanide Na (CN)	143339	Sodium cyanide	10	1,4	P106	A	10 (4.54)
Sodium dodecylbenzenesulfonate	25155300		1000	1		C	1000 (454)
Sodium fluoride	7681494		5000	1		C	1000 (454)
Sodium hydrosulfide	16721805		5000	1		D	5000 (2270)
Sodium hydroxide	1310732		1000	1		C	1000 (454)
Sodium hypochlorite	7681529		100	1		B	100 (45.4)
	10022705		100	1		B	100 (45.4)
Sodium methylate	124414		1000	1		C	1000 (454)
Sodium nitrite	7632000		100	1		B	100 (45.4)
Sodium phosphate, dibasic	7558794		5000	1		D	5000 (2270)
	10039324		5000	1		D	5000 (2270)
	10140655		5000	1		D	5000 (2270)
Sodium phosphate, tribasic	7601549		5000	1		D	5000 (2270)
	7758294		5000	1		D	5000 (2270)
	7785844		5000	1		D	5000 (2270)
	10101890		5000	1		D	5000 (2270)
	10124568		5000	1		D	5000 (2270)
	10361894		5000	1		D	5000 (2270)
Sodium selenite	10102188		1000	1		B	100 (45.4)
	7782823						
Streptozotocin	18883664	D-Glucose, 2-deoxy-2-[[(methylnitrosoamino)-carbonyl]amino]-Glucopyranose, 2-deoxy-2-(3-methyl-3-nitrosoureido)-	1*	4	U206	X	1 (0.454)
Strontium chromate	7789062		1000	1		A	10 (4.54)
Strychnidin-10-one	57249	Strychnine, & salts	10	1,4	P108	A	10 (4.54)
Strychnidin-10-one, 2,3-dimethoxy-	357573	Brucine	1*	4	P018	B	100 (45.4)
Strychnine, & salts	57249	Strychnidin-10-one	10	1,4	P108	A	10 (4.54)

Hazardous Substance	CASRN	Regulatory Synonyms	Statutory			Final RQ	
			RQ	Code †	RCRA Waste #	Category	Pounds (Kg)
Styrene	100425		1000	1		C	1000 (454)
Sulfur monochloride	12771083		1000	1		C	1000 (454)
Sulfur phosphide	1314803	Phosphorus pentasulfide Phosphorus sulfide	100	1,4	U189	B	100 (45.4)
Sulfuric acid	7664939		1000	1		C	1000 (454)
	8014957		1000	1		C	1000 (454)
Sulfuric acid, dithallium (1+) salt	7446186	Thallium (I) sulfate	1000	1,4	P115	B	100 (45.4)
	10031591		1000	1,4	P115	B	100 (45.4)
Sulfuric acid, dimethyl ester	77781	Dimethyl sulfate	1*	4	U103	B	100 (45.4)
2,4,5-T acid	93765	Acetic acid, (2,4,5-trichlorophenoxy) 2,4,5-T	100	1,4	U232	C	1000 (454)
2,4,5-T amines	2008460		100	1		D	5000 (2270)
	1319728		100	1		D	5000 (2270)
	3813147		100	1		D	5000 (2270)
	6369966		100	1		D	5000 (2270)
	6369977		100	1		D	5000 (2270)
2,4,5-T esters	93798		100	1		C	1000 (454)
	1928478		100	1		C	1000 (454)
	2545597		100	1		C	1000 (454)
	25168154		100	1		C	1000 (454)
	61792072		100	1		C	1000 (454)
2,4,5-T salts	13560991		100	1		C	1000 (454)
2,4,5-T	93765	Acetic acid, (2,4,5-trichlorophenoxy) 2,4,5-T acid	100	1,4	U232	C	1000 (454)
TDE	72548	Benzene, 1,1'-(2,2-dichloroethylidene)bis[4-chloro-DDD 4,4' DDD	1	1,2,4	U060	X	1 (0.454)
1,2,4,5-Tetrachlorobenzene	95943	Benzene, 1,2,4,5-tetrachloro-	1*	4	U207	D	5000 (2270)
2,3,7,8-Tetrachlorodibenzo-p-dioxin (TCDD)	1746016		1*	2		X	1 (0.454)
1,1,1,2-Tetrachloroethane	630206	Ethane, 1,1,1,2-tetrachloro-	1*	4	U208	B	100 (45.4)
1,1,2,2-Tetrachloroethane	79345	Ethane, 1,1,2,2-tetrachloro-	1*	2,4	U209	B	100 (45.4)

Hazardous Substance	CASRN	Regulatory Synonyms	Statutory			Final RQ	
			RQ	Code †	RCRA Waste #	Cate-gory	Pounds (Kg)
Tetrachloroethene	127184	Ethene, tetrachloro- Perchloroethylene Tetrachloroethylene	1 *	2,4	U210	B	100 (45.4)
Tetrachloroethylene	127184	Ethene, tetrachloro- Perchloroethylene Tetrachloroethene	1 *	2,4	U210	B	100 (45.4)
2,3,4,6-Tetrachlorophenol	58902	Phenol, 2,3,4,6-tetrachloro-	1 *	4	U212	A	10 (4.54)
Tetraethyl lead	78002	Plumbane, tetraethyl-	100	1,4	P110	A	10 (4.54)
Tetraethyl pyrophosphate	107493	Diphosphoric acid, tetraethyl ester	100	1,4	P111	A	10 (4.54)
Tetraethyldithiopyrophosphate	3689245	Thiodiphosphoric acid, tetraethyl ester	1 *	4	P109	B	100 (45.4)
Tetrahydrofuran	109999	Furan, tetrahydro-	1 *	4	U213	C	1000 (454)
Tetranitromethane	509148	Methane, tetranitro-	1 *	4	P112	A	10 (4.54)
Tetraphosphoric acid, hexaethyl ester	757584	Hexaethyl tetraphosphate	1 *	4	P062	B	100 (45.4)
Thallic oxide	1314325	Thallium oxide Tl2O3	1 *	4	P113	B	100 (45.4)
Thallium † †	7440280		1 *	2		C	1000 (454)
Thallium and compounds	N/A		1 *	2			* *
Thallium (I) acetate	563688	Acetic acid, thallium (1 +) salt	1 *	4	U214	B	100 (45.4)
Thallium (I) carbonate	6533739	Carbonic acid, dithallium (1 +) salt	1 *	4	U215	B	100 (45.4)
Thallium (I) chloride	7791120	Thallium chlorice TlCl	1 *	4	U216	B	100 (45.4)
Thallium chloride TlCl	7791120	Thallium (I) chloride	1 *	4	U216	B	100 (45.4)
Thallium (I) nitrate	10102451	Nitric acid, thallium (1 +) salt	1 *	4	U217	B	100 (45.4)
Thallium oxide Tl2O3	1314325	Thallic oxide	1 *	4	P113	B	100 (45.4)
Thallium selenite	12039520	Selenious acid, dithallium (1 +) salt	1 *	4	P114	C	1000 (454)
Thallium (I) sulfate	7446186	Sulfuric acid, dithallium (1 +) salt	1000	1,4	P115	B	100 (45.4)
	10031591		1000	1,4	P115	B	100 (45.4)
Thioacetamide	62555	Ethanethioamide	1 *	4	U218	A	10 (4.54)
Thiodiphosphoric acid, tetraethyl ester	3689245	Tetraethyldithiopyrophosphate	1 *	4	P109	B	100 (45.4)
Thiofanox	39196184	2-Butanone, 3,3-dimethyl-1-(methylthio)-, O[(methylamino) carbonyl] oxime	1 *	4	P045	B	100 (45.4)
Thioimidodicarbonic diamide [(H2N)C(S)] 2NH	541537	Dithiobiuret	1 *	4	P049	B	100 (45.4)
Thiomethanol	74931	Methanethiol Methylmercaptan	100	1,4	U153	B	100 (45.4)

Hazardous Substance	CASRN	Regulatory Synonyms	Statutory			Final RQ	
			RQ	Code†	RCRA Waste #	Category	Pounds (Kg)
Thioperoxydicarbonic diamide [(H2N)C(S)] 2S2, tetramethyl-	137268	Thiram	1*	4	U244	A	10 (4.54)
Thiophenol	108985	Benzenethiol	1*	4	P014	B	100 (45.4)
Thiosemicarbazide	79196	Hydrazinecarbothioamide	1*	4	P116	B	100 (45.4)
Thiourea	62566		1*	4	U219	A	10 (4.54)
Thiourea, (2-chlorophenyl)-	5344821	1-(o-Chlorophenyl)thiourea	1*	4	P026	B	100 (45.4)
Thiourea, 1-naphthalenyl-	86884	alpha-Naphthylthiourea	1*	4	P072	B	100 (45.4)
Thiourea, phenyl-	103855	Phenylthiourea	1*	4	P093	B	100 (45.4)
Thiram	137268	Thioperoxydicarbonic diamide [(H2N)C(S)] 2S2, tetramethyl-	1*	4	U244	A	10 (4.54)
Toluene	108883	Benzene, methyl-	1000	1,2,4	U220	C	1000 (454)
Toluenediamine	95807	Benzenediamine, ar-methyl-	1*	4	U221	A	10 (4.54)
	496720		1*	4	U221	A	10 (4.54)
	823405		1*	4	U221	A	10 (4.54)
	25376458		1*	4	U221	A	10 (4.54)
Toluene diisocyanate	584849	Benzene, 1,3-diisocyanatomethyl-	1*	4	U223	B	100 (45.4)
	91087		1*	4	U223	B	100 (45.4)
	26471625		1*	4	U223	B	100 (45.4)
o-Toluidine	95534	Benzenamine, 2-methyl-	1*	4	U328	B	100 (45.4)
p-Toluidine	106490	Benzenamine, 4-methyl-	1*	4	U353	B	100 (45.4)
o-Toluidine hydrochloride	636215	Benzenamine, 2-methyl-, hydrochloride	1*	4	U222	B	100 (45.4)
Toxaphene	8001352	Camphene, octachloro-	1*	1,2,4	P123	X	1 (0.454)
2,4,5-TP acid	93721	Propionic acid 2-(2,4,5-trichlorophenoxy)- Silvex (2,4,5-TP)	100	1,4	U233	B	100 (45.4)
2,4,5-TP esters	32534955		100	1		B	100 (45.4)
1H-1,2,4-Triazol-3-amine	61825	Amitrole	1*	4	U011	A	10 (4.54)
Trichlorfon	52686		1000	1		B	100 (45.4)
1,2,4-Trichlorobenzene	120821		1*	2		B	100 (45.4)
1,1,1-Trichloroethane	71556	Ethane, 1,1,1-trichloro- Methyl chloroform	1*	2,4	U226	C	1000 (454)
1,1,2-Trichloroethane	79005	Ethane, 1,1,2-trichloro-	1*	2,4	U227	B	100 (45.4)
Trichloroethene	79016	Ethene, trichloro- Trichloroethylene	1000	1,2,4	U228	B	100 (45.4)

Hazardous Substance	CASRN	Regulatory Synonyms	Statutory			Final RQ	
			RQ	Code †	RCRA Waste #	Category	Pounds (Kg)
Trichloroethylene	79016	Ethene, trichloro- Trichloroethene	1000	1,2,4	U228	B	100 (45.4)
Trichloromethanesulfenyl chloride	594423	Methanesulfenyl chloride, trichloro-	1*	4	P118	B	100 (45.4)
Trichloromonofluoromethane	75694	Methane, trichlorofluoro-	1*	4	U121	D	5000 (2270)
Trichlorophenol	25167822		10	1		A	10 (4.54)
2,3,4-Trichlorophenol	15950660		10	1		A	10 (4.54)
2,3,5-Trichlorophenol	933788		10	1		A	10 (4.54)
2,3,6-Trichlorophenol	933755		10	1		A	10 (4.54)
2,4,5-Trichlorophenol	95954	Phenol, 2,4,5-trichloro-	10*	1,4	U230	A	10 (4.54)
2,4,6-Trichlorophenol	88062	Phenol, 2,4,6-trichloro-	10*	1,2,4	U231	A	10 (4.54)
3,4,5-Trichlorophenol	609198						
2,4,5-Trichlorophenol	95954	Phenol, 2,4,5-trichloro-	10*	1,4	U230	A	10 (4.54)
2,4,6-Trichlorophenol	88062	Phenol, 2,4,6-trichloro-	10	1,2,4	U231	A	10 (4.54)
Triethanolamine dodecylbenzenesulfonate	27323417		1000	1		C	1000 (454)
Triethylamine	121448		5000	1		D	5000 (2270)
Trimethylamine	75503		1000	1		B	100 (45.4)
1,3,5-Trinitrobenzene	99354	Benzene, 1,3,5-trinitro-	1*	4	U234	A	10 (4.54)
1,3,5-Trioxane, 2,4,6-trimethyl-	123637	Paraldehyde	1*	4	U182	C	1000 (454)
Tris(2,3-dibromopropyl) phosphate	126727	1-Propanol, 2,3-dibromo-, phosphate [(3:1)	1*	4	U235	A	10 (4.54)
Trypan blue	72571	2,7-Naphthalenedisulfonic acid, 3,3'-3,3'-dimethyl-(1,1'-biphenyl)-4,4'-diyl)-bis(azo)]bis(5-amino-4-hydroxy)-tetrasodium salt	1*	4	U236	A	10 (4.54)
Unlisted Hazardous Wastes Characteristic of Corrosivity	N/A		1*	4	D002	B	100 (45.4)
Unlisted Hazardous Wastes Characteristics: Characteristic of Toxicity:	N/A		1*	4			
Arsenic (D004)	N/A		*1	4	D004	X	1 (0.454)
Barium (D005)	N/A		*1	4	D005	C	1000 (454)
Benzene (D018)	N/A		1000	1,2, 3,4	D018	A	10 (4.54)
Cadmium (D006)	N/A		*1	4	D006	A	10 (4.54)
Carbon tetrachloride (D019)	N/A		5000	1,2,4	D019	A	10 (4.54)

Hazardous Substance	CASRN	Regulatory Synonyms	Statutory			Final RQ	
			RQ	Code †	RCRA Waste #	Category	Pounds (Kg)
Chlordane (D020)	N/A		1	1,2,4	D020	X	1 (0.454)
Chlorobenzene (D021)	N/A		100	1,2,4	D021	B	100 (45.4)
Chloroform (D022)	N/A		5000	1,2,4	D022	A	10 (4.54)
Chromium (D007)	N/A		*1	4	D007	A	10 (4.54)
o-Cresol (D023)	N/A		1000	1,4	D023	C	1000 (454)
m-Cresol (D024)	N/A		1000	1,4	D024	C	1000 (454)
p-Cresol (D025)	N/A		1000	1,4	D025	C	1000 (454)
Cresol (D026)	N/A		1000	1,4	D026	C	1000 (454)
2,4-D (D016)	N/A		100	1,4	D016	B	100 (45.4)
1,4-Dichlorobenzene (D027)	N/A		100	1,2,4	D027	B	100 (45.4)
1,2-Dichloroethane (D028)	N/A		5000	1,2,4	D028	B	100 (45.4)
1,1-Dichloroethylene (D029)	N/A		5000	1,2,4	D029	B	100 (45.4)
2,4-Dinitrotoluene (D030)	N/A		1000	1,2,4	D030	A	10 (4.54)
Endrin (D012)	N/A		1	1,4	D012	X	1 (0.454)
Heptachlor (and epoxide) (D031)	N/A		1	1,2,4	D031	X	1 (0.454)
Hexachlorobenzene (D032)	N/A		*1	2,4	D032	A	10 (4.54)
Hexachlorobutadiene (D033)	N/A		*1	2,4	D033	X	1 (0.454)
Hexachloroethane (D034)	N/A		*1	2,4	D034	B	100 (45.4)
Lead (D008)	N/A		*1	4	D008		(#)
Lindane (D013)	N/A		1	1,4	D013	X	1 (0.454)
Mercury (D009)	N/A		*1	4	D009	X	1 (0.454)
Methoxychlor (D014)	N/A		1	1,4	D014	X	1 (0.454)
Methyl ethyl ketone (D035)	N/A		*1	4	D035	D	5000 (2270)
Nitrobenzene (D036)	N/A		1000	1,2,4	D036	C	1000 (454)
Pentachlorophenol (D037)	N/A		10	1,2,4	D037	A	10 (4.54)
Pyridine (D038)	N/A		*1	4	D038	C	1000 (454)
Selenium (D010)	N/A		*1	4	D010	A	10 (4.54)
Silver (D011)	N/A		*1	4	D011	X	1 (0.454)
Tetrachloroethylene (D039)	N/A		*1	2,4	D039	B	100 (45.4)
Toxaphene (D015)	N/A		1	1,4	D015	X	1 (0.454)
Trichloroethylene (D040)	N/A		1000	1,2,4	D040	B	100 (45.4)
2,4,5-Trichlorophenol (D041)	N/A		10	1,4	D041	A	10 (4.54)

Hazardous Substance	CASRN	Regulatory Synonyms	Statutory			Final RQ	
			RQ	Code†	RCRA Waste #	Category	Pounds (Kg)
2,4,6-Trichlorophenol (D042)	N/A		10	1,2,4	D042	A	10 (4.54)
2,4,5-TP (D017)	N/A		100	1,4	D017	B	100 (45.4)
Vinyl chloride (D043)	N/A		*1	2,3,4	D043	X	1 (0.454)
Unlisted Hazardous Wastes Characteristic of Ignitability	N/A		1*	4	D001	B	100 (45.4)
Unlisted Hazardous Wastes Characteristic of Reactivity	N/A		1*	4	D003	B	100 (45.4)
Uracil mustard	66751	2,4-(1H,3H)-Pyrimidinedione, 5-[bis(2-chloroethyl)amino]-	1*	4	U237	A	10 (4.54)
Uranyl acetate	5411093		5000	1		B	100 (45.4)
Uranyl nitrate	10102064		5000	1		B	100 (45.4)
	36478769					B	
Urea, N-ethyl-N-nitroso-	759739	N-Nitroso-N-ethylurea	1*	4	U176	X	1 (0.454)
Urea, N-methyl-N-nitroso	684935	N-Nitroso-N-methylurea	1*	4	U177	X	1 (0.454)
Vanadic acid, ammonium salt	7803556	Ammonium vanadate	1*	4	P119	C	1000 (454)
Vanadium oxide V2O5	1314621	Vanadium pentoxide	1000	1,4	P120	C	1000 (454)
Vanadium pentoxide	1314621	Vanadium oxide V2O5	1000	1,4	P120	C	1000 (454)
Vanadyl sulfate	27774136		1000	1		C	1000 (454)
Vinyl chloride	75014	Ethene, chloro-	1*	2,3,4	U043	X	1 (0.454)
Vinyl acetate	108054	Vinyl acetate monomer	1000	1		D	5000 (2270)
Vinyl acetate monomer	108054	Vinyl acetate	1000	1		D	5000 (2270)
Vinylamine, N-methyl-N-nitroso-	4549400	N-Nitrosomethylvinylamine	1*	4	P084	A	10 (4.54)
Vinylidene chloride	75354	Ethene, 1,1-dichloro-1,1-Dichloroethylene	5000	1,2,4	U078	B	100 (45.4)
Warfarin, & salts, when present at concentrations greater than 0.3%	81812	2H-1-Benzopyran-2-one, 4-hydroxy-3-(3-oxo-1-phenyl-butyl)-, & salts, when present at concentrations greater than 0.3%	1*	4	P001	B	100 (45.4)
Xylene (mixed)	1330207	Benzene, dimethyl	1000	1,4	U239	C	1000 (454)
m-Benzene, dimethyl	108383	m-Xylene	1000	1,4	U239	C	1000 (454)
o-Benzene, dimethyl	95476	o-Xylene	1000	1,4	U239	C	1000 (454)
p-Benzene, dimethyl	106423	p-Xylene	1000	1,4	U239	C	1000 (454)

Hazardous Substance	CASRN	Regulatory Synonyms	Statutory			Final RQ	
			RQ	Code†	RCRA Waste #	Category	Pounds (Kg)
Xylenol	1300716		1000	1		C	1000 (454)
Yohimban-16-carboxylic acid, 11,17-dimethoxy-18-((3,4,5-trimethoxybenzoyl)oxy)-, methyl ester (3beta,16beta,17alpha,18beta,20alpha)-	50555	Reserpine	1*	4	U200	D	5000 (2270)
Zinc††	7440666		1*	2		C	1000 (454)
ZINC AND COMPOUNDS	N/A		1*	2			**
Zinc acetate	557346		1000	1		C	1000 (454)
Zinc ammonium chloride	52628258		5000	1		C	1000 (454)
	14639975		5000	1		C	1000 (454)
	14639986		5000	1		C	1000 (454)
Zinc borate	1332076		1000	1		C	1000 (454)
Zinc bromide	7699458		5000	1		C	1000 (454)
Zinc carbonate	3486359		1000	1		C	1000 (454)
Zinc chloride	7646857		5000	1		C	1000 (454)
Zinc cyanide	557211	Zinc cyanide Zn(CN)2	10	1,4	P121	A	10 (4.54)
Zinc cyanide Zn(CN)2	557211	Zinc cyanide	10	1,4	P121	A	10 (4.54)
Zinc fluoride	7783495		1000	1		C	1000 (454)
Zinc formate	557415		1000	1		C	1000 (454)
Zinc hydrosulfite	7779864		1000	1		C	1000 (454)
Zinc nitrate	7779886		5000	1		C	1000 (454)
Zinc phenolsulfonate	127822		5000	1		D	5000 (2270)
Zinc phosphide	1314847	Zinc phosphide Zn3P2, when present at concentrations greater than 10%	1000	1,4	P122	B	100 (45.4)
Zinc phosphide Zn3P2, when present at concentrations greater than 10%	1314847	Zinc phosphide	1000	1,4	P122	B	100 (45.4)
Zinc silicofluoride	16871719		5000	1		D	5000 (2270)
Zinc sulfate	7733020		1000	1		C	1000 (454)
Zirconium nitrate	13746899		5000	1		D	5000 (2270)
Zirconium potassium fluoride	16923958		5000	1		C	1000 (454)
Zirconium sulfate	14644612		5000	1		D	5000 (2270)
Zirconium tetrachloride	10026116		5000	1		D	5000 (2270)

Hazardous Substance	CASRN	Regulatory Synonyms	Statutory			Final RQ	
			RQ	Code†	RCRA Waste #	Cate-gory	Pounds (Kg)
F001			1*	4	F001	A	10 (4.54)
The following spent halogenated solvents used in degreasing; all spent solvent mixtures/blends used in degreasing containing, before use, a total of ten percent or more (by volume) of one or more of the above halogenated solvents or those solvents listed in F002, F004, and F005; and still bottoms from the recovery of these spent solvents and spent solvent mixtures.							
(a) Tetrachloroethylene	127184		1*	2,4	U210	B	100 (45.4)
(b) Trichloroethylene	79016		1000	1,2,4	U228	B	100 (45.4)
(c) Methylene chloride	75092		1*	2,4	U080	C	1000 (454)
(d) 1,1,1-Trichloroethane	71556		1*	2,4	U226	C	1000 (454)
(e) Carbon tetrachloride	56235		5000	1,2,4	U211	A	10 (4.54)
(f) Chlorinated fluorocarbons	N/A					D	5000 (2270)
F002			1*	2,4	F002	A	10 (4.54)
The following spent halogenated solvents; all spent solvent mixtures/blends containing, before use, a total of ten percent or more (by volume) of one or more of the above halogenated solvents or those solvents listed in F002,F004, and F005; and still bottoms from the recovery of these spent solvents and spent solvent mixtures.							
(a) Tetracholoroethylene	127184		1*	4	U210	B	100 (45.4)
(b) Methylene chloride	75092		1*	2,4	U080	C	1000 (454)
(c) Trichloroethylene	79016		1000	1,2,4	U228	B	100 (45.4)
(d) 1,1,1-Trichloroethane	71556		1*	2,4	U226	C	1000 (454)
(e) Chlorobenzene	108907		100	1,2,4	U037	B	100 (45.4)
(f) 1,1,2-Trichloro-1,2,2-trifluoroethane	76131					D	5000 (2270)
(g) o-Dischlorobenzene	95501		100	1,2,4	U070	B	100 (45.4)
(h) Trichlorofluoromethane	75694		1*	4	U121	D	5000 (2270)
(i) 1,1,2-Trichloroethane	79005		1*	2,4	U227	B	100 (45.4)
F003			1*	4	F003	B	100 (45.4)

Hazardous Substance	CASRN	Regulatory Synonyms	Statutory			Final RQ	
			RQ	Code†	RCRA Waste #	Cate-gory	Pounds (Kg)
The following spent non-halogenated solvents and the still bottoms from the recovery of these solvents:							
(a) Xylene	1330207					C	1000 (454)
(b) Acetone	67641					D	5000 (2270)
(c) Ethyl acetate	141786					D	5000 (2270)
(d) Ethylbenzene	100414					C	1000 (454)
(e) Ethyl ether	60297					B	100 (45.4)
(f) Methyl isobutyl ketone	108101					D	5000 (2270)
(g) n-Butyl alcohol	71363					D	5000 (2270)
(h) Cyclohexanone	108941					D	5000 (2270)
(i) Methanol	67561					D	5000 (2270)
F004			1*	4	F004	C	1000 (454)
The following spent non-halogenated solvents and the still bottoms from the recovery of these solvents:							
(a) Cresols/Cresylic acid	1319773		1000	1,4	U052	C	1000 (454)
(b) Nitrobenzene	98953		1000	1,2,4	U169	C	1000 (454)
F005			1*	4	F005	B	100 (45.4)
The following spent non-halogenated solvents and the still bottoms from the recovery of these solvents:							
(a) Toluene	108883		1000	1,2,4	U220	C	1000 (454)
(b) Methyl ethyl ketone	78933		1*	4	U159	D	5000 (2270)
(c) Carbon disulfide	75150		5000	1,4	P022	B	100 (45.4)
(d) Isobutanol	78831		1*	4	U140	D	5000 (2270)
(e) Pyridine	110861		1*	4	U196	C	1000 (454)
F006			1*	4	F006	A	10 (4.54)

Hazardous Substance	CASRN	Regulatory Synonyms	Statutory			Final RQ	
			RQ	Code †	RCRA Waste #	Category	Pounds (Kg)
Wastewater treatment sludges from electroplating operations except from the following processes: (1) sulfuric acid anodizing of aluminum, (2) tin plating on carbon steel, (3) zinc plating (segregated basis) on carbon steel, (4) aluminum or zinc-aluminum plating on carbon steel, (5) cleaning/stripping associated with tin, zinc and aluminum plating on carbon steel, and (6) chemical etching and milling of aluminum.							
F007			1*	4	F007	A	10 (4.54)
Spent cyanide plating bath solutions from electroplating operations.							
F008			1*	4	F008	A	10 (4.54)
Plating bath residues from the bottom of plating baths from electroplating operations where cyanides are used in the process.							
F009			1*	4	F009	A	10 (4.54)
Spent stripping and cleaning bath solutions from electroplating operations where cyanides are used in the process.							
F010			1*	4	F010	A	10 (4.54)
Quenching bath residues from oil baths from metal heat treating operations where cyanides are used in the process.							
F011			1*	4	F011	A	10 (4.54)
Spent cyanide solution from salt bath pot cleaning from metal heat treating operations.							
F012			1*	4	F012	A	10 (4.54)
Quenching wastewater treatment sludges from metal heat treating operations where cyanides are used in the process.							
F019			1	4	F019	A	10 (4.54)

Hazardous Substance	CASRN	Regulatory Synonyms	Statutory			Final RQ	
			RQ	Code †	RCRA Waste #	Category	Pounds (Kg)
Wastewater treatment sludges from the chemical conversion coating of aluminum except from zirconium phosphating in aluminum can washing when such phosphating is an exclusive conversion coating process.							
F020			1 *	4	F020	X	1 (0.454)
Wastes (except wastewater and spent carbon from hydrogen chloride purification) from the production or manufacturing use (as a reactant, chemical intermediate, or component in a formulating process) of tri-or-tetrachlorophenol, or of intermediates used to produce their pesticide derivatives. (This listing does not include wastes from the production of hexachlorophene from highly purified 2,4,5-trichlorophenol.)							
FO21			1 *	4	F021	X	1 (0.454)
Wastes (except wastewater and spent carbon from hydrogen chloride purification) from the production or manufacturing use (as a reactant, chemical intermediate, or component in a formulating process) of pentachlorophenol, or of intermediates used to produce its derivatives.							
F022			1 *	4	F022	X	1 (0.454)
Wastes (except wastewater and spent carbon from hydrogen chloride purification) from the manufacturing use (as a reactant, chemical intermediate, or component in a formulating process) of tetra-, penta-, or hexachlorobenzenes under alkaline conditions.							
F023			1 *	4	F023	X	1 (0.454)

Hazardous Substance	CASRN	Regulatory Synonyms	Statutory			Final RQ	
			RQ	Code†	RCRA Waste #	Category	Pounds (Kg)
Wastes (except wastewater and spent carbon from hydrogen chloride purification) from the production of materials on equipment previously used for the production or manufacturing use (as a reactant, chemical intermediate, or component in a formulating process) of tri- and tetrachlorophenols. (This listing does not include wastes from equipment used only for the production or use of hexachlorophene from highly purified 2,4,5-tri-chlorophenol.)							
F024			1*	4	F024	X	1 (0.454)
Wastes, including but not limited to distillation residues, heavy ends, tars, and reactor cleanout wastes, from the production of chlorinated aliphatic hydrocarbons, having carbon content from one to five, utilizing free radical catalyzed processes. (This listing does not include light ends, spent filters and filter aids, spent dessicants(sic), wastewater, wastewater treatment sludges, spent catalysts, and wastes listed in Section 261.32.)							
F025			1*	4	F025	X	##1 (0.454)
Condensed light ends, spent filters and filter aids, and spent dessicant wastes from the production of certain chlorinated aliphatic hydrocarbons, by free radical catalyzed processes. These chlorinated aliphatic hydrocarbons are those having carbon chain lengths ranging from one to and including five, with varying amounts and positions of chlorine substitution.							

Hazardous Substance	CASRN	Regulatory Synonyms	Statutory			Final RQ	
			RQ	Code†	RCRA Waste #	Category	Pounds (Kg)
F026			1*	4	F026	X	1 (0.454)
Wastes (except wastewater and spent carbon from hydrogen chloride purification) from the production of materials on equipment previously used for the manufacturing use (as a reactant, chemical intermediate, or component in a formulating process) of tetra-, penta-, or hexachlorobenzene under alkaline conditions.							
F027			1*	4	F027	X	1 (0.454)
Discarded unused formulations containing tri-, tetra-, or pentachlorophenol or discarded unused formulations containing compounds derived from these chlorophenols. (This listing does not include formulations containing hexachlorophene synthesized from prepurified 2,4,5-tri-chlorophenol as the sole component.)							
F028			1*	4	F028	X	1 (0.454)
Residues resulting from the incineration or thermal treatment of soil contaminated with EPA Hazardous Waste Nos. F020, F021, F022, F023, F026, and F027.							
F032			1*	4	F032	X	1 (0.454)
Wastewaters, process residuals, preservative drippage, and spent formulations from wood preserving processes generated at plants that currently use or have previously used chlorophenolic formulations (except wastes from processes that have had the F032 waste code deleted in accordance with §261.35 and do not resume or initiate use of chlorophenolic formulations). This listing does not include K001 bottom sediment sludge from the treatment of wastewater from wood preserving processes that use creosote and/or pentachlorophenol.							
F034			1*	4	F034	X	1 (0.454)

Hazardous Substance	CASRN	Regulatory Synonyms	Statutory			Final RQ	
			RQ	Code†	RCRA Waste #	Cate-gory	Pounds (Kg)
Wastewaters, process residuals, preservative drippage, and spent formulations from wood preserving processes generated at plants that use creosote formulations. This listing does not include K001 bottom sediment sludge from the treatment of wastewater from wood preserving processes that use creosote and/or pentachlorophenol.							
F035			1 *	4	F035	X	1 (0.454)
Wastewaters, process residuals, preservative drippage, and spent formulations from wood preserving processes generated at plants that use inorganic preservatives containing arsenic or chromium. This listing does not include K001 bottom sediment sludge from the treatment of wastewater from wood preserving processes that use creosote and/or penetachlorophenol.							
F037			1 *	4	F037	X	1 (0.454)

Hazardous Substance	CASRN	Regulatory Synonyms	Statutory			Final RQ	
			RQ	Code†	RCRA Waste #	Cate-gory	Pounds (Kg)
Petroleum refinery primary oil/water/solids separation sludge-- Any sludge generated from the gravitational separation of oil/water/solids during the storage or treatment of process wastewaters and oily cooling wastewaters from petroleum refineries. Such sludges include, but are not limited to, those generated in: oil/water/solids separators; tanks and impoundments; ditches and other conveyances; sumps; and stormwater units receiving dry weather flow. Sludge generated in stormwater units that do not receive dry weather flow, sludges generated from non-contact once-through cooling waters segregated for treatment from other process or oily cooling waters, sludges generated in aggressive biological treatment units as defined in §261.31(b)(2) (including sludges generated in one or more additional units after wastewaters have been treated in aggressive biological treatment units) and K051 wastes are not included in this listing.							
F038			1*	4	F038	X	1 (0.454)

Hazardous Substance	CASRN	Regulatory Synonyms	Statutory			Final RQ	
			RQ	Code†	RCRA Waste #	Category	Pounds (Kg)
Petroleum refinery secondary (emulsified) oil/water/solids separation sludge--Any sludge and/or float generated from the physical and/or chemical separation of oil/water/solids in process wastewaters and oily cooling wastewaters from petroleum refineries. Such wastes include, but are not limited to, all sludges and floats generated in: induced air flotation (IAF) units, tanks and impoundments, and all sludges generated in DAF units. Sludges generated in stormwater units that do not receive dry weather flow, sludges generated from once-through non-contact cooling waters segregated for treatment from other process or oil cooling wastes, sludges and floats generated in aggressive biological treatment units as defined in §261.31(b)(2) (including sludges and floats generated in one or more additional units after wastewaters have been treated in aggressive biological treatment units) and F037, K048, and K051 wastes are not included in this listing.							
K001			1*	4	K001	X	1 (0.454)
Bottom sediment sludge from the treatment of wastewaters from wood preserving processes that use creosote and/or pentachlorophenol.							
K002			1*	4	K002		#
Wastewater treatment sludge from the production of chrome yellow and organge pigments.							
K003			1*	4	K003		#
Wastewater treatment sludge from the production of molybdate orange pigments.							
K004			1*	4	K004	A	10 (4.54)
Wastewater treatment sludge from the production of zinc yellow pigments.							
K005			1*	4	K005		#

Hazardous Substance	CASRN	Regulatory Synonyms	Statutory			Final RQ	
			RQ	Code†	RCRA Waste #	Category	Pounds (Kg)
Wastewater treatment sludge from the production of chrome green pigments.							
K006			1*	4	K006	A	10 (4.54)
Wastewater treatment sludge from the production of chrome oxide green pigments (anhydrous and hydrated).							
K007			1*	4	K007	A	10 (4.54)
Wastewater treatment sludge from the production or iron blue pigments.							
K008			1*	4	K008	A	10 (4.54)
Oven residue from the production of chrome oxide green pigments.							
K009			1*	4	K009	A	10 (4.54)
Distillation bottoms from the production of acetaldehyde from ethylene.							
K010			1*	4	K010	A	10 (4.54)
Distillation side cuts from the production of acetaldehyde from ethylene.							
K011			1*	4	K011	A	10 (4.54)
Bottom stream from the wastewater stripper inthe production of acrylonitrile.							
K013			1*	4	K013	A	10 (4.54)
Bottom stream from the acetonitrile column in the production of acrylonitrile.							
K014			1*	4	K014	D	5000 (2270)
Bottoms from the acetonitrile purification column in the production of acrylonitrile.							
K015			1*	4	K015	A	10 (4.54)
Still bottoms from the distillation of benzyl chloride.							
K016			1*	4	K016	X	1 (0.454)

Hazardous Substance	CASRN	Regulatory Synonyms	Statutory			Final RQ	
			RQ	Code†	RCRA Waste #	Cate-gory	Pounds (Kg)
Heavy ends or distillation residues from the production of carbon tetrachloride.							
K017			1*	4	K017	A	10 (4.54)
Heavy ends (still bottoms) from the purification column in the production of epi-chlorohydrin.							
K018			1*	4	K018	X	1 (0.454)
Heavy ends from the fractionation column in ethyl chloride production.							
K019			1*	4	K019	X	1 (0.454)
Heavy ends from the distillation of ethylene dichloride in ethylene dichloride production.							
K020			1*	4	K020	X	1 (0.454)
Heavy ends from the distillation of vinyl chloride in vinyl chloride monomer production.							
K021			1*	4	K021	A	10 (4.54)
Aqueous spent antimony catalyst waste from fluoromethanes production.							
K022			1*	4	K022	X	1 (0.454)
Distillation bottom tars from the production of phenol/acetone from cumene.							
K023			1*	4	K023	D	5000 (2270)
Distillation light ends from the production of phthalic anhydride from naphthalene.							
K024			1*	4	K024	D	5000 (2270)
Distillation bottoms from the production of phthalic anhydride from naphthalene.							
K025			1*	4	K025	A	10 (4.54)
Distillation bottoms from the production of nitrobenzene by the nitration of benzene.							
K026			1*	4	K026	C	1000 (454)

Hazardous Substance	CASRN	Regulatory Synonyms	Statutory			Final RQ	
			RQ	Code †	RCRA Waste #	Category	Pounds (Kg)
Stripping still tails from the production of methyl ethyl pyridines.							
K027			1*	4	K027	A	10 (4,54)
Centrifuge and distillation residues from tolune diisocyanate production.							
K028			1*	4	K028	X	1 (0.454)
Spent catalyst from the hydrochlorinator reactor in the production of 1,1,1-trichloroethane.							
K029			1*	4	K029	X	1 (0.454)
Waste from the product steam stripper in the production of 1,1,1-trichloroethane.							
K030			1*	4	K030	X	1 (0.454)
Column bottoms or heavy ends from the combined production of trichloroethylene and perchloroethylene.							
K031			1*	4	K031	X	1 (0.454)
By-product salts generated in the production of MSMA and cacodylic acid.							
K032			1*	4	K032	A	10 (4,54)
Wastewater treatment sludge from the production of chlordane.							
K033			1*	4	K033	A	10 (4.54)
Wastewater and scrub water from the chlorination of cyclopentadiene in the production of chlordane.							
K034			1*	4	K034	A	10 (4.54)
Filter solids from the filtration of hexachlorocyclo-pentadiene in the production of chlordane.							
K035			1*	4	K035	X	1 (0.454)
Wastewater treatment sludges generated in the production of creosote.							
K036			1*	4	K036	X	1 (0.454)

Hazardous Substance	CASRN	Regulatory Synonyms	Statutory			Final RQ	
			RQ	Code†	RCRA Waste #	Category	Pounds (Kg)
Still bottoms from toluene reclamation distillation in the production of disulfoton.							
K037			1*	4	K037	X	1 (0.454)
Wastewater treatment sludges from the production of disulfoton.							
K038			1*	4	K038	A	10 (4.54)
Wastewater from the washing and stripping of phorate production.							
K039			1*	4	K039	A	10 (4.54)
Filter cake from the filtration of diethylphosphorodithioic acid in the production of phorate.							
K040			1*	4	K040	A	10 (4.54)
Wastewater treatment sludge from the production of phorate.							
K041			1*	4	K041	X	1 (0.454)
Wastewater treatment sludge from the production of toxaphene.							
K042			1*	4	K042	A	10 (4.54)
Heavy ends or distillation residues from the distillation of tetrachlorobenzene in the production of 2,4,5-T.							
K043			1*	4	K043	A	10 (4.54)
2,6-Dichlorophenol waste from the production of 2,4-D.							
K044			1*	4	K044	A	10 (4.54)
Wastewater treatment sludges from the manufacturing and processing of explosives.							
K045			1*	4	K045	A	10 (4.54)
Spent carbon from the treatment of wastewater containing explosives.							
K046			1*	4	K046	B	100 (45.4)
Wastewater treatment sludges from the manufacturing, formulation and loading of lead-based initiating compounds.							

Hazardous Substance	CASRN	Regulatory Synonyms	Statutory			Final RQ	
			RQ	Code †	RCRA Waste #	Category	Pounds (Kg)
K047			1 *	4	K047	A	10 (4,54)
Pink/red water from TNT operations.							
K048			1 *	4	K048		#
Dissolved air flotation (DAF) float from the petroleum refining industry.							
K049			1 *	4	K049		#
Slop oil emulsion solids from the petroleum refining industry.							
K050			1 *	4	K050	A	10 (4.54)
Heat exchanger bundle cleaning sludge from the petroleum refining industry.							
K051			1 *	4	K051		#
API separator sludge from the petroleum refining industry.							
K052			1 *	4	K052	A	10 (4.54)
Tank bottoms (leaded) from the petroleum refining industry.							
K060			1 *	4	K060	X	1 (0,454)
Ammonia still lime sludge coking operations.							
K061			1 *	4	K061		#
Emission control dust/sludge from the primary production of steel in electric furnances.							
K062			1 *	4	K062		#
Spent pickle liquor generated by steel finishing operations of facilities within the iron and steel industry (SIC Codes 331 and 332).							
K064			1 *	4	K064		##
Acid plant blowdown slurry/sludge resulting from thickening of blowdown slurry from primary copper production.							
K065			1 *	4	K065		##

Hazardous Substance	CASRN	Regulatory Synonyms	Statutory			Final RQ	
			RQ	Code †	RCRA Waste #	Cate-gory	Pounds (Kg)
Surace impoundment solids contained in and dredged from surface impoundments at primary lead smelting facilities.							
K066			1 *	4	K066		##
Sludge from treatment of process wastewater and/or acid plant blowdown from primary zinc production.							
K069			1 *	4	K069		#
Emission control dust/sludge from secondary lead smelting.							
K071			1 *	4	K071	X	1 (0.454)
Brine purification muds from the mercury cell process in chlorine production, where separately prepurified brine is not used.							
K073			1 *	4	K073	A	10 (4.54)
Chlorinated hydrocarbon waste from the purification step of the diaphragm cell process using graphite anodes in chlorine production.							
K083			1 *	4	K083	B	100 (45.4)
Distillatin bottoms from aniline extraction.							
K084			1 *	4	K084	X	1 (0.454)
Wastewater treatment sludges generated during the production of veterinary pharmaceuticals from arsenic or organo-arsenic compounds.							
K085			1 *	4	K085	A	10 (4.54)
Distillation or fractionation column bottoms from the production of chlorobenzenes.							
K086			1 *	4	K086		#

Hazardous Substance	CASRN	Regulatory Synonyms	Statutory			Final RQ	
			RQ	Code †	RCRA Waste #	Cate-gory	Pounds (Kg)
Solvent washes and sludges, caustic washes and sludges, or water washes and sludges from cleaning tubs and equipment used in the formulation of ink from pigments, driers, soaps, and stabilizers containing chromium and lead.							
K087			1*	4	K087	B	100 (45.4)
Decanter tank tar sludge from coking operations.							
K088			1*	4	K088		
Spent potliners from primary aluminum reduction.							
K090			1*	4	K090		
Emission control dust or sludge from ferrochromiumsilicon production.							
K091			1	4	K091		
Emission control dust or sludge from ferrochromium production.							
K093			1*	4	K093	D	5000 (2270)
Distillation light ends from the production of phthalic anhydride from ortho-xylene.							
K094			1*	4	K094	D	5000 (2270)
Distillation bottoms from the production of phthalic anhydride from ortho-xylene.							
K095			1*	4	K095	B	100 (45.4)
Distillation bottoms from the production of 1,1,1-trichloroethane.							
K096			1*	4	K096	B	100 (45.4)
Heavy ends from the heavy ends column from the production of 1,1,1-trichloroethane.							
K097			1*	4	K097	X	1 (0.454)
Vacuum stripper discharge from the chlordane chlorinator in the production of chlordane.							
K098			1*	4	K098	X	1 (0.454)

Hazardous Substance	CASRN	Regulatory Synonyms	Statutory			Final RQ	
			RQ	Code†	RCRA Waste #	Cate-gory	Pounds (Kg)
Untreated process wastewater from the production of toxaphene.							
K099			1*	4	K099	A	10 (4.54)
Untreated wastewater from the production of 2,4-D.							
K100			1*	4	K100		#
Waste leaching solution from acid leaching of emission control dust/sludge from secondary lead smelting.							
K101			1*	4	K101	X	1 (0.454)
Distillation tar residues from the distillation of aniline-based compounds in the production of veterinary pharmaceuticals from arsenic or organo-arsenic compounds.							
K102			1*	4	K102	X	1 (0.454)
Residue from the use of activated carbon for decolorization in the production of veterinary pharmaceuticals from arsenic or organo-arsenic compounds.							
K103			1*	4	K103	B	100 (45.4)
Process residues from aniline extraction from the production of aniline.							
K104			1*	4	K104	A	10 (4.54)
Combined wastewater streams generated from nitrobenzene/aniline production.							
K105			1*	4	K105	A	10 (4.54)
Separated aqueous stream from the reactor product washing step in the production of chlorobenzenes.							
K106			1*	4	K106	X	1 (0.454)
Wastewater treatment sludge from the mercury cell process in chlorine production.							
K107			10	4	K107	X	10 (4.54)

Hazardous Substance	CASRN	Regulatory Synonyms	Statutory			Final RQ	
			RQ	Code†	RCRA Waste #	Category	Pounds (Kg)
Column bottoms from product separation from the production of 1,1-dimethylhydrazine (UDMH) from carboxylic acid hydrazines.							
K108			10	4	K108	X	10 (4.54)
Condensed column overheads from product separation and condensed reactor vent gases from the production of 1,1-dimethylhydrazine (UDMH) from carboxylic acid hydrazides.							
K109			10	4	K109	X	10 (4.54)
Spent filter cartridges from product purification from the production of 1,1-dimethylhydrazine (UDMH) from carboxylic acid hydrazides.							
K110			10	4	K110	X	10 (4.54)
Condensed column overheads from intermediate separation from the production of 1,1-dimethylhydrazine (UDMH) from carboxylic acid hydrazides.							
K111			1*	4	K111	A	10 (4.54)
Product washwaters from the production of dinitrotoluene via nitration of toluene.							
K112			1*	4	K112	A	10 (4.54)
Reaction by-product water from the drying column in the production of toluenediamine via hydrogenation of dinitrotoluene.							
K113			1*	4	K113	A	10 (4.54)
Condensed liquid light ends from the purification of toluenediamine in the production of toluenediamine via hydrogenation of dinitrotoluene.							
K114			1*	4	K114	A	10 (4.54)
Vicinals from the purification of toluenediamine in the production of toluenediamine via hydrogenation of dinitrotoluene.							
K115			1*	4	K115	A	10 (4.54)

Hazardous Substance	CASRN	Regulatory Synonyms	Statutory			Final RQ	
			RQ	Code†	RCRA Waste #	Cate-gory	Pounds (Kg)
Heavy ends from the purification of toluenediamine in the production of toluenediamine via hydrogenation of dinitrotoluene.							
K116			1*	4	K116	A	10 (4.54)
Organic condensate from the solvent recovery column in the production of toluene diisocyanate via phosgenation of toluenediamine.							
K117			1*	4	K117	X	1 (0.454)
Wastewater from the reaction vent gas scrubber in the production of ethylene bromide via bromination of ethene.							
K118			1*	4	K118	X	1 (0.454)
Spent absorbent solids from purification of ethylene dibromide in the production of ethylene dibromide.							
K123			1*	4	K123	A	10 (4.54)
Process wastewater (including supernates, filtrates, and washwaters) from the production of ethylene-bisdithiocarbamic acid and its salts.							
K124			1*	4	K124	A	10 (4.54)
Reactor vent scrubber water from the production of ethylenebisdithiocarbamic acid and its salts.							
K125			1*	4	K125	A	10 (4.54)
Filtration, evaporation, and centrifugation solids from the production of ethylenebisdithiocarbamic acid and its salts.							
K126			1*	4	K126	A	10 (4.54)
Baghouse dust and floor sweepings in milling and packaging operations from the production or formulation of ethylenebisdithiocarbamic acid and its salts.							
K131			100	4	K131	X	100 (45.4)

Hazardous Substance	CASRN	Regulatory Synonyms	Statutory			Final RQ	
			RQ	Code†	RCRA Waste #	Cate-gory	Pounds (Kg)
Wastewater from the reactor and spent sulfuric acid from the acid dryer in the production of methyl bromide.							
K132			1000	4	K132	X	1000 (454)
Spent absorbent and wastewater solids from the production of methyl bromide.							
K136			1*	4	K136	X	1 (0.454)
Still bottoms from the purification of ethylene dibromide in the production of ethylene dibromide via bromination of ethene.							

†Indicates the statutory source as defined by 1,2,3, and 4 below.
††No reporting of releases of this hazardous substance is required if the diameter of the pieces of the solid metal released is equal to or exceeds 100 micrometers (0.004 inches).
†††The RQ for asbestos is limited to friable forms only.
1--Indicates that the statutory source for designation of this hazardous substance under CERCLA is CWA Section 311(b)(4).
2--Indicates that the statutory source for designation of this hazardous substance under CERCLA is CWA Section 307(a).
3--Indicates that the statutory source for designation of this hazardous substance under CERCLA is CAA Section 112.
4--Indicates that the statutory source for designation of this hazardous substance under CERCLA is RCRA Section 3001.
1*--Indicates that the 1-pound RQ is a CERCLA statutory RQ.
#Indicates that the RQ is subject to change when the assessment of potential carcinogenicity is completed.
##The Agency may adjust the statutory RQ for this hazardous substance in a future rulemaking; until then the statutory RQ applies.
§--The adjusted RQs for radionuclides may be found in Appendix 2 to this table.
**--Indicates that no RQ is being assigned to the generic or broad class.

APPENDIX 1 TO § 302.4—SEQUENTIAL CAS REGISTRY NUMBER LIST OF CERCLA HAZARDOUS SUBSTANCES

CASRN	Hazardous substance
50000	Formaldehyde.
50077	Azirino[2',3':3,4]pyrrolo[1,2-a]indole-4,7-dione,6-amino-8-[[(aminocarbonyl)oxy]methyl]-1,1a,2,8,8a,8b-hexahydro-8a-methoxy-5-methyl-, [1aS-(1aalpha, 8beta,8aalpha,8balpha)]-
	Mitomycin C.
50180	Cyclophosphamide.
	2H-1,3,2-Oxazaphosphorin-2-amine, N,N-bis(2-chloroethyl)tetrahydro-, 2-oxide.
50293	Benzene, 1,1'-(2,2,2-trichloroethylidene)bis[4-chloro-.
	DDT.
	4,4'DDT.
50328	Benzo[a]pyrene.
	3,4-Benzopyrene.
50555	Reserpine.
	Yohimban-16-carboxylic acid,11,17-dimethoxy-18-[(3 ,4,5-trimethoxybenzoyl)oxy]-, methyl ester (3beta,16beta,17alpha,18beta,20alpha)-
51285	Phenol, 2,4-dinitro-.
	2,4-Dinitrophenol.
51434	Epinephrine.
	1,2-Benzenediol,4-[1-hydroxy-2-(methylamino) ethyl]-.
51796	Carbamic acid, ethyl ester.
	Ethyl carbamate (urethane).
52686	Trichlorfon.
52857	Famphur.
	Phosphorothioic acid, O,[4-[(dimethyl- amino) sulfonyl]phenyl]O,O-dimethyl ester.
53703	Dibenz[a,h]anthracene.
	Dibenzo[a,h]anthracene.
	1,2:5,6-Dibenzanthracene.
53963	Acetamide, N-9H-fluoren-2-yl-.
	2-Acetylaminofluorene.
54115	Nicotine, & salts.
	Pyridine, 3-(1-methyl-2-pyrrolidinyl)-, (S)-.
55185	Ethanamine, N-ethyl-N-nitroso-.
	N-Nitrosodiethylamine.
55630	Nitroglycerine.
	1,2,3-Propanetriol, trinitrate-.
55914	Diisopropylfluorophosphate.
	Phosphorofluoridic acid, bis(1-methyl- ethyl) ester.
56042	Methylthiouracil.
	4(1H)-Pyrimidinone, 2,3-dihydro-6-methyl-2-thioxo-.
56235	Carbon tetrachloride.
	Methane, tetrachloro-.
56382	Parathion.
	Phosphorothioic acid, O,O-diethyl O-(4-nitrophenyl) ester.
56495	Benz[j]aceanthrylene, 1,2-dihydro-3-methyl-.
	3-Methylcholanthrene.
56531	Diethylstilbestrol.
	Phenol, 4,4'-(1,2-diethyl-1,2-ethenediyl)bis-, (E).
56553	Benz[a]anthracene.
	Benzo[a]anthracene.
	1,2-Benzanthracene.
56724	Coumaphos.
57125	Cyanides (soluble salts and complexes) not otherwise specified.
57147	Hydrazine, 1,1-dimethyl-.
	1,1-Dimethylhydrazine.
57249	Strychnidin-10-one.

APPENDIX 1 TO § 302.4—SEQUENTIAL CAS REGISTRY NUMBER LIST OF CERCLA HAZARDOUS SUBSTANCES—Continued

CASRN	Hazardous substance
	Strychnine, & salts.
57749	Chlordane.
	Chlordane, alpha & gamma isomers.
	Chlordane, technical.
	4,7-Methano-1H-indene, 1,2,4,5,6,7,8,8-octachloro- 2,3,3a,4,7,7a-hexahydro-.
57976	1,2-Benzanthracene, 7,12-dimethyl-.
	7,12-Dimethylbenz[a]anthracene.
58899	Cyclohexane, 1,2,3,4,5,6-hexachloro-, (1alpha,2alpha,3beta,4alpha,5alpha, 6beta)-.
	gamma - BHC.
	Hexachlorocyclohexane (gamma isomer).
	Lindane.
58902	Phenol, 2,3,4,6-tetrachloro-.
	2,3,4,6-Tetrachlorophenol.
59507	p-Chloro-m-cresol.
	Phenol, 4-chloro-3-methyl-.
	4-Chloro-m-cresol.
60004	Ethylenediamine-tetraacetic acid (EDTA).
60117	Benzenamine, N,N-dimethyl-4-(phenylazo-).
	p-Dimethylaminoazobenzene.
60297	Ethane, 1,1'-oxybis-.
	Ethyl ether.
60344	Hydrazine, methyl-.
	Methyl hydrazine.
60515	Dimethoate.
	Phosphorodithioic acid, O,O-dimethyl S-[2(methylamino)-2-oxoethyl] ester.
60571	Dieldrin.
	2,7:3,6-Dimethanonaphth[2,3-b]oxirene, 3,4,5,6,9,9-hexachloro-1a,2, 2a,3,6,6a,7,7a-octahydro-, (1aalpha,2beta,2aalpha,3beta,6beta, 6aalpha,7beta, 7aalpha)-.
61825	Amitrole.
	1H-1,2,4-Triazol-3-amine.
62384	Mercury, (acetato-O)phenyl-.
	Phenylmercury acetate.
62442	Acetamide, N-(4-ethoxyphenyl)-.
	Phenacetin.
62500	Ethyl methanesulfonate.
	Methanesulfonic acid, ethyl ester.
62533	Aniline.
	Benzenamine.
62555	Ethanethioamide.
	Thioacetamide.
62566	Thiourea.
62737	Dichlorvos.
62748	Acetic acid, fluoro-, sodium salt.
	Fluoroacetic acid, sodium salt.
62759	Methanamine, N-methyl-N-nitroso-.
	N-Nitrosodimethylamine.
63252	Carbaryl.
64186	Formic acid.
64197	Acetic acid.
65850	Benzoic acid.
66751	Uracil mustard.
	2,4-(1H,3H)-Pyrimidinedione, 5-[bis(2-chloroethyl) amino]-.
67561	Methanol.
	Methyl alcohol.
67641	Acetone.
	2-Propanone.
67663	Chloroform.
	Methane, trichloro-.
67721	Ethane, hexachloro-.
	Hexachloroethane.
70257	Guanidine, N-methyl-N'-nitro-N-nitroso-.

APPENDIX 1 TO § 302.4—SEQUENTIAL CAS REGISTRY NUMBER LIST OF CERCLA HAZARDOUS SUBSTANCES—Continued

APPENDIX 1 TO § 302.4—SEQUENTIAL CAS REGISTRY NUMBER LIST OF CERCLA HAZARDOUS SUBSTANCES—Continued

CASRN	Hazardous substance
	MNNG.
70304	Hexachlorophene.
	Phenol, 2,2'-methylenebis[3,4,6-tri- chloro-.
71363	n-Butyl alcohol.
	1-Butanol.
71432	Benzene.
71556	Ethane, 1,1,1-trichloro-.
	Methyl chloroform.
	1,1,1-Trichloroethane.
72208	Endrin.
	Endrin, & metabolites.
	2,7:3,6-Dimethanonaphth[2,3-b]oxirene, 3,4,5,6,9,9-hexachloro-1a,2,2a,3,6,6a,7,7a-octa-hydro-, (1aalpha,2beta,2abeta,3alpha,6alpha, 6abeta,7beta,7aalpha)-.
72435	Benzene, 1,1'-(2,2,2-trichloroethylidene)bis[4-methoxy-.
	Methoxychlor.
72548	Benzene, 1,1'-(2,2-dichloroethylidene)bis[4-chloro-.
	DDD.
	TDE.
	4,4' DDD.
72559	DDE.
	4,4' DDE.
72571	Trypan blue.
	2,7-Naphthalenedisulfonic acid, 3,3'-[(3,3'-dimethyl-(1,1'-biphenyl)-4,4'-diyl)-bis(azo)]bis(5-amino-4-hydroxy)-tetrasodium salt.
74839	Methane, bromo-.
	Methyl bromide.
74873	Methane, chloro-.
	Methyl chloride.
74884	Methane, iodo-.
	Methyl iodide.
74895	Monomethylamine.
74908	Hydrocyanic acid.
	Hydrogen cyanide.
74931	Methanethiol.
	Methylmercaptan.
	Thiomethanol.
74953	Methane, dibromo-.
	Methylene bromide.
75003	Chloroethane.
75014	Ethene, chloro-.
	Vinyl chloride.
75047	Monoethylamine.
75058	Acetonitrile.
75070	Acetaldehyde.
	Ethanal.
75092	Methane, dichloro-.
	Methylene chloride.
75150	Carbon disulfide.
75207	Calcium carbide.
75218	Ethylene oxide.
	Oxirane.
75252	Bromoform.
	Methane, tribromo-.
75274	Dichlorobromomethane.
75343	Ethane, 1,1-dichloro-.
	Ethylidene dichloride.
	1,1-Dichloroethane.
75354	Ethene, 1,1-dichloro-.
	Vinylidene chloride.
	1,1-Dichloroethylene.
75365	Acetyl chloride.

CASRN	Hazardous substance
75445	Carbonic dichloride.
	Phosgene.
75503	Trimethylamine.
75558	Aziridine, 2-methyl-.
	1,2-Propylenimine.
75569	Propylene oxide.
75605	Arsinic acid, dimethyl-.
	Cacodylic acid.
75649	tert-Butylamine.
75694	Methane, trichlorofluoro-.
	Trichloromonofluoromethane.
75718	Dichlorodifluoromethane.
	Methane, dichlorodifluoro-.
75865	Acetone cyanohydrin.
	Propanenitrile, 2-hydroxy-2-methyl-.
	2-Methyllactonitrile.
75876	Acetaldehyde, trichloro-.
	Chloral.
75990	2,2-Dichloropropionic acid.
76017	Ethane, pentachloro-.
	Pentachloroethane.
76448	Heptachlor.
	4,7-Methano-1H-indene, 1,4,5,6,7,8,8-heptachloro-3a,4,7,7a-tetrahydro-.
77474	Hexachlorocyclopentadiene.
	1,3-Cyclopentadiene, 1,2,3,4,5,5-hexa- chloro-.
77781	Dimethyl sulfate.
	Sulfuric acid, dimethyl ester.
78002	Plumbane, tetraethyl-.
	Tetraethyl lead.
78591	Isophorone.
78795	Isoprene.
78819	iso-Butylamine.
78831	Isobutyl alcohol.
	1-Propanol, 2-methyl-.
78875	Propane, 1,2-dichloro-.
	Propylene dichloride.
	1,2-Dichloropropane.
78886	2,3-Dichloropropene.
78933	Methyl ethyl ketone (MEK).
	2-Butanone.
78999	1,1-Dichloropropane.
79005	Ethane, 1,1,2-trichloro-.
	1,1,2-Trichloroethane.
79016	Ethene, trichloro-.
	Trichloroethene.
	Trichloroethylene-.
79061	Acrylamide.
	2-Propenamide.
79094	Propionic acid.
79107	Acrylic acid.
	2-Propenoic acid.
79196	Hydrazinecarbothioamide.
	Thiosemicarbazide.
79221	Carbonochloridic acid, methyl ester.
	Methyl chlorocarbonate.
	Methyl chloroformate.
79312	iso-Butyric acid.
79345	Ethane, 1,1,2,2-tetrachloro-.
	1,1,2,2-Tetrachloroethane.
79447	Carbamic chloride, dimethyl-.
	Dimethylcarbamoyl chloride.
79469	Propane, 2-nitro-.
	2-Nitropropane.
80159	alpha,alpha-Dimethylbenzylhydroperoxide.
	Hydroperoxide, 1-methyl-1-phenylethyl-.
80626	Methyl methacrylate.

CASRN	Hazardous substance
	2-Propenoic acid, 2-methyl-, methyl ester.
81072	Saccharin and salts.
	1,2-Benzisothiazol-3(2H)-one, 1,1-dioxide.
81812	Warfarin, & salts, when present at concentrations greater than 0.3%.
	2H-1-Benzopyran-2-one, 4-hydroxy-3-(3-oxo-1-phenyl -butyl)-, & salts, when present at concentrations greater than 0.3%.
82688	Benzene, pentachloronitro-.
	Pentachloronitrobenzene (PCNB).
83329	Acenaphthene.
84662	Diethyl phthalate.
	1,2-Benzenedicarboxylic acid, diethyl ester.
84742	Di-n-butyl phthalate.
	Dibutyl phthalate.
	n-Butyl phthalate.
	1,2-Benzenedicarboxylic acid, dibutyl ester.
85007	Diquat.
85018	Phenanthrene.
85449	Phthalic anhydride.
	1,3-Isobenzofurandione.
85687	Butyl benzyl phthalate.
86306	N-Nitrosodiphenylamine.
86500	Guthion.
86737	Fluorene.
86884	alpha-Naphthylthiourea.
	Thiourea, 1-naphthalenyl-.
87650	Phenol, 2,6-dichloro-.
	2,6-Dichlorophenol.
87683	Hexachlorobutadiene.
	1,3-Butadiene, 1,1,2,3,4,4-hexachloro-.
87865	Pentachlorophenol.
	Phenol, pentachloro-.
88062	Phenol, 2,4,6-trichloro-.
	2,4,6-Trichlorophenol.
88722	o-Nitrotoluene.
88755	o-Nitrophenol.
	2-Nitrophenol.
88857	Dinoseb.
	Phenol, 2-(1-methylpropyl)-4,6-dinitro.
91087	Benzene, 1,3-diisocyanatomethyl-.
	Toluene diisocyanate.
91203	Naphthalene.
91225	Quinoline.
91587	beta-Chloronaphthalene.
	Naphthalene, 2-chloro-.
	2-Chloronaphthalene.
91598	beta-Naphthylamine.
	2-Naphthalenamine.
91805	Methapyrilene.
	1,2-Ethanediamine, N,N-dimethyl-N'-2-pyridinyl-N'- (2-thienylmethyl)-.
91941	[1,1'-Biphenyl]-4,4'diamine,3,3'dichloro-.
	3,3'-Dichlorobenzidine.
92875	(1,1'-Biphenyl)-4,4'diamine.
	Benzidine.
93721	Propionic acid, 2-(2,4,5-trichlorophenoxy)-.
	Silvex (2,4,5-TP).
	2,4,5-TP acid.
93765	Acetic acid, (2,4,5-trichlorophenoxy).
	2,4,5-T.
	2,4,5-T acid.
93798	2,4,5-T esters.
94111	2,4-D Ester.
94586	Dihydrosafrole.
	1,3-Benzodioxole, 5-propyl-.
94597	Safrole.

CASRN	Hazardous substance
	1,3-Benzodioxole, 5-(2-propenyl)-.
94757	Acetic acid (2,4-dichlorophenoxy)-.
	2,4-D Acid.
	2,4-D, salts and esters.
94791	2,4-D Ester.
94804	2,4-D Ester.
95476	o-Benzene, dimethyl.
	o-Xylene.
95487	o-Cresol.
	o-Cresylic acid.
95501	Benzene, 1,2-dichloro-.
	o-Dichlorobenzene.
	1,2-Dichlorobenzene.
95534	Benzenamine, 2-methyl-.
	o-Toluidine.
95578	o-Chlorophenol.
	Phenol, 2-chloro-.
	2-Chlorophenol.
95807	Benzenediamine, ar-methyl-.
	Toluenediamine.
95943	Benzene, 1,2,4,5-tetrachloro-.
	1,2,4,5-Tetrachlorobenzene.
95954	Phenol, 2,4,5-trichloro-.
	2,4,5-Trichlorophenol.
96128	Propane, 1,2-dibromo-3-chloro-.
	1,2-Dibromo-3-chloropropane.
96184	1,2,3-Trichloropropane.
96457	Ethylenethiourea.
	2-Imidazolidinethione.
97632	Ethyl methacrylate.
	2-Propenoic acid, 2-methyl-, ethyl ester.
98011	Furfural.
	2-Furancarboxaldehyde.
98077	Benzene, (trichloromethyl)-.
	Benzotrichloride.
98099	Benzenesulfonic acid chloride.
	Benzenesulfonyl chloride.
98828	Benzene, 1-methylethyl-.
	Cumene.
98862	Acetophenone.
	Ethanone, 1-phenyl-.
98873	Benzal chloride.
	Benzene, dichloromethyl-.
98884	Benzoyl chloride.
98953	Benzene, nitro-.
	Nitrobenzene.
99081	m-Nitrotoluene.
99354	Benzene, 1,3,5-trinitro-.
	1,3,5-Trinitrobenzene.
99558	Benzenamine, 2-methyl-5-nitro-.
	5-Nitro-o-toluidine.
99650	m-Dinitrobenzene.
99990	p-Nitrotoluene.
100016	Benzenamine, 4-nitro-.
	p-Nitroaniline.
100027	p-Nitrophenol.
	Phenol, 4-nitro-.
	4-Nitrophenol.
100254	p-Dinitrobenzene.
100414	Ethylbenzene.
100425	Styrene.
100447	Benzene, chloromethyl-.
	Benzyl chloride.
100470	Benzonitrile.
100754	N-Nitrosopiperidine.
	Piperidine, 1-nitroso-.
101144	Benzenamine, 4,4'-methylenebis(2-chloro-.

CASRN	Hazardous substance
	4,4'-Methylenebis(2-chloroaniline).
101553	Benzene, 1-bromo-4-phenoxy-.
	4-Bromophenyl phenyl ether.
103855	Phenylthiourea.
	Thiourea, phenyl-.
105464	sec-Butyl acetate.
105679	Phenol, 2,4-dimethyl-.
	2,4-Dimethylphenol.
106423	p-Benzene, dimethyl.
	p-Xylene.
106445	p-Cresol.
	p-Cresylic acid.
106467	Benzene, 1,4-dichloro-.
	p-Dichlorobenzene.
	1,4-Dichlorobenzene.
106478	Benzenamine, 4-chloro-.
	p-Chloroaniline.
106490	Benzenamine, 4-methyl-.
	p-Toluidine.
106503	Phenylenediamine (para-isomer).
106514	p-Benzoquinone.
	2,5-Cyclohexadiene-1,4-dione.
106898	Epichlorohydrin.
	Oxirane, (chloromethyl)-.
106934	Ethane, 1,2-dibromo-.
	Ethylene dibromide.
107028	Acrolein.
	2-Propenal.
107051	Allyl chloride.
107062	Ethane, 1,2-dichloro-.
	Ethylene dichloride.
	1,2-Dichloroethane.
107108	n-Propylamine.
	1-Propanamine.
107120	Ethyl cyanide.
	Propanenitrile.
107131	Acrylonitrile.
	2-Propenenitrile.
107153	Ethylenediamine.
107186	Allyl alcohol.
	2-Propen-1-ol.
107197	Propargyl alcohol.
	2-Propyn-1-ol.
107200	Acetaldehyde, chloro-.
	Chloroacetaldehyde.
107302	Chloromethyl methyl ether.
	Methane, chloromethoxy-.
107493	Diphosphoric acid, tetraethyl ester.
	Tetraethyl pyrophosphate.
107926	Butyric acid.
108054	Vinyl acetate.
	Vinyl acetate monomer.
108101	Methyl isobutyl ketone.
	4-Methyl-2-pentanone.
108247	Acetic anhydride.
108316	Maleic anhydride.
	2,5-Furandione.
108383	m-Benzene, dimethyl.
	m-Xylene.
108394	m-Cresol.
	m-Cresylic acid.
108463	Resorcinol.
	1,3-Benzenediol.
108601	Dichloroisopropyl ether.
	Propane, 2,2'-oxybis[2-chloro-.
108883	Benzene, methyl-.
	Toluene.

CASRN	Hazardous substance
108907	Benzene, chloro-.
	Chlorobenzene.
108941	Cyclohexanone.
108952	Benzene, hydroxy-.
	Phenol.
108985	Benzenethiol.
	Thiophenol.
109068	Pyridine, 2-methyl-.
	2-Picoline.
109739	Butylamine.
109773	Malononitrile.
	Propanedinitrile.
109897	Diethylamine.
109999	Furan, tetrahydro-.
	Tetrahydrofuran.
110009	Furan.
	Furfuran.
110167	Maleic acid.
110178	Fumaric acid.
110190	iso-Butyl acetate.
110758	Ethene, 2-chloroethoxy-.
	2-Chloroethyl vinyl ether.
110805	Ethanol, 2-ethoxy-.
	Ethylene glycol monoethyl ether.
110827	Benzene, hexahydro-.
	Cyclohexane.
110861	Pyridine.
111444	Bis (2-chloroethyl) ether.
	Dichloroethyl ether.
	Ethane, 1,1'-oxybis[2-chloro-.
111546	Carbamodithioic acid, 1,2-ethanediylbis, salts & esters.
	Ethylenebisdithiocarbamic acid, salts & esters.
111911	Bis(2-chloroethoxy) methane.
	Dichloromethoxy ethane.
	Ethane, 1,1'-[methylenebis(oxy)]bis(2-chloro-.
115026	Azaserine.
	L-Serine, diazoacetate (ester).
115297	Endosulfan.
	6,9-Methano-2,4,3-benzodioxathiepin, 6,7,8,9,10,10-hexachloro-1,5,5a,6,9,9a- hexahydro-, 3-oxide.
115322	Dicofol.
116063	Aldicarb.
	Propanal, 2-methyl-2-(methylthio)-, 0-[(methylamino)carbonyl]oxime.
117806	Dichlone.
117817	Bis (2-ethylhexyl)phthalate.
	Diethylhexyl phthalate.
	1,2-Benzenedicarboxylic acid, [bis(2-ethylhexyl)]ester.
117840	Di-n-octyl phthalate.
	1,2-Benzenedicarboxylic acid, dioctyl ester.
118741	Benzene, hexachloro-.
	Hexachlorobenzene.
119904	[1,1'-Biphenyl]-4,4'diamine,3,3'dimethoxy-.
	3,3'-Dimethoxybenzidine.
119937	[1,1'Biphenyl]-4,4'-diamine,3,3'-dimethyl-.
	3,3'-Dimethylbenzidine.
120127	Anthracene.
120581	Isosafrole.
	1,3-Benzodioxole, 5-)1-propenyl)-.
120821	1,2,4-Trichlorobenzene.
120832	Phenol, 2,4-dichloro-.
	2,4-Dichlorophenol.
121142	Benzene, 1-methyl-2,4-dinitro-.
	2,4-Dinitrotoluene.

CASRN	Hazardous substance
121211	Pyrethrins.
121299	Pyrethrins.
121448	Triethylamine.
121755	Malathion.
122098	alpha,alpha-Dimethylphenethylamine.
	Benzeneethanamine, alpha,alpha-dimethyl-.
122394	Diphenylamine.
·122667	Hydrazine, 1,2-diphenyl-.
	1,2-Diphenylhydrazine.
123331	Maleic hydrazide.
	3,6-Pyridazinedione, 1,2-dihydro-.
123626	Propionic anhydride.
123637	Paraldehyde.
	1,3,5-Trioxane, 2,4,6-trimethyl-.
123739	Crotonaldehyde.
	2-Butenal.
123864	Butyl acetate.
123911	1,4-Diethylenedioxide.
	1,4-Dioxane.
123922	iso-Amyl acetate.
124049	Adipic acid.
124403	Dimethylamine.
	Methanamine, N-methyl-.
124414	Sodium methylate.
124481	Chlorodibromomethane.
126727	Tris(2,3-dibromopropyl) phosphate.
	1-Propanol, 2,3-dibromo-, phosphate (3:1).
126987	Methacrylonitrile.
	2-Propenenitrile, 2-methyl-.
126998	2-Chloro-1,3-butadiene.
127184	Ethene, tetrachloro-.
	Perchloroethylene.
	Tetrachloroethene.
	Tetrachloroethylene.
127822	Zinc phenolsulfonate.
129000	Pyrene.
130154	1,4-Naphthalenedione.
	1,4-Naphthoquinone.
131113	Dimethyl phthalate.
	1,2-Benzenedicarboxylic acid, dimethyl ester.
131748	Ammonium picrate.
	Phenol, 2,4,6-trinitro-, ammonium salt.
131895	Phenol, 2-cyclohexyl-4,6-dinitro-.
	2-Cyclohexyl-4,6-dinitrophenol.
133062	Captan.
134327	alpha-Naphthylamine.
	1-Naphthalenamine.
137268	Thioperoxydicarbonic diamide ([H2N)C(S)]2S2, tetramethyl-.
	Thiram.
140885	Ethyl acrylate.
	2-Propenoic acid, ethyl ester.
141786	Acetic acid, ethyl ester.
	Ethyl acetate.
142289	1,3-Dichloropropane.
142712	Cupric acetate.
142847	Dipropylamine.
	1-Propanamine, N-propyl-.
143339	Sodium cyanide.
	Sodium cyanide Na(CN).
143500	Kepone.
	1,3,4-Metheno-2H-cyclobuta[cd]pentalen-2-one,
	1,1a,3,3a,4,5,5,5a,5b,6-decachlorooctahydro-.
145733	Endothall.
	7-Oxabicyclo[2.2.1]heptane-2,3-dicarboxylic acid.

CASRN	Hazardous substance
148823	L-Phenylalanine, 4-[bis(2-chloroethyl) aminol].
	Melphalan.
151508	Potassium cyanide.
	Potassium cyanide K(CN).
151564	Aziridine.
	Ethylenimine.
152169	Diphosphoramide, octamethyl-.
	Octamethylpyrophosphoramide.
156605	Ethene, 1,2-dichloro- (E).
	1,2-Dichloroethylene.
189559	Benzo [rst]pentaphene.
	Dibenz[a,i]pyrene.
191242	Benzo[ghi]perylene.
193395	Indeno(1,2,3-cd)pyrene.
	1,10-(1,2-Phenylene)pyrene.
205992	Benzo[b]fluoranthene.
206440	Benzo[j,k]fluorene.
	Fluoranthene.
207089	Benzo(k)fluoranthene.
208968	Acenaphthylene.
218019	Chrysene.
	1,2-Benzphenanthrene.
225514	Benz[c]acridine.
297972	O,O-Diethyl O-pyrazinyl phosphorothioate.
	Phosphorothioic acid, O,O-diethyl O-pyrazinyl ester.
298000	Methyl parathion.
	Phosphorothioic acid, O,O-dimethyl O-(4-nitrophenyl) ester.
298022	Phorate.
	Phosphorodithioic acid, O,O-diethyl S-(ethylthio), methyl ester.
298044	Disulfoton.
	Phosphorodithioic acid, O,O-diethyl S-[2-(ethylthio)ethyl]ester.
300765	Naled.
301042	Acetic acid, lead(2+) salt.
	Lead acetate.
302012	Hydrazine.
303344	Lasiocarpine.
	2-Butenoic acid, 2-methyl-, 7[[2,3-dihydroxy-2-(1-methoxyethyl)-3- methyl-1-oxobutoxy]methyl] ·2,3,5,7a-tetrahydro-1H-pyrrolizin-1-yl ester, [1S-[1alpha(Z),7(2S*,3R*),7aalpha]]-.
305033	Benzenebutanoic acid, 4-[bis(2-chloroethyl)amino]-.
	Chlorambucil.
309002	Aldrin.
	1,4,5,8-Dimethanonaphthalene, 1,2,3,4,10,10-10-hexachloro-1, 4,4a,5,8,8a-hexahydro-(1alpha,4 alpha,4abeta,5alpha,8alpha,8abeta)-
311455	Diethyl-p-nitrophenyl phosphate.
	Phosphoric acid, diethyl 4-nitrophenyl ester.
315184	Mexacarbate.
319846	alpha—BHC.
319857	beta—BHC.
319868	delta—BHC.
329715	2,5-Dinitrophenol.
330541	Diuron.
333415	Diazinon.
353504	Carbon oxyfluoride.
	Carbonic difluoride.
357573	Brucine.
	Strychnidin-10-one, 2,3-dimethoxy-.

CASRN	Hazardous substance
460195	Cyanogen.
	Ethanedinitrile.
465736	Isodrin.
	1,4,5,8-Dimethanonaphthalene, 1,2,3,4,10,10-hexachloro-1,4,4a,5 8,8a-hexahydro (1alpha, 4alpha,4abeta,5beta,8beta,8abeta)-.
492808	Auramine.
	Benzenamine, 4,4'-carbonimidoylbis (N,N-dimethyl(N,N-D,methyl-)-.
494031	Chlornaphazine.
	Naphthalenamine, N,N'-bis(2-chloro-ethyl)-.
496720	Benzenediamine, ar-methyl-.
	Toluenediamine.
504245	4-Aminopyridine.
	4-Pyridinamine.
504609	1-Methylbutadiene.
	1,3-Pentadiene.
506616	Argentate(1-), bis(cyano-C)- ,potassium.
	Potassium silver cyanide.
506649	Silver cyanide.
	Silver cyanide Ag(CN).
506683	Cyanogen bromide.
	Cyanogen bromide (CN)Br.
506774	Cyanogen chloride.
	Cyanogen chloride (CN)Cl.
506876	Ammonium carbonate.
506967	Acetyl bromide.
509148	Methane, tetranitro-.
	Tetranitromethane.
510156	Benzeneacetic acid,4-chloro-alpha-(4-chlorophenyl)-alpha-hydroxy-, ethyl ester.
	Chlorobenzilate.
513495	sec-Butylamine.
528290	o-Dinitrobenzene.
534521	Phenol, 2-methyl-4,6 dinitro-.
	4,6-Dinitro-o-cresol and salts.
540738	Hydrazine, 1,2-dimethyl-.
	1,2-Dimethylhydrazine.
540885	tert-Butyl acetate.
541093	Uranyl acetate.
541537	Dithiobiuret.
	Thiolmidodicarbonic diamide [(H2N)C(S)]2NH.
541731	Benzene, 1,3-dichloro-.
	m-Dichlorobenzene.
	1,3-Dichlorobenzene.
542621	Barium cyanide.
542756	1-Propene, 1,3-dichloro-.
	1,3-Dichloropropene
542767	Propanenitrile, 3-chloro-.
	3-Chloropropionitrile
542881	Dichloromethyl ether
	Methane, oxybis(chloro)-.
543908	Cadmium acetate.
544183	Cobaltous formate.
544923	Copper cyanide CuCN.
	Copper cyanide.
554847	m-Nitrophenol.
557197	Nickel cyanide.
	Nickel cyanide Ni(CN)2.
557211	Zinc cyanide.
	Zinc cyanide Zn(CN)2.
557346	Zinc acetate.
557415	Zinc formate.
563122	Ethion.
563688	Acetic acid, thallium(1+) salt.

CASRN	Hazardous substance
	Thallium(I) acetate.
573568	2,6-Dinitrophenol.
584849	Benzene, 1,3-diisocyanatomethyl-.
	Toluene diisocyanate.
591082	Acetamide, N-(aminothioxomethyl)-.
	1-Acetyl-2-thiourea.
592018	Calcium cyanide.
	Calcium cyanide Ca(CN)2.
592041	Mercuric cyanide.
592858	Mercuric thiocyanate.
592870	Lead thiocyanate.
594423	Methanesulfenyl chloride, trichloro-.
	Trichloromethanesulfenyl chloride.
598312	Bromoacetone.
	2-Propanone, 1-bromo-.
606202	Benzene, 1-methyl-1,3-dinitro-.
	2,6-Dinitrotoluene.
608731	HEXACHLOROCYCLOHEXANE (all isomers).
608935	Benzene, pentachloro-.
	Pentachlorobenzene.
609198	3,4,5-Trichlorophenol.
610399	3,4-Dinitrotoluene.
615532	Carbamic acid, methylnitroso-, ethyl ester.
	N-Nitroso-N-methylurethane.
616239	n-,2,3 Dichloropropanol.
621647	Di-n-propylnitrosamine.
	1-Propanamine, N-nitroso-N-propyl-.
624839	Methane, isocyanato-.
	Methyl isocyanate.
625161	tert-Amyl acetate.
626380	sec-Amyl acetate.
628637	Amyl acetate.
628864	Fulminic acid, mercury(2+)salt.
	Mercury fulminate.
630104	Selenourea.
630206	Ethane, 1,1,1,2-tetrachloro-.
	1,1,1,2-Tetrachloroethane.
631618	Ammonium acetate.
636215	Benzenamine, 2-methyl-, hydrochloride.
	o-Toluidine hydrochloride.
640197	Acetamide, 2-fluoro-.
	Fluoroacetamide.
684935	N-Nitroso-N-methylurea.
	Urea, N-methyl-N-nitroso-.
692422	Arsine, diethyl-.
	Diethylarsine.
696286	Arsonous dichloride, phenyl-.
	Dichlorophenylarsine.
757584	Hexaethyl tetraphosphate.
	Tetraphosphoric acid, hexaethyl ester.
759739	N-Nitroso-N-ethylurea.
	Urea, N-ethyl-N-nitroso-.
764410	1,4-Dichloro-2-butene.
	2-Butene, 1,4-dichloro-.
765344	Glycidylaldehyde.
	Oxiranecarboxyaldehyde.
815827	Cupric tartrate.
823405	Benzenediamine, ar-methyl-.
	Toluenediamine.
924163	N-Nitrosodi-n-butylamine.
	1-Butanamine, N-butyl-N-nitroso-.
930552	N-Nitrosopyrrolidine.
	Pyrrolidine, 1-nitroso-.
933755	2,3,6-Trichlorophenol.
933788	2,3,5-Trichlorophenol.
959988	alpha-Endosulfan.
1024573	Heptachlor epoxide.

APPENDIX 1 TO § 302.4—SEQUENTIAL CAS REGISTRY NUMBER LIST OF CERCLA HAZARDOUS SUBSTANCES—Continued

CASRN	Hazardous substance
1031078	Endosulfan sulfate.
1066304	Chromic acetate.
1066337	Ammonium bicarbonate.
1072351	Lead stearate.
1111780	Ammonium carbamate.
1116547	Ethanol, 2,2'-(nitrosoimino)bis-.
	N-Nitrosodiethanolamine.
1120714	1,2-Oxathiolane, 2,2-dioxide.
	1,3-Propane sultone.
1185575	Ferric ammonium citrate.
1194656	Dichlobenil.
1300716	Xylenol.
1303282	Arsenic oxide As2O5.
	Arsenic pentoxide.
1303328	Arsenic disulfide.
1303339	Arsenic trisulfide.
1309644	Antimony trioxide.
1310583	Potassium hydroxide.
1310732	Sodium hydroxide.
1314325	Thallic oxide.
	Thallium oxide Tl2O3.
1314621	Vanadium oxide V2O5.
	Vanadium pentoxide.
1314803	Phosphorus pentasulfide.
	Phosphorus sulfide.
	Sulfur phosphide.
1314847	Zinc phosphide.
	Zinc phosphide Zn3P2, when present at concentrations greater than 10%.
1314870	Lead sulfide.
1319728	2,4,5-T amines.
1319773	Cresol(s).
	Cresylic acid.
	Phenol, methyl-.
1320189	2,4-D Ester.
1321126	Nitrotoluene.
1327522	Arsenic acid.
	Arsenic acid H3AsO4.
1327533	Arsenic oxide As2O3.
	Arsenic trioxide.
1330207	Benzene, dimethyl.
	Xylene (mixed).
1332076	Zinc borate.
1332214	Asbestos.
1333831	Sodium bifluoride.
1335326	Lead subacetate.
	Lead, bis(acetato-O)tetrahydroxytri.
1336216	Ammonium hydroxide.
1336363	Polychlorinated Biphenyls (PCBs).
1338234	Methyl ethyl ketone peroxide.
	2-Butanone peroxide.
1338245	Naphthenic acid.
1341497	Ammonium bifluoride.
1464535	1,2:3,4-Diepoxybutane.
	2,2'-Bioxirane.
1563662	Carbofuran.
1615801	Hydrazine, 1,2-diethyl-.
	N,N'-Diethylhydrazine.
1746016	2,3,7,8-Tetrachlorodibenzo-p-dioxin (TCDD).
1762954	Ammonium thiocyanate.
1863634	Ammonium benzoate.
1888717	Hexachloropropene.
	1-Propene, 1,1,2,3,3,3-hexachloro-.
1918009	Dicamba.
1928387	2,4-D Ester.
1928478	2,4,5-T esters.
1928616	2,4-D Ester.

APPENDIX 1 TO § 302.4—SEQUENTIAL CAS REGISTRY NUMBER LIST OF CERCLA HAZARDOUS SUBSTANCES—Continued

CASRN	Hazardous substance
1929733	2,4-D Ester.
2008460	2,4,5-T amines.
2032657	Mercaptodimethur.
2303164	Carbamothioic acid, bis(1-methylethyl)-, S-(2,3-dichloro-2-propenyl) ester.
	Diallate.
2312358	Propargite.
2545597	2,4,5-T esters
2763964	Muscimol.
	3(2H)-Isoxazolone, 5-(aminomethyl)-.
	5-(Aminomethyl)-3-isoxazolol.
2764729	Diquat
2921882	Chlorpyrifos.
2944674	Ferric ammonium oxalate.
2971382	2,4-D Ester.
3012655	Ammonium citrate, dibasic.
3164292	Ammonium tartrate.
3165933	Benzenamine, 4-chloro-2-methyl-, hydrochloride.
	4-Chloro-o-toluidine, hydrochloride.
3251238	Cupric nitrate.
3288582	O,O-Diethyl S-methyl dithiophosphate.
	Phosphorodithioic acid, O,O-diethyl S-methyl ester.
3486359	Zinc carbonate.
3689245	Tetraethyldithiopyrophosphate.
	Thiodiphosphoric acid, tetraethyl ester.
3813147	2,4,5-T amines.
4170303	Crotonaldehyde.
	2-Butenal.
4549400	N-Nitrosomethylvinylamine.
	Vinylamine, N-methyl-N-nitroso-.
5344821	Thiourea, (2-chlorophenyl)-.
	1-(o-Chlorophenyl)thiourea.
5893663	Cupric oxalate.
5972736	Ammonium oxalate.
6009707	Ammonium oxalate.
6369966	2,4,5-T amines.
6369977	2,4,5-T amines.
6533739	Carbonic acid, dithallium(1+) salt.
	Thallium(I) carbonate.
7005723	4-Chlorophenyl phenyl ether.
7421934	Endrin aldehyde.
7428480	Lead stearate.
7439921	Lead.
7439976	Mercury.
7440020	Nickel.
7440224	Silver.
7440235	Sodium.
7440280	Thallium.
7440360	Antimony.
7440382	Arsenic.
7440417	Beryllium.
	Beryllium dust.
7440439	Cadmium.
7440473	Chromium.
7440508	Copper.
7440666	Zinc.
7446084	Selenium dioxide.
	Selenium oxide.
7446142	Lead sulfate.
7446186	Sulfuric acid dithallium(1+) salt.
	Thallium(I) sulfate.
7446277	Lead phosphate.
	Phosphoric acid, lead(2+) salt (2:3).
7447394	Cupric chloride.
7488564	Selenium sulfide.

CASRN	Hazardous substance
	Selenium sulfide SeS2.
7558794	Sodium phosphate, dibasic.
7601549	Sodium phosphate, tribasic.
7631892	Sodium arsenate.
7631905	Sodium bisulfite.
7632000	Sodium nitrite.
7645252	Lead arsenate.
7646857	Zinc chloride.
7647010	Hydrochloric acid.
	Hydrogen chloride.
7647189	Antimony pentachloride.
7664382	Phosphoric acid.
7664393	Hydrofluoric acid.
	Hydrogen fluoride.
7664417	Ammonia.
7664939	Sulfuric acid.
7681494	Sodium fluoride.
7681529	Sodium hypochlorite.
7697372	Nitric acid.
7699458	Zinc bromide.
7705080	Ferric chloride.
7718549	Nickel chloride.
7719122	Phosphorus trichloride.
7720787	Ferrous sulfate.
7722647	Potassium permanganate.
7723140	Phosphorus.
7733020	Zinc sulfate.
7738945	Chromic acid.
7758294	Sodium phosphate, tribasic.
7758943	Ferrous chloride.
7758954	Lead chloride.
7758987	Cupric sulfate.
7761888	Silver nitrate.
7773060	Ammonium sulfamate
7775113	Sodium chromate.
7778394	Arsenic acid.
	Arsenic acid H3AsO4.
7778441	Calcium arsenate.
7778509	Potassium bichromate.
7778543	Calcium hypochlorite.
7779864	Zinc hydrosulfite.
7779886	Zinc nitrate.
7782414	Fluorine.
7782492	Selenium.
7782505	Chlorine.
7782630	Ferrous sulfate.
7782823	Sodium selenite.
7782867	Mercurous nitrate.
7783008	Selenious acid.
7783064	Hydrogen sulfide.
	Hydrogen sulfide H2S.
7783359	Mercuric sulfate.
7783462	Lead fluoride.
7783495	Zinc fluoride.
7783508	Ferric fluoride.
7783564	Antimony trifluoride.
7784341	Arsenic trichloride.
7784409	Lead arsenate.
7784410	Potassium arsenate.
7784465	Sodium arsenite.
7785844	Sodium phosphate, tribasic.
7786347	Mevinphos.
7786814	Nickel sulfate.
7787475	Beryllium chloride.
7787497	Beryllium fluoride.
7787555	Beryllium nitrate.
7788989	Ammonium chromate.

CASRN	Hazardous substance
7789006	Potassium chromate.
7789062	Strontium chromate.
7789095	Ammonium bichromate.
7789426	Cadmium bromide.
7789437	Cobaltous bromide.
7789619	Antimony tribromide.
7790945	Chlorosulfonic acid.
7791120	Thallium chloride TlCl.
	Thallium(I) chloride.
7803512	Phosphine.
7803556	Ammonium vanadate.
	Vanadic acid, ammonium salt.
8001352	Camphene, octachloro-
	Toxaphene.
8001589	Creosote.
8003198	Dichloropropane—Dichloropropene (mixture).
8003347	Pyrethrins.
8014957	Sulfuric acid.
10022705	Sodium hypochlorite.
10025873	Phosphorus oxychloride.
10025919	Antimony trichloride.
10026116	Zirconium tetrachloride.
10028225	Ferric sulfate.
10031591	Sulfuric acid, dithallium(1+) salt.
	Thallium(I) sulfate.
10039324	Sodium phosphate, dibasic.
10043013	Aluminum sulfate.
10045893	Ferrous ammonium sulfate.
10045940	Mercuric nitrate.
10049055	Chromous chloride.
10099748	Lead nitrate.
10101538	Chromic sulfate.
10101630	Lead iodide.
10101890	Sodium phosphate, tribasic.
10102064	Uranyl nitrate.
10102188	Sodium selenite.
10102439	Nitric oxide.
	Nitrogen oxide NO.
10102440	Nitrogen dioxide.
	Nitrogen oxide NO2.
10102451	Nitric acid, thallium(1+) salt.
	Thallium(I) nitrate.
10102484	Lead arsenate.
10108642	Cadmium chloride.
10124502	Potassium arsenite.
10124568	Sodium phosphate, tribasic.
10140655	Sodium phosphate, dibasic.
10192300	Ammonium bisulfite.
10196040	Ammonium sulfite.
10361894	Sodium phosphate, tribasic.
10380297	Cupric sulfate, ammoniated.
10415755	Mercurous nitrate.
10421484	Ferric nitrate.
10544726	Nitrogen dioxide.
	Nitrogen oxide NO2.
10588019	Sodium bichromate.
11096825	Aroclor 1260.
	Polychlorinated Biphenyls (PCBs).
11097691	Aroclor 1254.
	Polychlorinated Biphenyls (PCBs).
11104282	Aroclor 1221.
	Polychlorinated Biphenyls (PCBs).
11115745	Chromic acid.

CASRN	Hazardous substance
11141165	Aroclor 1232.
	Polychlorinated Biphenyls (PCBs).
12002038	Cupric acetoarsenite.
12039520	Selenious acid, dithallium(1+) salt.
	Thallium selenite.
12054487	Nickel hydroxide.
12125018	Ammonium fluoride.
12125029	Ammonium chloride.
12135761	Ammonium sulfide.
12672296	Aroclor 1248.
	Polychlorinated Biphenyls (PCBs).
12674112	Aroclor 1016.
	Polychlorinated Biphenyls (PCBs).
12771083	Sulfur monochloride.
13463393	Nickel carbonyl.
	Nickel carbonyl Ni(CO)4, (T-4)-.
13560991	2,4,5-T salts.
13597994	Beryllium nitrate.
13746899	Zirconium nitrate.
13765190	Calcium chromate.
	Chromic acid H2CrO4, calcium salt.
13814965	Lead fluoborate.
13826830	Ammonium fluoborate.
13952846	sec-Butylamine.
14017415	Cobaltous sulfamate.
14216752	Nickel nitrate.
14258492	Ammonium oxalate.
14307358	Lithium chromate.
14307438	Ammonium tartrate.
14639975	Zinc ammonium chloride.
14639986	Zinc ammonium chloride.
14644612	Zirconium sulfate.
15699180	Nickel ammonium sulfate.
15739807	Lead sulfate.
15950660	2,3,4-Trichlorophenol.
16721805	Sodium hydrosulfide.
16752775	Ethanimidothioic acid, N-[[(methyl-amino)carbonyl] oxy]-, methyl ester.
	Methomyl.
16871719	Zinc silicofluoride.
16919190	Ammonium silicofluoride.
16923958	Zirconium potassium fluoride.
18883664	D-Glucose, 2-deoxy-2-[[(methylnitrosoamino)-carbonyl]amino]-.
	Glucopyranose, 2-deoxy-2-(3-methyl-3-nitrosour-eido)-.
	Streptozotocin.
20816120	Osmium oxide OsO4 (T-4)-.
	Osmium tetroxide.
20830813	Daunomycin.
	5,12-Naphthacenedione, 8-acetyl-10-[3-amino-2,3,6-trideoxy-alpha-L-lyxo-hexopyranosyl)oxy]-7,8,9,10-tetrahydro-6,8,11-trihydroxy-1-methoxy-,(8S-cis)-.
20859738	Aluminum phosphide.
23950585	Benzamide, 3,5-dichloro-N-(1,1- dimethyl-2-pro-pynyl)-.
	Pronamide.
25154545	Dinitrobenzene (mixed).
25154556	Nitrophenol (mixed).
25155300	Sodium dodecylbenzenesulfonate.
25167822	Trichlorophenol.

CASRN	Hazardous substance
25168154	2,4,5-T esters.
25168267	2,4-D Ester.
25321146	Dinitrotoluene.
25321226	Dichlorobenzene.
25376458	Benzenediamine, ar-methyl-.
	Toluenediamine.
25550587	Dinitrophenol.
26264062	Calcium dodecylbenzenesulfonate.
26471625	Benzene, 1,3-diisocyanatomethyl-.
	Toluene diisocyanate
26628228	Sodium azide.
26638197	Dichloropropane.
26952238	Dichloropropene.
27176870	Dodecylbenzenesulfonic acid.
27323417	Triethanolamine dodecylbenzene sulfonate.
27774136	Vanadyl sulfate.
28300745	Antimony potassium tartrate.
30525894	Paraformaldehyde.
32534955	2,4,5-TP esters.
33213659	beta - Endosulfan.
36478769	Uranyl nitrate.
37211055	Nickel chloride.
39196184	Thiofanox
	2-Butanone, 3,3-dimethyl-1-(methyl-thio)-, O[(methylamino)carbonyl] oxime.
42504461	Isopropanolamine dodecylbenzenesulfonate.
52628258	Zinc ammonium chloride.
52652592	Lead stearate.
52740166	Calcium arsenite.
53467111	2,4-D Ester.
53469219	Aroclor 1242.
	Polychlorinated Biphenyls (PCBs)
55488874	Ferric ammonium oxalate.
56189094	Lead stearate.
61792072	2,4,5-T esters.

APPENDIX **2** TO § 302.4—RADIONUCLIDES

Radionuclide	Atomic Number	Final RQ Ci (Bq)
Radionuclides[e]		1& (3.7E 10)
Actinium-224	89	100 (3.7E 12)
Actinium-225	89	1 (3.7E 10)
Actinium-226	89	10 (3.7E 11)
Actinium-227	89	0.001 (3.7E 7)
Actinium-228	89	10 (3.7E 11)
Aluminum-26	13	10 (3.7E 11)
Americium-237	95	1000 (3.7E 13)
Americium-238	95	100 (3.7E 12)
Americium-239	95	100 (3.7E 12)
Americium-240	95	10 (3.7E 11)
Americium-241	95	0.01 (3.7E 8)
Americium-242m	95	0.01 (3.7E 8)
Americium-242	95	100 (3.7E 12)
Americium-243	95	0.01 (3.7E 8)
Americium-244m	95	1000 (3.7E 13)
Americium-244	95	10 (3.7E 11)
Americium-245	95	1000 (3.7E 13)
Americium-246m	95	1000 (3.7E 13)
Americium-246	95	1000 (3.7E 13)
Antimony-115	51	1000 (3.7E 13)
Antimony-116m	51	100 (3.7E 12)
Antimony-116	51	1000 (3.7E 13)

APPENDIX 2 TO § 302.4—RADIONUCLIDES— Continued

APPENDIX 2 TO § 302.4—RADIONUCLIDES— Continued

Radionuclide	Atomic Number	Final RQ Ci (Bq)
Antimony-117	51	1000 (3.7E 13)
Antimony-118m	51	10 (3.7E 11)
Antimony-119	51	1000 (3.7E 13)
Antimony-120 (16 min)	51	1000 (3.7E 13)
Antimony-120 (5.76 day)	51	10 (3.7E 11)
Antimony-122	51	10 (3.7E 11)
Antimony-124m	51	1000 (3.7E 13)
Antimony-124	51	10 (3.7E 11)
Antimony-125	51	10 (3.7E 11)
Antimony-126m	51	1000 (3.7E 13)
Antimony-126	51	10 (3.7E 11)
Antimony-127	51	10 (3.7E 11)
Antimony-128 (10.4 min)	51	1000 (3.7E 13)
Antimony-128 (9.01 hr)	51	10 (3.7E 11)
Antimony-129	51	100 (3.7E 12)
Antimony-130	51	100 (3.7E 12)
Antimony-131	51	1000 (3.7E 13)
Argon-39	18	1000 (3.7E 13)
Argon-41	18	10 (3.7E 11)
Arsenic-69	33	1000 (3.7E 13)
Arsenic-70	33	100 (3.7E 12)
Arsenic-71	33	100 (3.7E 12)
Arsenic-72	33	10 (3.7E 11)
Arsenic-73	33	100 (3.7E 12)
Arsenic-74	33	10 (3.7E 11)
Arsenic-76	33	100 (3.7E 12)
Arsenic-77	33	1000 (3.7E 13)
Arsenic-78	33	100 (3.7E 12)
Astatine-207	85	100 (3.7E 12)
Astatine-211	85	100 (3.7E 12)
Barium-126	56	1000 (3.7E 13)
Barium-128	56	10 (3.7E 11)
Barium-131m	56	1000 (3.7E 13)
Barium-131	56	10 (3.7E 11)
Barium-133m	56	100 (3.7E 12)
Barium-133	56	10 (3.7E 11)
Barium-135m	56	1000 (3.7E 13)
Barium-139	56	1000 (3.7E 13)
Barium-140	56	10 (3.7E 11)
Barium-141	56	1000 (3.7E 13)
Barium-142	56	1000 (3.7E 13)
Berkelium-245	97	100 (3.7E 12)
Berkelium-246	97	10 (3.7E 11)
Berkelium-247	97	0.01 (3.7E 8)
Berkelium-249	97	1 (3.7E 10)
Berkelium-250	97	100 (3.7E 12)
Beryllium-7	4	100 (3.7E 12)
Beryllium-10	4	1 (3.7E 10)
Bismuth-200	83	100 (3.7E 12)
Bismuth-201	83	100 (3.7E 12)
Bismuth-202	83	1000 (3.7E 13)
Bismuth-203	83	10 (3.7E 11)
Bismuth-205	83	10 (3.7E 11)
Bismuth-206	83	10 (3.7E 11)
Bismuth-207	83	10 (3.7E 11)
Bismuth-210m	83	0.1 (3.7E 9)
Bismuth-210	83	10 (3.7E 11)
Bismuth-212	83	100 (3.7E 12)
Bismuth-213	83	100 (3.7E 12)
Bismuth-214	83	100 (3.7E 12)
Bromine-74m	35	100 (3.7E 12)
Bromine-74	35	100 (3.7E 12)
Bromine-75	35	100 (3.7E 12)
Bromine-76	35	10 (3.7E 11)
Bromine-77	35	100 (3.7E 12)
Bromine-80m	35	1000 (3.7E 13)
Bromine-80	35	1000 (3.7E 13)
Bromine-82	35	10 (3.7E 11)

Radionuclide	Atomic Number	Final RQ Ci (Bq)
Bromine-83	35	1000 (3.7E 13)
Bromine-84	35	100 (3.7E 12)
Cadmium-104	48	1000 (3.7E 13)
Cadmium-107	48	1000 (3.7E 13)
Cadmium-109	48	1 (3.7E 10)
Cadmium-113m	48	0.1 (3.7E 9)
Cadmium-113	48	0.1 (3.7E 9)
Cadmium-115m	48	10 (3.7E 11)
Cadmium-115	48	100 (3.7E 12)
Cadmium-117m	48	10 (3.7E 11)
Cadmium-117	48	100 (3.7E 12)
Calcium-41	20	10 (3.7E 11)
Calcium-45	20	10 (3.7E 11)
Calcium-47	20	10 (3.7E 11)
Californium-244	98	1000 (3.7E 13)
Californium-246	98	10 (3.7E 11)
Californium-248	98	0.1 (3.7E 9)
Californium-249	98	0.01 (3.7E 8)
Californium-250	98	0.01 (3.7E 8)
Californium-251	98	0.01 (3.7E 8)
Californium-252	98	0.1 (3.7E 9)
Californium-253	98	10 (3.7E 11)
Californium-254	98	0.1 (3.7E 9)
Carbon-11	6	1000 (3.7E 13)
Carbon-14	6	10 (3.7E 11)
Cerium-134	58	10 (3.7E 11)
Cerium-135	58	10 (3.7E 11)
Cerium-137m	58	100 (3.7E 12)
Cerium-137	58	1000 (3.7E 13)
Cerium-139	58	100 (3.7E 12)
Cerium-141	58	10 (3.7E 11)
Cerium-143	58	100 (3.7E 12)
Cerium-144	58	1 (3.7E 10)
Cesium-125	55	1000 (3.7E 13)
Cesium-127	55	100 (3.7E 12)
Cesium-129	55	100 (3.7E 12)
Cesium-130	55	1000 (3.7E 13)
Cesium-131	55	1000 (3.7E 13)
Cesium-132	55	10 (3.7E 11)
Cesium-134m	55	1000 (3.7E 13)
Cesium-134	55	1 (3.7E 10)
Cesium-135m	55	100 (3.7E 12)
Cesium-135	55	10 (3.7E 11)
Cesium-136	55	10 (3.7E 11)
Cesium-137	55	1 (3.7E 10)
Cesium-138	55	100 (3.7E 12)
Chlorine-36	17	10 (3.7E 11)
Chlorine-38	17	100 (3.7E 12)
Chlorine-39	17	100 (3.7E 12)
Chromium-48	24	100 (3.7E 12)
Chromium-49	24	1000 (3.7E 13)
Chromium-51	24	1000 (3.7E 13)
Cobalt-55	27	10 (3.7E 11)
Cobalt-56	27	10 (3.7E 11)
Cobalt-57	27	100 (3.7E 12)
Cobalt-58m	27	1000 (3.7E 13)
Cobalt-58	27	10 (3.7E 11)
Cobalt-60m	27	1000 (3.7E 13)
Cobalt-60	27	10 (3.7E 11)
Cobalt-61	27	1000 (3.7E 13)
Cobalt-62m	27	1000 (3.7E 13)
Copper-60	29	100 (3.7E 12)
Copper-61	29	100 (3.7E 12)
Copper-64	29	1000 (3.7E 13)
Copper-67	29	100 (3.7E 12)
Curium-238	96	1000 (3.7E 13)
Curium-240	96	1 (3.7E 10)
Curium-241	96	10 (3.7E 11)

APPENDIX **2** TO § 302.4—RADIONUCLIDES— Continued

Radionuclide	Atomic Number	Final RQ Ci (Bq)
Curium-242	96	1 (3.7E 10)
Curium-243	96	0.01 (3.7E 8)
Curium-244	96	0.01 (3.7E 8)
Curium-245	96	0.01 (3.7E 8)
Curium-246	96	0.01 (3.7E 8)
Curium-247	96	0.01 (3.7E 8)
Curium-248	96	0.001 (3.7E 7)
Curium-249	96	1000 (3.7E 13)
Dysprosium-155	66	100 (3.7E 12)
Dysprosium-157	66	100 (3.7E 12)
Dysprosium-159	66	100 (3.7E 12)
Dysprosium-165	66	1000 (3.7E 13)
Dysprosium-166	66	10 (3.7E 11)
Einsteinium-250	99	10 (3.7E 11)
Einsteinium-251	99	1000 (3.7E 13)
Einsteinium-253	99	10 (3.7E 11)
Einsteinium-254m	99	1 (3.7E 10)
Einsteinium-254	99	0.1 (3.7E 9)
Erbium-161	68	100 (3.7E 12)
Erbium-165	68	1000 (3.7E 13)
Erbium-169	68	100 (3.7E 12)
Erbium-171	68	100 (3.7E 12)
Erbium-172	68	10 (3.7E 11)
Europium-145	63	10 (3.7E 11)
Europium-146	63	10 (3.7E 11)
Europium-147	63	10 (3.7E 11)
Europium-148	63	10 (3.7E 11)
Europium-149	63	100 (3.7E 12)
Europium-150 (12.6 hr)	63	1000 (3.7E 13)
Europium-150 (34.2 yr)	63	10 (3.7E 11)
Europium-152m	63	100 (3.7E 12)
Europium-152	63	10 (3.7E 11)
Europium-154	63	10 (3.7E 11)
Europium-155	63	10 (3.7E 11)
Europium-156	63	10 (3.7E 11)
Europium-157	63	10 (3.7E 11)
Europium-158	63	1000 (3.7E 13)
Fermium-252	100	10 (3.7E 11)
Fermium-253	100	10 (3.7E 11)
Fermium-254	100	100 (3.7E 12)
Fermium-255	100	100 (3.7E 12)
Fermium-257	100	1 (3.7E 10)
Fluorine-18	9	1000 (3.7E 13)
Francium-222	87	100 (3.7E 12)
Francium-223	87	100 (3.7E 12)
Gadolinium-145	64	100 (3.7E 12)
Gadolinium-146	64	10 (3.7E 11)
Gadolinium-147	64	10 (3.7E 11)
Gadolinium-148	64	0.001 (3.7E7)
Gadolinium-149	64	100 (3.7E 12)
Gadolinium-151	64	100 (3.7E 12)
Gadolinium-152	64	0.001 (3.7E 7)
Gadolinium-153	64	10 (3.7E 11)
Gadolinium-159	64	1000 (3.7E 13)
Gallium-65	31	1000 (3.7E 13)
Gallium-66	31	10 (3.7E 11)
Gallium-67	31	100 (3.7E 12)
Gallium-68	31	1000 (3.7E 13)
Gallium-70	31	1000 (3.7E 13)
Gallium-72	31	10 (3.7E 11)
Gallium-73	31	100 (3.7E 12)
Germanium-66	32	100 (3.7E 12)
Germanium-67	32	1000 (3.7E 13)
Germanium-68	32	10 (3.7E 11)
Germanium-69	32	10 (3.7E 11)
Germanium-71	32	1000 (3.7E 13)
Germanium-75	32	1000 (3.7E 13)
Germanium-77	32	10 (3.7E 11)

APPENDIX **2** TO § 302.4—RADIONUCLIDES— Continued

Radionuclide	Atomic Number	Final RQ Ci (Bq)
Germanium-78	32	1000 (3.7E 13)
Gold-193	79	100 (3.7E 12)
Gold-194	79	10 (3.7E 11)
Gold-195	79	100 (3.7E 12)
Gold-198m	79	10 (3.7E 11)
Gold-198	79	100 (3.7E 12)
Gold-199	79	100 (3.7E 12)
Gold-200m	79	10 (3.7E 11)
Gold-200	79	1000 (3.7E 13)
Gold-201	79	1000 (3.7E 13)
Hafnium-170	72	100 (3.7E 12)
Hafnium-172	72	1 (3.7E 10)
Hafnium-173	72	100 (3.7E 12)
Hafnium-175	72	100 (3.7E 12)
Hafnium-177m	72	1000 (3.7E 13)
Hafnium-178m	72	0.1 (3.7E 9)
Hafnium-179m	72	100 (3.7E 12)
Hafnium-180m	72	100 (3.7E 12)
Hafnium-181	72	10 (3.7E 11)
Hafnium-182m	72	100 (3.7E 12)
Hafnium-182	72	0.1 (3.7E 9)
Hafnium-183	72	100 (3.7E 12)
Hafnium-184	72	100 (3.7E 12)
Holmium-155	67	1000 (3.7E 13)
Holmium-157	67	1000 (3.7E 13)
Holmium-159	67	1000 (3.7E 13)
Holmium-161	67	1000 (3.7E 13)
Holmium-162m	67	1000 (3.7E 13)
Holmium-162	67	1000 (3.7E 13)
Holmium-164m	67	1000 (3.7E 13)
Holmium-164	67	1000 (3.7E 13)
Holmium-166m	67	1 (3.7E 10)
Holmium-166	67	100 (3.7E 12)
Holmium-167	67	100 (3.7E 12)
Hydrogen-3	1	100 (3.7E 12)
Indium-109	49	100 (3.7E 12)
Indium-110 (69.1 min)	49	100 (3.7E 12)
Indium-110 (4.9 hr)	49	10 (3.7E 11)
Indium-111	49	100 (3.7E 12)
Indium-112	49	1000 (3.7E 13)
Indium-113m	49	1000 (3.7E 13)
Indium-114m	49	10 (3.7E 11)
Indium-115m	49	100 (3.7E 12)
Indium-115	49	0.1 (3.7E 9)
Indium-116m	49	100 (3.7E 12)
Indium-117m	49	100 (3.7E 12)
Indium-117	49	1000 (3.7E 13)
Indium-119m	49	1000 (3.7E 13)
Iodine-120m	53	100 (3.7E 12)
Iodine-120	53	10 (3.7E 11)
Iodine-121	53	100 (3.7E 12)
Iodine-123	53	10 (3.7E 11)
Iodine-124	53	0.1 (3.7E 9)
Iodine-125	53	0.01 (3.7E 8)
Iodine-126	53	0.01 (3.7E 8)
Iodine-128	53	1000 (3.7E 13)
Iodine-129	53	0.001 (3.7E 7)
Iodine-130	53	1 (3.7E 10)
Iodine-131	53	0.01 (3.7E 8)
Iodine-132m	53	10 (3.7E 11)
Iodine-132	53	10 (3.7E 11)
Iodine-133	53	0.1 (3.7E 9)
Iodine-134	53	100 (3.7E 12)
Iodine-135	53	10 (3.7E 11)
Iridium-182	77	1000 (3.7E 13)
Iridium-184	77	100 (3.7E 12)
Iridium-185	77	100 (3.7E 12)
Iridium-186	77	10 (3.7E 11)

APPENDIX 2 TO § 302.4—RADIONUCLIDES—
Continued

APPENDIX 2 TO § 302.4—RADIONUCLIDES—
Continued

Radionuclide	Atomic Number	Final RQ Ci (Bq)
Iridium-187	77	100 (3.7E 12)
Iridium-188	77	10 (3.7E 11)
Iridium-189	77	100 (3.7E 12)
Iridium-190m	77	1000 (3.7E 13)
Iridium-190	77	10 (3.7E 11)
Iridium-192m	77	100 (3.7E 12)
Iridium-192	77	10 (3.7E 11)
Iridium-194m	77	10 (3.7E 11)
Iridium-194	77	100 (3.7E 12)
Iridium-195m	77	100 (3.7E 12)
Iridium-195	77	1000 (3.7E 13)
Iron-52	26	100 (3.7E 12)
Iron-55	26	100 (3.7E 12)
Iron-59	26	10 (3.7E 11)
Iron-60	26	0.1 (3.7E 9)
Krypton-74	36	10 (3.7E 11)
Krypton-76	36	10 (3.7E 11)
Krypton-77	36	10 (3.7E 11)
Krypton-79	36	100 (3.7E 12)
Krypton-81	36	1000 (3.7E 13)
Krypton-83m	36	1000 (3.7E 13)
Krypton-85m	36	100 (3.7E 12)
Krypton-85	36	1000 (3.7E 13)
Krypton-87	36	10 (3.7E 11)
Krypton-88	36	10 (3.7E 11)
Lanthanum-131	57	1000 (3.7E 13)
Lanthanum-132	57	100 (3.7E 12)
Lanthanum-135	57	1000 (3.7E 13)
Lanthanum-137	57	10 (3.7E 11)
Lanthanum-138	57	1 (3.7E 10)
Lanthanum-140	57	10 (3.7E 11)
Lanthanum-141	57	1000 (3.7E 13)
Lanthanum-142	57	100 (3.7E 12)
Lanthanum-143	57	1000 (3.7E 13)
Lead-195m	82	1000 (3.7E 13)
Lead-198	82	100 (3.7E 12)
Lead-199	82	100 (3.7E 12)
Lead-200	82	100 (3.7E 12)
Lead-201	82	100 (3.7E 12)
Lead-202m	82	10 (3.7E 11)
Lead-202	82	1 (3.7E 10)
Lead-203	82	100 (3.7E 12)
Lead-205	82	100 (3.7E 12)
Lead-209	82	1000 (3.7E 13)
Lead-210	82	0.01 (3.7E 8)
Lead-211	82	100 (3.7E 12)
Lead-212	82	10 (3.7E 11)
Lead-214	82	100 (3.7E 12)
Lutetium-169	71	10 (3.7E 11)
Lutetium-170	71	10 (3.7E 11)
Lutetium-171	71	10 (3.7E 11)
Lutetium-172	71	10 (3.7E 11)
Lutetium-173	71	100 (3.7E 12)
Lutetium-174m	71	10 (3.7E 11)
Lutetium-174	71	10 (3.7E 11)
Lutetium-176m	71	1000 (3.7E 13)
Lutetium-176	71	1 (3.7E 10)
Lutetium-177m	71	10 (3.7E 11)
Lutetium-177	71	100 (3.7E 12)
Lutetium-178m	71	1000 (3.7E 13)
Lutetium-178	71	1000 (3.7E 13)
Lutetium-179	71	1000 (3.7E 13)
Magnesium-28	12	10 (3.7E 11)
Manganese-51	25	1000 (3.7E 13)
Manganese-52m	25	1000 (3.7E 13)
Manganese-52	25	10 (3.7E 11)
Manganese-53	25	1000 (3.7E 13)
Manganese-54	25	10 (3.7E 11)

Radionuclide	Atomic Number	Final RQ Ci (Bq)
Manganese-56	25	100 (3.7E 12)
Mendelevium-257	101	100 (3.7E 12)
Mendelevium-258	101	1 (3.7E 10)
Mercury-193m	80	10 (3.7E 11)
Mercury-193	80	100 (3.7E 12)
Mercury-194	80	0.1 (3.7E 9)
Mercury-195m	80	100 (3.7E 12)
Mercury-195	80	100 (3.7E 12)
Mercury-197m	80	1000 (3.7E 13)
Mercury-197	80	1000 (3.7E 13)
Mercury-199m	80	1000 (3.7E 13)
Mercury-203	80	10 (3.7E 11)
Molybdenum-90	42	100 (3.7E 12)
Molybdenum-93m	42	10 (3.7E 11)
Molybdenum-93	42	100 (3.7E 12)
Molybdenum-99	42	100 (3.7E 12)
Molybdenum-101	42	1000 (3.7E 13)
Neodymium-136	60	1000 (3.7E 13)
Neodymium-138	60	1000 (3.7E 13)
Neodymium-139m	60	100 (3.7E 12)
Neodymium-139	60	1000 (3.7E 13)
Neodymium-141	60	1000 (3.7E 13)
Neodymium-147	60	10 (3.7E 11)
Neodymium-149	60	100 (3.7E 12)
Neodymium-151	60	1000 (3.7E 13)
Neptunium-232	93	1000 (3.7E 13)
Neptunium-233	93	1000 (3.7E 13)
Neptunium-234	93	10 (3.7E 11)
Neptunium-235	93	1000 (3.7E 13)
Neptunium-236 (1.2 E 5 yr)	93	0.1 (3.7E 9)
Neptunium-236 (22.5 hr)	93	100 (3.7E 12)
Neptunium-237	93	0.01 (3.7E 8)
Neptunium-238	93	10 (3.7E 11)
Neptunium-239	93	100 (3.7E 12)
Neptunium-240	93	100 (3.7E 12)
Nickel-56	28	10 (3.7E 11)
Nickel-57	28	10 (3.7E 11)
Nickel-59	28	100 (3.7E 12)
Nickel-63	28	100 (3.7E 12)
Nickel-65	28	100 (3.7E 12)
Nickel-66	28	10 (3.7E 11)
Niobium-88	41	100 (3.7E 12)
Niobium-89 (66 min)	41	100 (3.7E 12)
Niobium-89 (122 min)	41	100 (3.7E 12)
Niobium-90	41	10 (3.7E 11)
Niobium-93m	41	100 (3.7E 12)
Niobium-94	41	10 (3.7E 11)
Niobium-95m	41	100 (3.7E 12)
Niobium-95	41	10 (3.7E 11)
Niobium-96	41	10 (3.7E 11)
Niobium-97	41	100 (3.7E 12)
Niobium-98	41	1000 (3.7E 13)
Osmium-180	76	1000 (3.7E 13)
Osmium-181	76	100 (3.7E 12)
Osmium-182	76	100 (3.7E 12)
Osmium-185	76	10 (3.7E 11)
Osmium-189m	76	1000 (3.7E 13)
Osmium-191m	76	1000 (3.7E 13)
Osmium-191	76	100 (3.7E 12)
Osmium-193	76	100 (3.7E 12)
Osmium-194	76	1 (3.7E 10)
Palladium-100	46	100 (3.7E 12)
Palladium-101	46	100 (3.7E 12)
Palladium-103	46	100 (3.7E 12)
Palladium-107	46	100 (3.7E 12)
Palladium-109	46	1000 (3.7E 13)
Phosphorus-32	15	0.1 (3.7E 9)
Phosphorus-33	15	1 (3.7E 10)

APPENDIX 2 TO § 302.4—RADIONUCLIDES—
Continued

Radionuclide	Atomic Number	Final RQ Ci (Bq)
Platinum-186	78	100 (3.7E 12)
Platinum-188	78	100 (3.7E 12)
Platinum-189	78	100 (3.7E 12)
Platinum-191	78	100 (3.7E 12)
Platinum-193m	78	100 (3.7E 12)
Platinum-193	78	1000 (3.7E 13)
Platinum-195m	78	100 (3.7E 12)
Platinum-197m	78	1000 (3.7E 13)
Platinum-197	78	1000 (3.7E 13)
Platinum-199	78	1000 (3.7E 13)
Platinum-200	78	100 (3.7E 12)
Plutonium-234	94	1000 (3.7E 13)
Plutonium-235	94	1000 (3.7E 13)
Plutonium-236	94	0.1 (3.7E 9)
Plutonium-237	94	1000 (3.7E 13)
Plutonium-238	94	0.01 (3.7E 8)
Plutonium-239	94	0.01 (3.7E 8)
Plutonium-240	94	0.01 (3.7E 8)
Plutonium-241	94	1 (3.7E 10)
Plutonium-242	94	0.01 (3.7E 8)
Plutonium-243	94	1000 (3.7E 13)
Plutonium-244	94	0.01 (3.7E 8)
Plutonium-245	94	100 (3.7E 12)
Polonium-203	84	100 (3.7E 12)
Polonium-205	84	100 (3.7E 12)
Polonium-207	84	10 (3.7E 11)
Polonium-210	84	0.01 (3.7E 8)
Potassium-40	19	1 (3.7E 10)
Potassium-42	19	100 (3.7E 12)
Potassium-43	19	10 (3.7E 11)
Potassium-44	19	100 (3.7E 12)
Potassium-45	19	1000 (3.7E 13)
Praseodymium-136	59	1000 (3.7E 13)
Praseodymium-137	59	1000 (3.7E 13)
Praseodymium-138m	59	100 (3.7E 12)
Praseodymium-139	59	1000 (3.7E 13)
Praseodymium-142m	59	1000 (3.7E 13)
Praseodymium-142	59	100 (3.7E 12)
Praseodymium-143	59	10 (3.7E 11)
Praseodymium-144	59	1000 (3.7E 13)
Praseodymium-145	59	1000 (3.7E 13)
Praseodymium-147	59	1000 (3.7E 13)
Promethium-141	61	1000 (3.7E 13)
Promethium-143	61	100 (3.7E 12)
Promethium-144	61	10 (3.7E 11)
Promethium-145	61	100 (3.7E 12)
Promethium-146	61	10 (3.7E 11)
Promethium-147	61	10 (3.7E 11)
Promethium-148m	61	10 (3.7E 11)
Promethium-148	61	10 (3.7E 11)
Promethium-149	61	100 (3.7E 12)
Promethium-150	61	100 (3.7E 12)
Promethium-151	61	100 (3.7E 12)
Protactinium-227	91	100 (3.7E 12)
Protactinium-228	91	10 (3.7E 11)
Protactinium-230	91	10 (3.7E 11)
Protactinium-231	91	0.01 (3.7E 8)
Protactinium-232	91	10 (3.7E 11)
Protactinium-233	91	100 (3.7E 12)
Protactinium-234	91	10 (3.7E 11)
Radium-223	88	1 (3.7E 10)
Radium-224	88	10 (3.7E 11)
Radium-225	88	1 (3.7E 10)
Radium-226Φ	88	0.1 (3.7E 9)
Radium-227	88	1000 (3.7E 13)
Radium-228	88	0.1 (3.7E 9)
Radon-220	86	0.1 (3.7E 9)
Radon-222	86	0.1 (3.7E 9)

APPENDIX 2 TO § 302.4—RADIONUCLIDES—
Continued

Radionuclide	Atomic Number	Final RQ Ci (Bq)
Rhenium-177	75	1000 (3.7E 13)
Rhenium-178	75	1000 (3.7E 13)
Rhenium-181	75	100 (3.7E 12)
Rhenium-182 (12.7 hr)	75	10 (3.7E 11)
Rhenium-182 (64.0 hr)	75	10 (3.7E 11)
Rhenium-184m	75	10 (3.7E 11)
Rhenium-184	75	10 (3.7E 11)
Rhenium-186m	75	10 (3.7E 11)
Rhenium-186	75	100 (3.7E 12)
Rhenium-187	75	1000 (3.7E 13)
Rhenium-188m	75	1000 (3.7E 13)
Rhenium-188	75	1000 (3.7E 13)
Rhenium-189	75	1000 (3.7E 13)
Rhodium-99m	45	100 (3.7E 12)
Rhodium-99	45	10 (3.7E 11)
Rhodium-100	45	10 (3.7E 11)
Rhodium-101m	45	100 (3.7E 12)
Rhodium-101	45	10 (3.7E 11)
Rhodium-102m	45	10 (3.7E 11)
Rhodium-102	45	10 (3.7E 11)
Rhodium-103m	45	1000 (3.7E 13)
Rhodium-105	45	100 (3.7E 12)
Rhodium-106m	45	10 (3.7E 11)
Rhodium-107	45	1000 (3.7E 13)
Rubidium-79	37	1000 (3.7E 13)
Rubidium-81m	37	1000 (3.7E 13)
Rubidium-81	37	100 (3.7E 12)
Rubidium-82m	37	10 (3.7E 11)
Rubidium-83	37	10 (3.7E 11)
Rubidium-84	37	10 (3.7E 11)
Rubidium-86	37	10 (3.7E 11)
Rubidium-88	37	1000 (3.7E 13)
Rubidium-89	37	1000 (3.7E 13)
Rubidium-87	37	10 (3.7E 11)
Ruthenium-94	44	1000 (3.7E 13)
Ruthenium-97	44	100 (3.7E 12)
Ruthenium-103	44	10 (3.7E 11)
Ruthenium-105	44	100 (3.7E 12)
Ruthenium-106	44	1 (3.7E 10)
Samarium-141m	62	1000 (3.7E 13)
Samarium-141	62	1000 (3.7E 13)
Samarium-142	62	1000 (3.7E 13)
Samarium-145	62	100 (3.7E 12)
Samarium-146	62	0.01 (3.7E 8)
Samarium-147	62	0.01 (3.7E 8)
Samarium-151	62	10 (3.7E 11)
Samarium-153	62	100 (3.7E 12)
Samarium-155	62	1000 (3.7E 13)
Samarium-156	62	100 (3.7E 12)
Scandium-43	21	1000 (3.7E 13)
Scandium-44m	21	10 (3.7E 11)
Scandium-44	21	100 (3.7E 12)
Scandium-46	21	10 (3.7E 11)
Scandium-47	21	100 (3.7E 12)
Scandium-48	21	10 (3.7E 11)
Scandium-49	21	1000 (3.7E 13)
Selenium-70	34	1000 (3.7E 13)
Selenium-73m	34	100 (3.7E 12)
Selenium-73	34	10 (3.7E 11)
Selenium-75	34	10 (3.7E 11)
Selenium-79	34	10 (3.7E 11)
Selenium-81m	34	1000 (3.7E 13)
Selenium-81	34	1000 (3.7E 13)
Selenium-83	34	1000 (3.7E 13)
Silicon-31	14	1000 (3.7E 13)
Silicon-32	14	1 (3.7E 10)
Silver-102	47	100 (3.7E 12)
Silver-103	47	1000 (3.7E 13)

APPENDIX 2 TO § 302.4—RADIONUCLIDES—
Continued

APPENDIX 2 TO § 302.4—RADIONUCLIDES—
Continued

Radionuclide	Atomic Number	Final RQ Ci (Bq)
Silver-104m	47	1000 (3.7E 13)
Silver-104	47	1000 (3.7E 13)
Silver-105	47	10 (3.7E 11)
Silver-106m	47	10 (3.7E 11)
Silver-106	47	1000 (3.7E 13)
Silver-108m	47	10 (3.7E 11)
Silver-110m	47	10 (3.7E 11)
Silver-111	47	10 (3.7E 11)
Silver-112	47	100 (3.7E 12)
Silver-115	47	1000 (3.7E 13)
Sodium-22	11	10 (3.7E 11)
Sodium-24	11	10 (3.7E 11)
Strontium-80	38	100 (3.7E 12)
Strontium-81	38	1000 (3.7E 13)
Strontium-83	38	100 (3.7E 12)
Strontium-85m	38	1000 (3.7E 13)
Strontium-85	38	10 (3.7E 11)
Strontium-87m	38	100 (3.7E 12)
Strontium-89	38	10 (3.7E 11)
Strontium-90	38	0.1 (3.7E 9)
Strontium-91	38	10 (3.7E 11)
Strontium-92	38	100 (3.7E 12)
Sulfur-35	16	1 (3.7E 10)
Tantalum-172	73	100 (3.7E 12)
Tantalum-173	73	100 (3.7E 12)
Tantalum-174	73	100 (3.7E 12)
Tantalum-175	73	100 (3.7E 12)
Tantalum-176	73	10 (3.7E 11)
Tantalum-177	73	1000 (3.7E 13)
Tantalum-178	73	1000 (3.7E 13)
Tantalum-179	73	1000 (3.7E 13)
Tantalum-180m	73	1000 (3.7E 13)
Tantalum-180	73	100 (3.7E 12)
Tantalum-182m	73	1000 (3.7E 13)
Tantalum-182	73	10 (3.7E 11)
Tantalum-183	73	100 (3.7E 12)
Tantalum-184	73	10 (3.7E 11)
Tantalum-185	73	1000 (3.7E 13)
Tantalum-186	73	1000 (3.7E 13)
Technetium-93m	43	1000 (3.7E 13)
Technetium-93	43	100 (3.7E 12)
Technetium-94m	43	100 (3.7E 12)
Technetium-94	43	10 (3.7E 11)
Technetium-96m	43	1000 (3.7E 13)
Technetium-96	43	10 (3.7E 11)
Technetium-97m	43	100 (3.7E 12)
Technetium-97	43	100 (3.7E 12)
Technetium-98	43	10 (3.7E 11)
Technetium-99m	43	100 (3.7E 12)
Technetium-99	43	10 (3.7E 11)
Technetium-101	43	1000 (3.7E 13)
Technetium-104	43	1000 (3.7E 13)
Tellurium-116	52	1000 (3.7E 13)
Tellurium-121m	52	10 (3.7E 11)
Tellurium-121	52	10 (3.7E 11)
Tellurium-123m	52	10 (3.7E 11)
Tellurium-123	52	10 (3.7E 11)
Tellurium-125m	52	10 (3.7E 11)
Tellurium-127m	52	10 (3.7E 11)
Tellurium-127	52	1000 (3.7E 13)
Tellurium-129m	52	10 (3.7E 11)
Tellurium-129	52	1000 (3.7E 13)
Tellurium-131m	52	10 (3.7E 11)
Tellurium-131	52	1000 (3.7E 13)
Tellurium-132	52	10 (3.7E 11)
Tellurium-133m	52	1000 (3.7E 13)
Tellurium-133	52	1000 (3.7E 13)
Tellurium-134	52	1000 (3.7E 13)

Radionuclide	Atomic Number	Final RQ Ci (Bq)
Terbium-147	65	100 (3.7E 12)
Terbium-149	65	100 (3.7E 12)
Terbium-150	65	100 (3.7E 12)
Terbium-151	65	10 (3.7E 11)
Terbium-153	65	100 (3.7E 12)
Terbium-154	65	10 (3.7E 11)
Terbium-155	65	100 (3.7E 12)
Terbium-156m (5.0 hr)	65	1000 (3.7E 13)
Terbium-156m (24.4 hr)	65	1000 (3.7E 13)
Terbium-156	65	10 (3.7E 11)
Terbium-157	65	100 (3.7E 12)
Terbium-158	65	10 (3.7E 11)
Terbium-160	65	10 (3.7E 11)
Terbium-161	65	100 (3.7E 12)
Thallium-194m	81	100 (3.7E 12)
Thallium-194	81	1000 (3.7E 13)
Thallium-195	81	100 (3.7E 12)
Thallium-197	81	100 (3.7E 12)
Thallium-198m	81	100 (3.7E 12)
Thallium-198	81	10 (3.7E 11)
Thallium-199	81	100 (3.7E 12)
Thallium-200	81	10 (3.7E 11)
Thallium-201	81	1000 (3.7E 13)
Thallium-202	81	10 (3.7E 11)
Thallium-204	81	10 (3.7E 11)
Thorium-226	90	100 (3.7E 12)
Thorium-227	90	1 (3.7E 10)
Thorium-228	90	0.01 (3.7E 8)
Thorium-229	90	0.001 (3.7E 7)
Thorium-230	90	0.01 (3.7E 8)
Thorium-231	90	100 (3.7E 12)
Thorium-232Φ	90	0.001 (3.7E 7)
Thorium-234	90	100 (3.7E 12)
Thulium-162	69	1000 (3.7E 13)
Thulium-166	69	10 (3.7E 11)
Thulium-167	69	100 (3.7E 12)
Thulium-170	69	10 (3.7E 11)
Thulium-171	69	100 (3.7E 12)
Thulium-172	69	100 (3.7E 12)
Thulium-173	69	100 (3.7E 12)
Thulium-175	69	1000 (3.7E 13)
Tin-110	50	100 (3.7E 12)
Tin-111	50	1000 (3.7E 13)
Tin-113	50	10 (3.7E 11)
Tin-117m	50	100 (3.7E 12)
Tin-119m	50	10 (3.7E 11)
Tin-121m	50	10 (3.7E 11)
Tin-121	50	1000 (3.7E 13)
Tin-123m	50	1000 (3.7E 13)
Tin-123	50	10 (3.7E 11)
Tin-125	50	10 (3.7E 11)
Tin-126	50	1 (3.7E 10)
Tin-127	50	100 (3.7E 12)
Tin-128	50	1000 (3.7E 13)
Titanium-44	22	1 (3.7E 10)
Titanium-45	22	1000 (3.7E 13)
Tungsten-176	74	1000 (3.7E 13)
Tungsten-177	74	100 (3.7E 12)
Tungsten-178	74	100 (3.7E 12)
Tungsten-179	74	1000 (3.7E 13)
Tungsten-181	74	100 (3.7E 12)
Tungsten-185	74	10 (3.7E 11)
Tungsten-187	74	100 (3.7E 12)
Tungsten-188	74	10 (3.7E 11)
Uranium-230	92	1 (3.7E 10)
Uranium-231	92	1000 (3.7E 13)
Uranium-232	92	0.01 (3.7E 8)
Uranium-233	92	0.1 (3.7E 9)

Radionuclide	Atomic Number	Final RQ Ci (Bq)
Uranium-234φ	92	0.1 (3.7E 9)
Uranium-235φ	92	0.1 (3.7E 9)
Uranium-236	92	0.1 (3.7E 9)
Uranium-237	92	100 (3.7E 12)
Uranium-238φ	92	0.1& (3.7E 9)
Uranium-239	92	1000 (3.7E 13)
Uranium-240	92	1000 (3.7E 13)
Vanadium-47	23	1000 (3.7E 13)
Vanadium-48	23	10 (3.7E 11)
Vanadium-49	23	1000 (3.7E 13)
Xenon-120	54	100 (3.7E 12)
Xenon-121	54	10 (3.7E 11)
Xenon-122	54	100 (3.7E 12)
Xenon-123	54	10 (3.7E 11)
Xenon-125	54	100 (3.7E 12)
Xenon-127	54	100 (3.7E 12)
Xenon-129m	54	1000 (3.7E 13)
Xenon-131m	54	1000 (3.7E 13)
Xenon-133m	54	1000 (3.7E 13)
Xenon-133	54	1000 (3.7E 13)
Xenon-135m	54	10 (3.7E 11)
Xenon-135	54	100 (3.7E 12)
Xenon-138	54	10 (3.7E 11)
Ytterbium-162	70	1000 (3.7E 13)
Ytterbium-166	70	10 (3.7E 11)
Ytterbium-167	70	1000 (3.7E 13)
Ytterbium-169	70	10 (3.7E 11)
Ytterbium-175	70	100 (3.7E 12)
Ytterbium-177	70	1000 (3.7E 13)
Ytterbium-178	70	1000 (3.7E 13)
Yttrium-86m	39	1000 (3.7E 13)
Yttrium-86	39	10 (3.7E 11)
Yttrium-87	39	10 (3.7E 11)
Yttrium-88	39	10 (3.7E 11)
Yttrium-90m	39	100 (3.7E 12)
Yttrium-90	39	10 (3.7E 11)
Yttrium-91m	39	1000 (3.7E 13)
Yttrium-91	39	10 (3.7E 11)
Yttrium-92	39	100 (3.7E 12)
Yttrium-93	39	100 (3.7E 12)
Yttrium-94	39	1000 (3.7E 13)
Yttrium-95	39	1000 (3.7E 13)
Zinc-62	30	100 (3.7E 12)
Zinc-63	30	1000 (3.7E 13)
Zinc-65	30	10 (3.7E 11)
Zinc-69m	30	100 (3.7E 12)
Zinc-69	30	1000 (3.7E 13)
Zinc-71m	30	100 (3.7E 12)
Zinc-72	30	100 (3.7E 12)
Zirconium-86	40	100 (3.7E 12)
Zirconium-88	40	10 (3.7E 11)
Zirconium-89	40	100 (3.7E 12)
Zirconium-93	40	1 (3.7E 10)
Zirconium-95	40	10 (3.7E 11)
Zirconium-97	40	10 (3.7E 11)

Ci—Curie. The curie represents a rate of radioactive decay. One curie is the quantity of any radioactive nuclide which undergoes 3.7E 10 disintegrations per second.

Bq—Becquerel. The becquerel represents a rate of radioactive decay. One becquerel is the quantity of any radioactive nuclide which undergoes one disintegration per second. One curie is equal to 3.7E 10 becquerel.

■—Final RQs for all radionuclides apply to chemical compounds containing the radionuclides and elemental forms regardless of the diameter of pieces of solid material.

&—The adjusted RQ of one curie applies to all radionuclides not otherwise listed. Whenever the RQs in Table 302.4 and this appendix to the table are in conflict, the lowest RQ shall apply. For example, uranyl acetate and uranyl nitrate have adjusted RQs shown in Table 302.4 of 100 pounds, equivalent to about one-tenth the RQ level for uranium-238 listed in this appendix.

E—Exponent to the base 10. For example, 1.3E 2 is equal to 130 while 1.3E 3 is equal to 1300.

m—Signifies a nuclear isomer which is a radionuclide in a higher energy metastable state relative to the parent isotope.

φ—Notification requirements for releases of mixtures or solutions of radionuclides can be found in Section 302.6(b) of this rule. Final RQs for the following four common radionuclide mixtures are provided: radium-226 in secular equilibrium with its daughters (0.053 curie); natural uranium (0.1 curie); natural uranium in secular equilibrium with its daughters (0.052 curie); and natural thorium in secular equilibrium with its daughters (0.011 curie).

[54 FR 33449, Aug. 14, 1989]

APPENDIX I

SECTION 313 WATER PRIORITY CHEMICALS

SECTION 313 WATER PRIORITY CHEMICALS	
CAS Number	Common Name
75-07-0	Acetaldehyde
75865	Acetane cynohydrin
107-02-8	Acrolein
107-13-1	Acrylonitrile
309-00-2	Aldrin[1,4:5,8-Dimethanonaphthalene, 1,2,3,4,10,10-hexachloro-1,4,4a,5,8,8a-hexahydro-(1.alpha.,4.alpha.,4a.beta.,5.alpha.,8.alpha., 8a.beta.)-]
107-05-1	Allyl Chloride
7429-90-5	Aluminum (fume or dust)
7664-41-7	Ammonia
62-53-3	Aniline
120-12-7	Anthracene
7440-36-0	Antimony
7647189	Antimony pentachloride
28300745	Antimony potassium tartrate
7789619	Antimony tribromide
10025919	Antimony trichloride
7783564	Antimony trifluoride
1309644	Antimony trioxide
7440-38-2	Arsenic
1303328	Arsenic disulfide
1303282	Arsenic pentoxide
7784341	Arsenic trichloride
1327533	Arsenic trioxide
1303339	Arsenic trisulfide
1332-21-4	Asbestos (friable)
542621	Barium cyanide
71-43-2	Benzene
92-87-5	Benzidine
100470	Benzonitrile
98-88-4	Benzoyl chloride

SECTION 313 WATER PRIORITY CHEMICALS	
CAS Number	**Common Name**
100-44-7	Benzyl chloride
7440-41-7	Beryllium
7787475	Beryllium chloride
7787497	Beryllium fluoride
7787555	Beryllium nitrate
111-44-4	Bis(2-chloroethyl) ether
75-25-2	Bromoform
74-83-9	Bromomethane (Methyl bromide)
85-68-7	Butyl benzyl phthalate
7440-43-9	Cadmium
543908	Cadmium acetate
7789426	Cadmium bromide
10108642	Cadmium chloride
7778441	Calcium arsenate
52740166	Calcium arsenite
13765190	Calcium chromate
592018	Calcium cyanide
133-06-2	Captan [1H-Isoindole-1,3(2H)-dione,3a,4,7,7a-tetrahydro-2-[(trichloromethyl)thio]-]
63-25-2	Carbaryl [1-Naphthalenol, methylcarbamate]
75-15-0	Carbon disulfide
56-23-5	Carbon tetrachloride
57-74-9	Chlordane [4,7-Methanoindan,1,2,4,5,6,7,8,8-octachloro-2,3,3a,4,7,7a-hexahydro-]
7782-50-5	Chlorine
59-50-7	Chloro-4-methyl-3-phenol p-Chloro-m-cresol
108-90-7	Chlorobenzene
75-00-3	Chloroethane (Ethyl chloride)
67-66-3	Chloroform
74-87-3	Chloromethane (Methyl chloride)
95-57-8	2-Chlorophenol

SECTION 313 WATER PRIORITY CHEMICALS	
CAS Number	Common Name
106-48-9	4-Chlorophenol
1066304	Chromic acetate
11115745	Chromic acid
10101538	Chromic sulfate
7440-47-3	Chromium
1308-14-1	Chromium (Tri)
10049055	Chromous chloride
7789437	Cobaltous bromide
544183	Cobaltous formate
14017415	Cobaltous sulfamate
7440-50-8	Copper
108-39-4	*m*-Cresol
9548-7	*o*-Cresol
106-44-5	*p*-Cresol
1319-77-3	Cresol (mixed isomers)
142712	Cupric acetate
12002038	Cupric acetoarsenite
7447394	Cupric chloride
3251238	Cupric nitrate
5893663	Cupric oxalate
7758987	Cupric sulfate
10380297	Cupric sulfate, ammoniated
815827	Cupric tartrate
57-12-5	Cyanide
506774	Cyanogen chloride
110-82-7	Cyclohexane
94-75-7	2,4-D [Acetic acid, (2,4-dichlorophenoxy)-]
106-93-4	1,2-Dibromoethane (Ethylene dibromide)
84-74-2	Dibutyl phthalate
25321-22-6	Dichlorobenzene (mixed isomers)

SECTION 313 WATER PRIORITY CHEMICALS	
CAS Number ·	Common Name
95-50-1	1,2-Dichlorobenzene
541-73-1	1,3-Dichlorobenzene
106-46-7	1,4-Dichlorobenzene
91-94-1	3,3'-Dichlorobenzidine
75-27-4	Dichlorobromomethane
107-06-2	1,2-Dichloroethane (Ethylene dichloride)
540-59-0	1,2-Dichloroethylene
120-83-2	2,4-Dichlorophenol
78-87-5	1,2-Dichloropropane
542-75-6	1,3-Dichloropropylene
62-73-7	Dichlorvos [Phosphoric acid, 2,2-dichloroethenyl dimethyl ester]
115-32-2	Dicofol [Benzenemethanol, 4-chloro-.alpha.-(4-chlorophenyl)-.alpha.-(trichloromethyl)-]
177-81-7	Di-(2-ethylhexyl phthalate (DEHP)
84-66-2	Diethyl phthalate
105-67-9	2,4-Dimethylphenol
131-11-3	Dimethyl phthalate
534-52-1	4,6-Dinitro-o-cresol
51-28-5	2,4-Dinitrophenol
121-14-2	2,4-Dinitrotoluene
606-20-2	2,6-Dinitrotoluene
117-84-0	n-Dioctyl phthalate
122-66-7	1,2-Diphenylhydrazine (Hydrazobenzene)
106-89-8	Epichlorohydrin
100-41-4	Ethylbenzene
106934	Ethylene dibromide
50-00-0	Formaldehyde
76-44-8	Heptachlor [1,4,5,6,7,8,8-Heptachloro-3a,4,7,7a-tetrahydro-4,7-methano-1H-indene]
118-74-1	Hexachlorobenzene
87-68-3	Hexachloro-1,3-butadiene

SECTION 313 WATER PRIORITY CHEMICALS	
CAS Number	Common Name
77-47-4	Hexachlorocyclopentadiene
67-72-1	Hexachloroethane
7647-01-0	Hydrochloric acid
74-90-8	Hydrogen cyanide
7664-39-3	Hydrogen fluoride
7439-92-1	Lead
301042	Lead acetate
7784409	Lead arsenate
7645252	" "
10102484	" "
7758954	Lead chloride
13814965	Lead fluoborate
7783462	Lead fluoride
10101630	Lead iodide
10099748	Lead nitrate
7428480	Lead stearate
1072351	" "
52652592	" "
7446142	Lead sulfate
1314870	Lead sulfide
592870	Lead thiocyanate
58-89-9	Lindane [Cyclohexane, 1,2,3,4,5,6-hexachloro- (1.alpha.,3.beta., 4.alpha.,5.alpha.,6.beta.)-]
14307358	Lithium chromate
108-31-6	Maleic anhydride
592041	Mercuric cyanide
10045940	Mercuric nitrate
7783359	Mercuric sulfate
592858	Mercuric thiocyanate
7782867	Mercurous nitrate
7439-97-6	Mercury

SECTION 313 WATER PRIORITY CHEMICALS	
CAS Number	Common Name
72-43-5	Methoxychlor [Benzene, 1,1'-(2,2,2-trichloroethylidene)bis[4- methoxy-]
80-62-6	Methyl methacrylate
91-20-3	Naphthalene
7440-02-0	Nickel
15699180	Nickel ammonium sulfate
37211055	Nickel chloride
7718549	" "
12054487	Nickel hydroxide
14216752	Nickel nitrate
7786814	Nickel sulfate
7697-37-2	Nitric acid
98-95-3	Nitrobenzene
88-75-5	2-Nitrophenol
100-02-7	4-Nitrophenol
62-75-9	N-Nitrosodimethylamine
86-30-6	N-Nitrosodiphenylamine
621-64-7	N-Nitrosodi-n-propylamine
56-38-2	Parathion [Phosphorothioic acid, O,O-diethyl-O-(4-nitrophenyl) ester]
87-86-5	Pentachlorophenol (PCP)
108-95-2	Phenol
75-44-5	Phosgene
7664-38-2	Phosphoric acid
7723-14-0	Phosphorus (yellow or white)
1336-36-3	Polychlorinated biphenyls (PCBs)
7784410	Potassium arsenate
10124502	Potassium arsenite
7778509	Potassium bichromate
7789006	Potassium chromate
151508	Potassium cyanide
75-56-9	Propylene oxide

SECTION 313 WATER PRIORITY CHEMICALS	
CAS Number	**Common Name**
91-22-5	Quinoline
7782-49-2	Selenium
7446084	Selenium oxide
7440-22-4	Silver
7761888	Silver nitrate
7631892	Sodium arsenate
7784465	Sodium arsenite
10588019	Sodium bichromate
7775113	Sodium chromate
143339	Sodium cyanide
10102188	Sodium selenite
7782823	▪ ▪
7789062	Strontium chromate
100-42-5	Styrene
7664-93-9	Sulfuric acid
79-34-5	1,1,2,2-Tetrachloroethane
127-18-4	Tetrachloroethylene (Perchloroethylene)
935-95-5	2,3,5,6-Tetrachlorophenol
78002	Tetraethyl lead
7440-28-0	Thallium
10031591	Thallium sulfate
108-88-3	Toluene
8001-35-2	Toxaphene
52-68-6	Trichlorfon [Phosphonic acid, (2,2,2-trichloro-1-hydroxyethyl)-dimethylester]
120-82-1	1,2,4-Trichlorobenzene
71-55-6	1,1,1-Trichloroethane (Methyl chloroform)
79-00-5	1,1,2-Trichloroethane
79-01-6	Trichloroethylene
95-95-4	2,4,5-Trichlorophenol
88-06-2	2,4,6-Trichlorophenol

SECTION 313 WATER PRIORITY CHEMICALS	
CAS Number	**Common Name**
7440-62-2	Vanadium (fume or dust)
108-05-4	Vinyl acetate
75-01-4	Vinyl chloride
75-35-4	Vinylidene chloride
108-38-3	m-Xylene
95-47-6	o-Xylene
106-42-3	p-Xylene
1330-20-7	Xylene (mixed isomers)
7440-66-6	Zinc (fume or dust)
557346	Zinc acetate
14639975	Zinc ammonium chloride
14639986	" " "
52628258	" " "
1332076	Zinc borate
7699458	Zinc bromide
3486359	Zinc carbonate
7646857	Zinc chloride
557211	Zinc cyanide
7783495	Zinc fluoride
557415	Zinc formate
7779864	Zinc hydrosulfite
7779886	Zinc nitrate
127822	Zinc phenolsulfonate
1314847	Zinc phosphide
16871719	Zinc silicofluoride
7733020	Zinc sulfate

APPENDIX J

TABLE OF MONITORING REQUIREMENTS IN EPA'S GENERAL PERMIT

EPA FINAL GENERAL PERMIT MONITORING REQUIREMENTS[1]				
Type of Facility	Type of Storm Water Discharge	Parameters	Monitoring Frequency	Reporting Frequency
EPCRA, Section 313 Facilities Subject to Reporting Requirements for Water Priority Chemicals	Storm water discharges that come into contact with any equipment, tank, container, or other vessel or area used for storage of a Section 313 water priority chemical, or located at a truck or rail car loading or unloading area where a Section 313 water priority chemical is handled	Oil and Grease, BOD5, COD, TSS, Total Kjeldahl Nitrogen, Total Phosphorus, pH, acute whole effluent toxicity[2], any Section 313 water priority chemical for which the facility reports	Semi-annual	Annual
Primary Metal Industries (SIC 33)	All storm water discharges associated with industrial activity	Oil and Grease, COD, TSS, pH, acute whole effluent toxicity[2], Total Recoverable Lead, Total Recoverable Cadmium, Total Recoverable Copper, Total Recoverable Arsenic, Total Recoverable Chromium, and any pollutant limited in an effluent guideline to which the facility is subject	Semi-annual	Annual
Land Disposal Units/ Incinerators/ BIFs	Storm water discharges from active or inactive land disposal units without a stabilized cover that have received any waste from industrial facilities other than construction sites; and storm water discharges from incinerators and BIFs that burn hazardous waste	Total Recoverable Magnesium, Magnesium (dissolved), Total Kjeldahl Nitrogen, COD, TDS, TOC, Oil and Grease, pH, Total Recoverable Arsenic, Total Recoverable Barium, Total Recoverable Cadmium, Total Recoverable Chromium, Total Cyanide, Total Recoverable Lead, Total Mercury, Total Recoverable Selenium, Total Recoverable Silver, acute whole effluent toxicity[2]	Semi-annual	Annual

EPA FINAL GENERAL PERMIT MONITORING REQUIREMENTS[1]				
Type of Facility	Type of Storm Water Discharge	Parameters	Monitoring Frequency	Reporting Frequency
Wood Treatment Facilities	Storm water discharges from areas that are used for wood treatment, wood surface application or storage of treated or surface protected wood	Oil and Grease, pH, COD, TSS	Semi-annual	Annual
	Facilities that use chlorophenolic formulations	Plus Pentachlorophenol and acute whole effluent toxicity[2]		
	Facilities that use creosote formulations	Plus acute whole effluent toxicity[2]		
	Facilities that use chromium-arsenic formulations	Plus Total Recoverable Arsenic, Total Recoverable Chromium, Total Recoverable Copper		
Industrial Facilities with Coal Piles	Storm water discharges from coal pile runoff	Oil and Grease, pH, TSS, Total Recoverable Copper, Total Recoverable Nickel, Total Recoverable Zinc	Semi-annual	Annual
Battery Reclaimers	Storm water discharges from areas for storage of lead acid batteries, reclamation products, or waste products, and areas used for lead acid battery reclamation	Oil and Grease, COD, TSS, pH, Total Recoverable Copper, Total Recoverable Lead	Semi-annual	Annual
Airports (with over 50,000 flight operations per year)	Storm water discharges from aircraft or airport deicing areas	Oil and Grease, BOD5, COD, TSS, pH, and the primary ingredient used in the deicing materials	Annual	Retain onsite
Coal-fired Steam Electric Facilities	Storm water discharges from coal handling sites (other than runoff from coal piles which is not eligible for coverage under this permit)	Oil and Grease, pH, TSS, Total Recoverable Copper, Total Recoverable Nickel, Total Recoverable Zinc	Annual	Retain onsite

EPA FINAL GENERAL PERMIT MONITORING REQUIREMENTS[1]				
Type of Facility	Type of Storm Water Discharge	Parameters	Monitoring Frequency	Reporting Frequency
Animal Handling/ Meat Packing Facilities	Storm water discharges from animal handling areas, manure management areas, production waste management areas exposed to precipitation at meat packing plants, poultry packing plants, facilities that manufacture animal and marine fats and oils	BOD5, Oil and Grease, COD, TSS, Total Kjeldahl Nitrogen (TKN), Total Phosphorus, pH, Fecal Coliform	Annual	Retain onsite
Chemical and Allied Product Manufacturers/ Rubber Manufacturers (SIC 28 and 30)	Storm water discharges that come into contact with solid chemical storage piles	Oil and Grease, COD, TSS, pH, any pollutant limited in an effluent guideline to which the facility is subject	Annual	Retain onsite
Automobile Junkyards	Storm water discharges exposed to: (a) over 250 auto/truck bodies with drivelines, 250 drivelines, or any combination thereof (b) over 500 auto/truck units (c) over 100 units dismantled per year where automotive fluids are drained or stored	Oil and Grease, COD, TSS, pH, any pollutant limited in an effluent guideline to which the facility is subject	Annual	Retain onsite
Lime Manufacturing Facilities	Storm water discharges that have come into contact with lime storage piles	Oil and Grease, COD, TSS, pH, any pollutant limited in an effluent guideline to which the facility is subject	Annual	Retain onsite
Oil-fired Steam Electric Power Generating Facilities	Storm water discharges from oil handling sites	Oil and Grease, COD, TSS, pH, any pollutant limited in an effluent guideline to which the facility is subject	Annual	Retain onsite
Cement Manufacturing Facilities and Cement Kilns	All storm water discharges associated with industrial activity (except those from material storage piles that are not eligible for coverage under this permit)	Oil and Grease, COD, TSS, pH, any pollutant limited in an effluent guideline to which the facility is subject	Annual	Retain onsite

EPA FINAL GENERAL PERMIT MONITORING REQUIREMENTS[1]				
Type of Facility	Type of Storm Water Discharge	Parameters	Monitoring Frequency	Reporting Frequency
Ready-mix Concrete Facilities	All storm water discharges associated with industrial activity	Oil and Grease, COD, TSS, pH, any pollutant limited in an effluent guideline to which the facility is subject	Annual	Retain onsite
Ship Building and Repairing Facilities	All storm water discharges associated with industrial activity	Oil and Grease, COD, TSS, pH, any pollutant limited in an effluent guideline to which the facility is subject	Annual	Retain onsite

[1]A discharger is not subject to the monitoring requirements provided the discharger makes a certification for a given outfall, on an annual basis, under penalty of law, that material handling equipment or activities, raw materials, intermediate products, final products, waste materials, by-products, industrial machinery or operations, significant materials from past industrial activities, or, in the case of airports, deicing activities, that are located in areas of the facility that are within the drainage area of the outfall are not presently exposed to storm water and will not be exposed to storm water for the certification period.

[2]A discharger may, in lieu of monitoring for acute whole effluent toxicity, monitor for pollutants identified in Tables II and III of Appendix D of 40 CFR Part 122 that the discharger knows or has reason to believe are present at the facility site. Such determinations are to be based on reasonable best efforts to identify significant quantities of materials or chemical present at the facility.

MUNICIPAL WASTEWATER TREATMENT TECHNOLOGY
Recent Developments

U.S. Environmental Protection Agency

This book presents recent developments in municipal wastewater treatment technology, based on a forum sponsored by the USEPA in 1991. The 25 presentations included cover the areas of land treatment, sand and gravel filters, operation and maintenance, biological nutrient removal, sludge, stormwater, disinfection, constructed wetlands, and municipal water use efficiency.

Several recent and/or upcoming changes in Federal regulations will affect all of those involved in wastewater technology development and transfer. These changes include impending sludge and stormwater regulations and the reauthorization of the Clean Water Act. Because of the technology transfer implications, the information presented here will be beneficial to those engineers, managers, and coordinators involved in treating municipal wastewaters.

The appendices to the book contain a summary of innovative and alternative technology projects by state.

CONTENTS

ISBN 0-8155-1309-7 (1993)   7" x 10"   250 pages

Other Noyes Publications

POLLUTION PREVENTION
TECHNOLOGY HANDBOOK

Edited by

Robert Noyes

This book presents technical information relating to current and potential pollution prevention and waste minimization techniques in 36 industries. Many of these industries have similar problems, and there are many opportunities for **cross-fertilization** in adopting pollution prevention techniques across industry boundaries.

In general each chapter provides for each industry: (1) description of manufacturing processes, (2) types of waste generated, and (3) specific pollution prevention and waste minimization opportunities.

There are a number of benefits involved in adopting pollution prevention techniques, the most important of which is economic. When wastes are reduced or eliminated, substantial cost savings can be realized by reduced expenditures for pollution control equipment, and lower treatment and disposal costs. In some firms, substantial source reduction activities have been implemented with minor capital expenditures, with resultant payback within six months.

Other considerations include lessened liability problems, and improved public image.

The thousands of items of technological advice in this book make it a valuable source of current and potential pollution prevention technology.

CONTENTS

ISBN 0-8155-1311-9 (1993)   7" x 10"   683 pages

Other Noyes Publications

RCRA REGULATORY COMPLIANCE GUIDE

by

Mark S. Dennison
Attorney at Law

The *RCRA Regulatory Compliance Guide* was written especially for companies that need an easy-to-read guide to the Resource Conservation and Recovery Act (RCRA), and information on how hazardous waste regulations impact day-to-day business operations.

This is the first book to explain this law, as well as other environmental laws and regulations, in practical terms. It provides step-by-step explanations of hazardous waste identification, storage, disposal and transportation. Useful checklists, sample forms, and various types of information sources have been included to assist those company officials involved in regulatory compliance.

The book details how to identify hazardous wastes at the company, what generator classification the company will fall under, how to safely handle the wastes, how to store hazardous wastes, and how to properly dispose of the wastes. The practical aspects of hazardous waste management are examined along with the body of laws and regulations the company needs to comply with in order to avoid costly liability and penalties.

Special issues concerning recordkeeping and reporting, worker safety, and environmental liability insurance are also closely examined.

The author has explained an extremely complex regulatory scheme as concisely as possible, using a practical, organized approach.

A condensed contents listing **chapter titles and selected subtitles** is given below.

ISBN 0-8155-1321-6 (1993)   7" x 10"   354 pages

HANDBOOK OF
LEAK, SPILL AND ACCIDENTAL RELEASE
PREVENTION TECHNIQUES

Edited by

Robert Noyes

Leaks, spills, and accidental releases are a major source of release of toxic and hazardous substances to the environment. For example, equipment leaks alone account for 35% of all VOC emissions in the chemical industry.

Leaks, spills, and accidental releases can take many forms including (1) gaseous emissions; (2) liquid releases that allow VOCs to enter the atmosphere, as well as the liquid portion entering the ground; (3) heavier liquid releases that do not vaporize, or have very low volatility; and (4) solid waste emissions mainly in the form of dust.

Fugitive emissions are also distinguished from process (point-source) emissions. The term *fugitive emissions* includes the loss of chemicals through sealing mechanisms separating process fluid from the atmosphere. Examples of fugitive emissions are equipment leaks that come from the hundreds or thousands of valves, pumps, compressors, pressure relief devices, open-ended valves or lines, sampling connection systems, and flanges and other connectors within a processing plant.

The techniques used to control fugitive VOC emissions are quite different from those used to control process emissions, due in large part to the fact that process emissions are generally vented from a definable point or stack, while fugitive emission sources are more diffuse.

Emission sources in the chemical industry can be divided into six source types:

1) Process vent emissions
2) Storage tank emissions and leaks
3) Equipment and piping leak emissions
4) Transfer emissions, leaks and spills
5) Wastewater collection and treatment emissions
6) Waste storage piles

This book is designed to provide technical guidance to prevent leaks, spills or other accidental releases of hazardous substances from fixed facilities that produce hazardous substances, store them, or transfer them to and from transportation terminals. The audience addressed includes managerial and supervisory personnel as well as "hands on" personnel associated with both large and small manufacturers. As an aid to plant engineers and managers, federal workers, fire marshalls, and fire and casualty insurance inspectors, this document is offered as a guide to prevention of leaks, spills, and accidental releases. A **condensed contents** is given below.

ISBN 0-8155-1296-1 (1992)   6" x 9"   487 pages

www.ingramcontent.com/pod-product-compliance
Lightning Source LLC
Chambersburg PA
CBHW082002190326
41458CB00010B/3046